Jürgen Brandes, Jan Czerniawski, Ludwig Neidhart

Special and General Theory of Relativity

for physicists and philosophers

Jürgen Brandes, Jan Czerniawski, Ludwig Neidhart

SPECIAL and **GENERAL THEORY OF RELATIVITY** for physicists and philosophers

Einstein and Lorentz Interpretation
Paradoxes
Space and Time
Experiments

Translation of the 5th German edition
augmented, with over 100 illustrations
Publishing House of Relativistic Interpretations - VRI
Karlsbad, Germany 2023

Bibliographic information of the German National Library. The German National Library lists this publication in the German National Bibliography; detailed bibliographic data is available on the Internet at http://dnb.d-nb.de.

ISBN 9783930879144

Danziger Str. 65, D-76307 Karlsbad, Germany, 2023

Translation of 5th German edition augmented with over 100 illustrations

ISBN 9783930879144

Cover design, Figure 7.3, 19.1: Stephanie BURGSTALLER,
A 5273 St. Veit, Austria, using the following sources:
Hendrik Antoon Lorentz, 1853 - 1928, Nobel Prize in Physics 1902. Portrait of Jan Veth 1899 - see Berends. [19.1]. Copyright: Instituut-Lorentz, Universiteit Leiden, Netherlands, Prof. Dr. Beenakker
Bundesarchiv, Picture 102-10447 Einstein / Photographer: o.Ang. / Licence CC-BY-SA 3.0

Our families

Foreword

Disappointed critics of relativity theory were and are not uncommon, and so this book is written "for those who studied EINSTEIN'S theory of relativity, became fascinated by it, but ultimately gave up trying to understand it" (quotation from a concerned person).

... and also for those who cannot comprehend this at all, because for them philosophical objections suggest a lack of understanding of Minkowski space and Riemann spaces, or because for them the solution of the paradoxes is a calculation experiment - perhaps not yet carried out by themselves - taking into account the relativity of simultaneity, or because for them unsolvable questions about the conservation of energy can only be temporarily unsolved. More details in the chapter "Three reasons (not) to read this book".

All explanations and attempts at explanations in the specialist literature, insofar as they have significance, are systematically discussed. This is surprisingly simple, since it is sufficient to examine the different proposed solutions of the Einstein and Lorentz interpretations. The main aim is to enable the reader to make his or her **own decisions** on the theory of relativity and to check those already been done, so the Lorentz interpretation is examined on an equal footing with the classical, Einsteinian theory. The contribution by Jan CZERNIAWSKI supports this goal and discusses not only NEWTON and KANT but the philosophy of LEIBNIZ, which is closer to general relativity.

A major advance in this and the previous edition is the understanding of the conservation of energy of general relativity, including the Newtonian limiting case. Thus, an object at rest in the Newtonian gravitational field has *negative* potential energy. For special relativity, this means negative mass times c^2, but negative masses do not exist. The Lorentz interpretation can be valuated by the way it solves this problem of understanding. The same applies to the formulas for the total energy of a particle at rest in the gravitational field, which seem contradictory at first glance.

Dr. habil. Ludwig NEIDHART, Augsburg, many thanks for his numerous suggestions for corrections and for allowing me to print his review.

Karlsbad, September 1994
June 1995
February 2001
December 2010
January 2022
June 2023 Jürgen Brandes

Table of contents

Table of contents

Foreword 7

1. Introduction 19

2. Einstein (EI) and Lorentz (LI) Interpretation - first definition 21

3. Three reasons to (not) read this book 25

3.1. Philosophical simultaneity 25
3.2. Twin paradox 25
3.3. Conservation of energy in free fall 27

I. The Paradoxes and the Space-Time Continuum of the Special Theory of Relativity 30

4. The statement of the special theory of relativity 30

4.1. Introduction 30
4.2. The Lorentz and Galileo Transformations 30
4.3. The relativity of simultaneity and the Einstein convention of clock synchronisation 33
4.4. Conclusions from the Lorentz transformations 35
4.5. Conservation of momentum, mass and energy in the special theory of relativity 38
4.6. The different shape of natural laws in special relativity and classical physics 43

5. The Experimental Evidence of Special Relativity 44

5.1. Introduction 44
5.2. Experimental proof of time dilation 44
5.3. The dependence of mass on velocity 45
5.4. On the experimental proof of the constancy of the speed of light 47
5.5. The Michelson-Morley and Kennedy-Thorndike experiments to prove relativistic length contraction. 48

6. Derivation of the Lorentz transformations 53

6.1. Introduction 53

6.2. Heuristic derivation of the Lorentz transformations from length contraction and time dilation. 54
6.3. Alternative derivation of the Lorentz transformations with the help of the reciprocity principle. 57
6.4. The Einstein and Lorentz Interpretation of the Lorentz Transformations 58
6.5. Derivation from the principle of relativity and the principle of constancy of the speed of light for both interpretations 60
6.6. Definition of EI and LI - somewhat more abstract 62

7. Thought experiments to differentiate between the Einstein and Lorentz interpretations 63

7.1. Introduction 63
7.2. Thought experiment on particle decay 63
7.3. A thought experiment on measuring the one-way speed of light using lasers 66
7.4. Statements on the Lorentz Interpretation - MITTELSTAEDT, TREDER, SEXL 68

8. JAN CZERNIAWSKI – On rods, clocks, and relativity 74

9. The concepts of length and distance in special relativity 107

9.1. Lengths and distances interpreted relativistically 107
9.2. Length measurements in the LI 109
9.3. Linearly accelerated bodies 111
9.4. Circularly accelerated bodies 113

10. Paradoxes 119

10.1. Introduction 119
10.2. Garage problem 120
10.3. Wood's paradox 121
10.4. Lid paradox 123
10.5. Ehrenfest paradox 127
10.6. Bell's spaceship paradox 132
10.7. Sagnac effect 132

10.8. Clock or twin paradox 137
10.8.1. Introduction 137
10.8.2. Preliminary remark on proper time 138
10.8.3. Proper times relative to *S*, S′, *S″* 140
10.8.4. Proper times with change of *S′* and *S″* 143
10.8.5. Contradictory conclusions in the clock paradox 144
10.8.6. The inertial system requirement and the first solution of the clock paradox 146
10.8.7. Second solution of the clock paradox 149
10.8.8. The general relativistic solution 151
10.8.9. The twin paradox (ZP) in WIKIPEDIA 157

11. Philosophical contributions to Einstein (EI) and Lorentz interpretation (LI) 159

11.1. The terms one-dimensional and multidimensional 159
11.2. A first philosophical argument on the space-time continuum 163
11.3. The four-dimensional Minkowski space of the EI 165
11.4. Alternative interpretation of the Minkowski space of the LI 172
11.5. Inertial systems with the speed of light 173
11.6. Spatial objects and space, time sequences and time 175
11.6.1. Geometric objects in mathematical spaces, bodies in physical space 175
11.6.2. Relative and philosophical simultaneity 178
11.6.3. Minkowski space and real space 183
11.7. Ether and quantum ether 186
11.8. Physical quantities in the EI and LI 189

12. Comparison of the Einstein and Lorentz interpretations of special relativity theory 194

13. An exact definition of the Lorentz interpretation of special relativity 195

II. The Curved Space of General Relativity 199

14. Overview 199

15. The Significance of General Relativity for the Special Relativistic Space-Time Theses 200

16. Schwarzschild metric (SM) 202

16.1. Introduction 202
16.2. The experimental evidence for the validity of the Schwarzschild metric 202
16.3. Fundamentals of the Schwarzschild metric 206
16.4. Geometric model of the Schwarzschild metric 209
16.5. Gaussian coordinate system for centrally symmetric gravitational fields 212
16.6. The Einstein and Lorentz Interpretation of the Schwarzschild metric 214
16.7. Experimentally verifiable conclusions from the Schwarzschild metric 216
16.8. The Schwarzschild metric in other coordinates and the objection by WEYL 218
16.9. The Singularity of the Schwarzschild Metric 222

17. The Poincaré thesis and the empty, matter-free space of general relativity 224

18. For the definition of curved space with the help of mathematics 227

18.1. Introduction 227
18.2. Two-dimensional spherical surfaces 227
18.3. Three-dimensional spherical surfaces 233
18.4. To construct a coordinate system for the three-dimensional spherical surface 235
18.5. Interpretation of metrics as scale distortion 238
18.6. Other curved surfaces 239

19. Cosmology and Robertson-Walker metric (RWM) 242

19.1. Introduction 242
19.2. The finite curved cosmological space from the point of view of the Schwarzschild metric 243
19.3. The cosmological principle and the co-moving coordinate system 243

19.4. The Robertson-Walker Metric (RWM) 246
19.5. The Lorentz interpretation of the Robertson-Walker metric as scale changes 249
19.6. Observations in the universe 250
19.7. The cosmological alternatives: The infinite Euclidean and infinite hyperbolic universe. 254
19.8. Deepening the Einstein and Lorentz interpretation of the RWM 256
19.8.1. Overview 256
19.8.2. Newton's law of gravity and the big bang 258
19.8.3. The Robertson-Walker metric and the gravitational collapse of stars 259
19.8.4. Expanding space and cosmological simultaneity for the special relativistic limiting case 260
19.8.5. The expansion of the universe as a function of t and τ 265

20. The Lorentz Interpretation (LI) of general relativity 267

20.1. Approach to an exact definition 267
20.2. Plausibility and verifiability of the additional assumptions 270

21. Black holes and conservation of energy of the SM 271

21.1. Conservation of energy for light 271
21.2. Fall laws 272
21.3. Law of conservation of energy for freely falling particles 276
21.4. Energy of a particle at rest in the gravitational field 279
21.5. Conservation of Energy in the Inner Schwarzschild Metric. 283
21.6. Black holes of the SM and their observability 284
21.7. Calculation of supermassive objects using TOV 290
21.8. Black holes when using the Inner SM and the TOV 293

22. Black holes and conservation of energy of the RWM 297

22.1. Formulas 297
22.2. Lorentz interpretation for exploding dust stars 298
22.3. Black holes (BH) and conservation of energy in the RWM 301

23. Comparison of the Einstein and Lorentz interpretations of the General Theory of Relativity 305

24. Notes, additions, comments on I. and II. 308

24.1. Absolute special relativistic effects 308
24.2. Wilhelm T. HERING: Discussion remarks on the lecture by J. BRANDES at the 64th DPG Conference in Dresden 312
24.3. Bell's spaceship paradox - addition 315
24.4. Historical note on the Lorentz interpretation 315
24.5. Critics of the Lorentz interpretation 316
24.6. Fermat's Principle in General Relativity 317
24.7. Weight force, gravitational force 318
24.8. The negative Newtonian gravitational potential 318
24.9. EINSTEIN's law of conservation of energy 319
24.10. Are we living in a black hole? 321
24.11. Astronomical observations 322
24.11.1 Observation options 322
24.11.2 Contradictory measurements of the Hubble parameter 324
24.11.3 Questionable predictions of Sgr A* characteristics 325
24.12. Cosmic Microwave Background (CMB) and the so-called WMAP cold spot 329
24.13. The smallness of the Cosmological Constant 330

25. LUDWIG NEIDHART - Review 331

26. Directories 334

26.1. Useful and hypothetical figures 334
26.2. Glossary: Special and General Theory of Relativity 335
26.3. Bibliography and references 340
26.4. The authors 366
26.5. Index 368

Special and General Theory of Relativity for physicists and philosophers

"Don't bite my finger, look where it's pointing."
Warren S. McCulloch

1. Introduction

If you are looking for your own way through this book, the first thing you should do is read chapter 3 so that you know what you are promised. Resolved are the three problem areas - philosophical objections, contradictions through paradoxes, conservation of energy in general relativity - by the Lorentz interpretation of special and general relativity or already by the Einstein interpretation. For the reader's own decision, both interpretations are discussed equally. Chapter (ch.) 2, 6.6, 13, 20 define more and more precisely what is meant by Einstein and Lorentz interpretation. The associated chapters 13, 23 summarise the differences between the two interpretations and what is experimentally verifiable.

Chapters 4 to 19 have endured since the first edition and describe in an elementary yet exact manner how the philosophical objections and paradoxes can be resolved with the help of the Lorentz interpretation **without denying the fundamental physical insights of relativity.** Chapters 21 and 22 deal with the conservation of energy in general relativity and concern the physicist. Here, the Lorentz interpretation finally becomes experimentally verifiable and thus a **physical** theory.

These points seem essential to the author:

- The proof that the special theory of relativity can be derived alternatively, ch. 6 and ch. 8
- The contradictory nature of the twin paradox, ch. 10.8
- The special-relativistic limiting case and cosmological simultaneity, ch. 19.8.4
- The explanation of what the negative Newtonian gravitational potential means, ch. 21.3, 24.8
- Contradictory energy formulas, ch. 21.3.

Already in 1999, in his first DPG lecture, the author pointed out that the highest-energy cosmic rays (UHECR) could have its origin in the galactic centers as the experiments of the Pierre Auger Observatory [266] and H.E.S.S. [155.1] now seem to confirm. Wherever black holes play a role (supernovae, gamma bursts, gravitational waves, jets from AGN), the different statements on the conservation of energy and the properties of black holes of the Einstein and Lorentz interpretations could be of importance. For suggestions on this topic, see ch. 21.3 and 24.11.

Some, hopefully excusable, shortcomings should be pointed out: Some formulas may contain spelling mistakes or be difficult to read, some chapters somewhat lengthy, the translation from German sometimes ambiguous although the translator Deepl was really helpful. The LI of SM and RWM in ch. 16 - 18 is a preliminary stage of the more general one in ch. 20, but is already called LI. Some ideas are not to be regarded as proven, but are intended to serve understanding, to stimulate, in any case merely to help. The author cannot exclude for each individual case that he may have overlooked one or the other argument in favour of EI or LI that appears to be essential.

For the correct approach to solving relativistic tasks, knowledge of both interpretations is helpful, and it is not decisive which of the two one prefers. **This is also the view of KIP S. THORNE, Nobel Prize 2017: "When working in the field of relativity, it is extremely useful to have both paradigms at hand."** [366.2] Since both should be well explained here, this text is suitable for use alongside the usual lectures, which on the one hand are more mathematically demanding, but on the other do not address the Lorentz interpretation. Although more is intended, the minimal aim remains a systematic presentation of the paradoxes of relativity together with suggestions to solve them.

The most important physical statement lies in the understanding of the *negative* Newtonian gravitational potential.

2. Einstein (EI) and Lorentz (LI) Interpretation - first definition

The Lorentz interpretation (LI) is an extension of classical special and general relativity. This means: For the LI, classical special and general relativity is neither wrong nor right, but incomplete. A wrong theory must be replaced, an incomplete one extended, the latter is what the LI wants. Einstein Interpretation (EI) is another, shorter name for classical special and general relativity, the emphasis being that the relativistic formulae are to be interpreted philosophically in EINSTEIN'S sense. Fig. 2.1 illustrates what is common to both interpretations and what distinguishes them. It follows somewhat more precisely: EI and LI use the same formulae of classical special and/or general relativity. EI interprets the relativistic formulae and measurement results as *philosophical* statements about space and time. The LI tries to deepen the relativistic formulae and measurement results *physically* (rod contraction, slower running clocks).

While the EI is a highly developed theory, the LI is in its infancy, but it is sufficiently well thought out to be experimentally testable. For example, the properties of black holes are verifiably different in both interpretations. For the LI, black holes are possible origins of the UHECR (cosmic particles of highest energy) and emit light, for the EI black holes are objects that can neither radiate like stars nor emit particles.

In his inauguration speech in Berlin in 1914, EINSTEIN had explained the principle of general relativity: all equations should apply independently of the respective coordinate system. The special theory of relativity "does not grant full satisfaction from the theoretical point of view, because [it] favors uniform motion". To this PLANCK replied: certain reference systems must have a priority, for example, the principle of relativity only applies in inertial systems and one can therefore "just as well be of the opposite opinion and see precisely in the preference for uniform motion a particularly important and valuable feature." [254]

Unintentionally, LI pays attention to the profound, seemingly contradictory natural philosophy of two famous physicists, for some laws of nature can be recognised by the independence of their equations from the respective coordinate system, others by examining them in excellent coordinate systems.

The ideas of Fig. 2.1 in a little more detail. For the classical special theory of relativity apply:

(a) the principle of the constancy of the speed of light. (alternatively: the speed of light is independent of the speed of the light source)

(b) the principle of relativity.

From this follows the Lorentz invariance of the laws of nature and the representability of these laws in Minkowski space. For the Lorentz interpretation (LI) of special relativity, the following applies: Physical objects change when they are moved. The same applies to the measuring instruments used to investigate such physical effects. If no distinction is made between the two effects and only the measurement results (adjusted for the usual measurement errors) are considered, the classical special theory of relativity applies, i.e. its formulae and principles.

Examples:

The measurement result for a moving rod measured with a moving meter stick does not change compared to a rod at rest. The Lorentz contraction only becomes apparent when the rod and the meter stick are moved relative to each other. The same applies to oscillations measured with a moving or stationary clock (time dilation). Since only the relative velocity of the measured object and the measuring instrument is included in the formula for the measurement result, there is no absolute velocity for the classical special theory of relativity, only a relative velocity. For LI, there are different concepts of velocity, although not that of an 'absolute' velocity through 'empty' space. The LI investigates what verifiable physical consequences this might have, in particular it does not reject such a thing as wrong from the outset.

The Maxwellian equations are Lorentz invariant. Moving electric charges have a magnetic field, but measuring instruments that move with them (e. g. a magnetic needle) cannot measure it. For the classical special theory of relativity, this means that electric and magnetic fields do not have an absolute character, for the LI only that this absolute character does not enter into the formulae for calculating the measurement results. Again, the LI asks whether physically verifiable or philosophically relevant conclusions could result from this in any other way.

For the classical general theory of relativity apply:

(a) the equivalence principle

(b) the covariance principle.

Gravitational fields similar to those of Newtonian gravitation theory, which exert forces on material objects, do not exist in classical general relativity. What takes their place are the pseudo-Riemannian spaces of index 1 (i.e. special curved four-dimensional spaces). If gravitational forces are spoken of often enough in the Einstein interpretation (EI), then only in analogy to the Newtonian gravitational fields.

Fig. 2.1 (opposite) LI and EI

The classical special and general theory of relativity has been proven not to be wrong, but possibly not right either, but incomplete.

The Lorentz interpretation (LI) extends both theories of relativity to include the idea of non-measurable relativistic effects, because for the LI, the measuring instruments are also subject to relativistic effects. Thus, differences arise in the assessment of paradoxes, for the law of conservation of energy, for black holes and in the space-time philosophy.

Physical assumptions about reality:
Special Rel.: Principle of relativity, principle of constancy of speed of light,
General Rel.: principle of equivalence, principle of covariance

These principles are only confirmed by experiment to a limited extent.
The following is considered experimentally confirmed:

Minkowski space, Riemannian space have *physical* meaning.
Relativistic laws correctly predict the *measurement results.*

Additional assumption of the LI:
There are not measurable relativistic effects, because also measuring instruments are subject to relativistic changes.

Additional assumption of the EI:
What is not measurable, does not exist or is without meaning for the relativity theory.

With these different additional assumptions, the result is:

LI:
Minkowski space, Riemannian space are *new physical spaces*. They describe the total effect from change of measuring object and measuring device. Kant's space and time, Newton's gravitational forces keep their justification.

EI:
Minkowski space, Riemannian space have *philosophical* meaning. They are the spaces in which we live. Gravitational forces do not exist, but only a curved space-time. Kant and Newton are in question.

For the Lorentz interpretation (LI) of general relativity applies (quite analogously to the LI of special relativity): Physical objects change in gravitational fields, as do the measuring instruments with which the properties of these objects are examined. If one does not ask about these two effects in detail, but only considers the overall effect, the formulae of the classical general theory of relativity apply. The measurement results (adjusted for the usual measurement errors) are what the general relativistic formulae, derived from the above principles, predict. The first difference: Riemannian spaces are for the LI mathematical spaces for the description of physical properties of gravitational fields and do not take their place.

Examples.

The comparison of lengths and periodic processes with a meter stick and a clock at the same place shows no difference to the measurement in gravity-free space. But e .g. the comparison of clocks inside and outside gravitational fields by clock transport shows that clocks in gravitational fields go slower (gravitational redshift). The classical general theory of relativity says that none of the clocks runs absolutely slower, but they move on different paths in the four-dimensional curved space-time, for which one clock needs more seconds than the other.

For classical general relativity, the formulas of the Schwarzschild metric (SM) and the Robertson-Walker metric (RWM) describe Riemannian spaces. Scales, clocks and physical objects are not changed in them as long as the dimensions of the objects are small and therefore the curvature of space may be considered constant (i.e. in the absence of tidal forces). For the LI, the formulas of the SM and the RWM describe scale contraction and clock dilation in gravitational fields; for the LI, the RWM is additionally the consequence of a new simultaneity definition (called 'cosmological simultaneity' in contrast to the special-relativistic simultaneity definition of an observer outside the gravitational field).

LI considers the gravitational field as something real like electric fields, something what exists. Classical general relativity considers the Riemannian space as the reality, there are no gravitational fields as real as electric fields.

What verifiable differences result from this for the LI? These are the differences in the space-time philosophy, the differences in the solution of paradoxes, different, experimentally verifiable predictions in special relativity, e.g. in the question: Are Lorentz contractions associated with forces or not? and in general relativity, e.g. from the other properties of black holes. All considerations are in agreement with previous relativistic experiments, in particular with the astronomical observations of Genzel et al [122] or the EHT group [105.1], but which are not sufficient to distinguish between the two variants of black holes, see ch. 21. In 2023 it is going to change, see ch. 24.11.

The following chapter lists three facts that speak in favour of the LI.

3. Three reasons to (not) read this book

The Einstein interpretation (EI, synonymous with classical special and general relativity) and the Lorentz interpretation (LI, same formulae but with different philosophical and physical conclusions) differ in their proposed solutions to a number of different questions. In this chapter, one philosophical, one scientific-theoretical and one physical question are selected as examples. Before doing so, it should be reiterated that the numerous experimental confirmations and the associated relativistic formulae are valid for both interpretations. It is a widespread prejudice that only EI can be consistent with this wealth of experiments.

Discussed are:

a.) a philosophical argument on absolute simultaneity
b.) a scientific-theoretical argument on the consistency of EI
c.) a physical argument - conservation of energy in free fall.

"All science is but a refinement of
the thinking of everyday life."
EINSTEIN 1936

3.1. Philosophical simultaneity

Every physicist knows it, in every physics book it can be read: There is no (absolute) simultaneity, but only a relative simultaneity, as can be deduced from the Lorentz transformations. In contradiction to this is the statement of the natural philosopher DESSAUER [68] and others: "Simultaneity is a legal ontological concept, also in the cases where it is not measurable." For these philosophers there is an (absolute or philosophical) simultaneity, but the relativity theory is not able to measure it.

To understand (absolute) simultaneity, one must visualize some simple concepts What means past, present, future? And: What is a 'real' event in contrast to an 'unreal' event? Each of us knows what is meant by this, so a few examples will suffice. Past, present, future are a statement about time. They indicate what has been, is happening or will be one day. A family celebration, a visit from friends can be past, both have happened. Both or one of both is happening now and is present or both is about to happen and is future. The statement, "My watch says 10 o'clock" can be true, if yes then it is real. Everyone who looks at the clock agrees here and he does not say, this is unreal. Another example: An attacking animal is a real danger, an only imagined attack on the other hand is not a real danger, it is unreal. As a future event, it could become real one day.

With these concept explanations one can understand the following statements: Every present event is a real event and vice versa. Past events were once real, future events will be once real. The above examples are an explanation. What was assumed to be present there was real at the same time and vice versa. One may equate "being present" and "being real".

This is the so-called 'present argument'. It contradicts the relativity of simultaneity and says: Everything what is present (no matter where in the space) is simultaneous. 'To be present' always means also 'to be real' and vice versa. Any process in the present is at the same time also a real process. On the other hand past and future are no more or not yet, in any case no reality. Simultaneity and 'being real' are equivalent. Whether a faraway star explosion is still reality or has passed and only the consequences can be dangerous, is certain for everyone who is there - completely independently of whether we know here on earth about it or not. *That, what is reality in the universe, is to our doing at the same time.*

So there are two things: The not measurable philosophical simultaneity, which each of us knows, and the measurable relative simultaneity of the relativity theory. In contrast to the EI, the LI has no difficulties with both terms. For a detailed discussion of simultaneity see chapter 11.6.2.

3.2. Twin paradox

The twin paradox must not be confused with the experimentally verifiable and verified fact that a space traveler is younger when he returns than a twin brother who stayed behind on Earth. This is a consequence of time dilation, the twin on Earth forms an inertial system, the space traveler moves in it with different speed and direction. The paradox is: when the space traveler moves away, he rests in an inertial system in which the Earth twin moves and therefore remains younger. Then the return travel follows and again the space traveler is resting in an inertial system in which the Earth twin moves and continues to remain younger. At least at first glance, this is a clear contradiction within classical special relativity.

A frequent solution approach discusses the different motion sequence. While the twin on earth rests unchanged in an inertial frame, the space traveler is accelerated during the reversal. But the time for this transition is negligible, because the duration of the space travel can be extended arbitrarily. The acceleration part of the total time can be neglected and a convincing solution of the twin paradox is not (yet) in sight.

Does it stay that way, although the main thing, the physics (the space traveler stays younger) is correctly predicted by the EI? Chapter 10.8 contains all the main solution attempts of the EI and the one of the LI, so that the reader can decide for himself. However, even if you find a particular solution convincing, the question remains why relativity is allowed to afford different solutions to pick and choose from without being criticised. Some solutions to pick and choose from means no solution - but that is precisely the situation of a theory that lacks an additional criterion leading to a choice, *a theory that is not wrong but incomplete*. So it is not a question of refuting the classical special theory of relativity, but of extending it, if necessary, in such a way that the twin paradox is solved. The once revolutionary, now self-evident insight of classical special relativity that the space traveler stays younger is not called into question with the LI solution approach.

"Anyone who wants to refute special relativity is a donkey;
it may be different with general relativity."
Ondit from TREDER, Potsdam, ca. 1970

3.3. Conservation of energy in free fall

By now, general relativity is well confirmed experimentally and one could extend the above quote, but that does not rule out flaws in the theory. Unlike special relativity and Newtonian gravitation theory, classical general relativity does not have a universal energy conservation theorem. Let us consider a simple, illustrative example, a free-falling stone. If you let it fall, it gains kinetic energy and therefore the total energy increases and, because of $E = mc^2$, the total mass of the stone also increases. Where does the growing proportion of the total energy come from?

(3.1) $$\Delta E = m_0 c^2 \left(\frac{1}{\sqrt{1-\beta^2}} - 1 \right)$$

E: energy, m: mass, m_0: rest mass, v : velocity, h: height of the stone, $\beta = v/c$.

In classical mechanics, the increase in energy corresponds to a decrease in potential energy mgh and something similar applies in Newtonian gravitation theory. Based on the special theory of relativity, the positive ΔE of the stone must be associated with a decrease in the total mass or total energy of the earth by the same amount, because the total energy of the earth and stone must remain constant. But in general relativity there is no convincing concept of potential energy, because one must not assume that the gravitational field transfers energy (and thus mass) to the falling stone. RINDLER [299] proves this for a special case. "By slowly lowering a particle of rest mass m_0 to the horizon [of a black hole] we could extract from it all energy $m_0 c^2$. A fair question: have we really extracted this energy from the particle, using the field as a mere catalyst, or have we extracted it from the field?" His answer: no energy is extracted from the field; the field only acts as a catalyst. TOLMAN [367] discusses the limits of the energy concept of general relativity with a similar result.

What could the solution for a free-falling particle, our stone, look like? A complete solution at this point is too much to ask. Therefore, the idea of LI and some objections of EI using the simple example of a particle at rest in the gravitational field.

Considering the general relativistic formulae for free fall relative to the standard coordinate system (*r, t coordinate system*) of the SM, the following assumption of the LI is justified: The rest energy of a freely falling particle decreases, but the total energy remains constant. For the mass of a particle at rest in the field, i.e. for the rest mass of a particle in the gravitational field $m_{0,field}$ applies quantitatively:

(3.2) $$m_{0,field} = m_0 \left(1 - {}^{2GM_{star}}/_{c^2 r}\right)^{\frac{1}{2}}$$

It decreases with decreasing r and has the old value for $r \to \infty$

M_{star}: mass of the star (or Earth), G : gravitational constant, r: distance from the center of the star (or Earth).

The EI does not take this approach. Because of TOLMAN and RINDLER see above, the alternative of demanding *a decrease in the gravitational mass of the star* is also omitted.

The gravitational force in this new conception is the effort of the gravitational field to reduce the rest mass of a particle. Thus, for example, the following applies to a bouncing ball - if the ball bounces upwards, the rest mass increases according to formula (3.2) (NEWTON potential energy increases) and the speed decreases and vice versa when falling. The total energy of the particle remains constant.

A first possible objection is the gravitational redshift of light. Light has no rest mass that could decrease and therefore experiences no gravitational forces, yet the frequency and thus the total energy is said to increase or decrease in gravitational fields. For details, see ch. 21. (In fact, it can be shown that light does not change its frequency and thus its energy when it propagates in a gravitational field. Instead, it changes its speed and wavelength, as in optically dense media).

The EI argues: The freely falling body experiences no forces, it moves on a geodesics (which corresponds to the straight line in curved space), the mass and total energy remain constant, because there are no forces acting ... and in view of this, completely incomprehensibly, the kinetic energy of the freely falling body increases relative to all resting local inertial systems and one does not know where this energy is taken from. Elsewhere, classical general relativity rightly recognises the difference between the constituent and gravitational mass of a star, see formula (22.18). In this sense, it is shown, see ch. 21, that *the energy concept of LI does not contradict that of EI, but extends it.*

Finally, once again the open question: The kinetic energy and thus the total energy of a freely falling stone increases. Where does the increase come from if RINDLER [299] is right that the gravitational field does not change and therefore cannot give off any energy to the stone?

And an outlook: The two chapters 4 and 16 bring basics, facts and partial solutions to be able to define the LI more abstractly and compare it with the EI. Much of this is general knowledge for a physicist, but these two chapters show more, namely also that the relativistic formulae can be derived not only from the assumptions of the EI, but also alternatively. If you already have a basic knowledge of general relativity, take a look at Chapter 21.4.

I. The Paradoxes and the Space-Time Continuum of Special Relativity

I. The Paradoxes and the Space-Time Continuum of the Special Theory of Relativity

4. The statement of the special theory of relativity

4.1. Introduction

The normal introduction to special relativity consists of "conjuring up" the Lorentz transformations on the basis of two general principles - the constancy of the speed of light and the principle of relativity. One of our goals, however, is to derive the special theory of relativity from different principles in different ways in order to better understand the space-time theories it contains. Therefore, we start with the result, the Lorentz transformations, and show by way of example how the special theory of relativity is built upon them. This firmly outlines what is to be derived from general principles and it is easier to see whether an alternative method of deriving relativistic formulas is complete.

The minimal scope of special relativity consists of the Lorentz transformations and the relativistic - as opposed to the classical - energy and momentum conservation laws. What goes beyond that - relativistic mechanics and the covariant formulation of Maxwell's electrodynamics - is not necessary, on the one hand, for the objective of this treatise and, on the other hand, also for many practical applications of relativistic formulae.

The basics of special relativity are presented in a generally understandable, sufficiently precise manner. Non-experts are recommended to use an introduction such as [sp1] as a supplement.

4.2. The Lorentz and Galileo Transformations

The core of special relativity is formed by the so-called Lorentz transformations. All relativistic phenomena are derived with or from them. That is why the Lorentz transformations are placed at the beginning - but they are simple enough for this: they merely describe the transition between inertial systems. Inertial systems are reference or coordinate systems with the property of either being at rest or moving uniformly in a straight line. Physically, this means they move without forces and are not accelerated. The bodies in them are of course subject to forces.

Reference- or coordinate systems assign coordinates to points in mathematics, in physics the points correspond to events or positions of bodies and the coordinates are the location information x, y, z together with the associated time t. This can be seen clearly in Fig. 4.1, the coordinate system S consists of three mutually perpendicular axes, the coordinate axes x, y, z (more precisely: x-, y- and z-axis) and any number of clocks, which are at rest in S. Relative to the coordinate system S, further coordinate systems S', S'' ... can move in any direction. The coordinate axes of S' are called x'-, y'- and z'-axes, and of S'' x''-axis, etc. Furthermore, x, y, z and x', y', z'

denote the location coordinates of the point P in S and in S' respectively and the same applies to S". The times measured in S are denoted by t, the times measured in S' and S'' by t' and t''. To express exactly which coordinates the point P should have, one also writes $P(x, y, z)$ or $P(x, y, z, t)$. Correspondingly, the coordinate axes of an inertial system can also be designated with $S(x, y, z)$ or $S(x, y, z, t)$. Whether with t, t' time itself is measured or whether they are only clock times, is not determined by this. With the help of the coordinate systems, it is merely specified when this or that body is located where, e. g. which length measurement describes its position at which time.

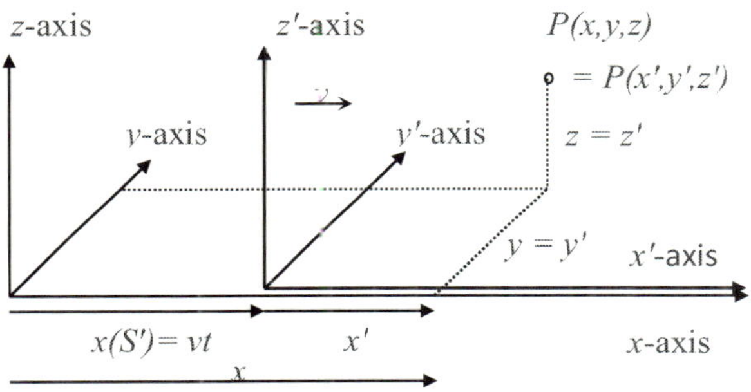

Fig. 4.1 The inertial systems $S = S(x, y, z, t)$ and $S' = S'(x', y', z', t')$
The point $P = P(x, y, z)$ resp. $P = P(x', y', z')$ has the (location) coordinates x, y, z in S and the (location-) coordinates x', y', z' in S'. S' moves with the speed v in S in the direction of the positive x-axis. The x-coordinate of the origin of S' has the value $x = vt$ in S. The x'-coordinate of the origin of S has the value $x' = -vt$ in S'. Not shown graphically are the time coordinate t in S and the time coordinate t' in S'.

How you measure lengths with scales and times with clocks is - apart from the accuracy - no different in relativity than in everyday life. One problem, however, is to set all clocks so that they show the same time "simultaneously". This is dealt with in ch. 4.3. First, we assume that the clocks are synchronised rational in accordance with the Lorentz transformations.

Different from fig. 4.1, two inertial systems may have different directions. Without limiting generality, rotate the coordinate axes of S and S' so that the x- and x' - axes coincide and point in a common direction and the y – and y' - as well as the z- and z'- axes are parallel to each other. The term "without limitation of generality" in this context means that everything that is physically important can be demonstrated on such inertial systems, fig. 4.1.

Since one can describe the same events in different inertial frames, the question arises as to how their coordinates are related to each other. If one is at rest in S, what coordinates does one have in S'? This is done by the - famous - Lorentz transformation (4.1) - (4.3). Because of their importance, these formulae, like others later, will be explained with the help of numerical examples.

(4.1)
$$x' = \gamma(x - vt)$$
$$y' = y$$
$$z' = z$$

(4.2)
$$t' = \gamma(t - vx/c^2)$$

where for γ :

(4-3)
$$\gamma = 1/\sqrt{1-v^2/c^2}$$

c is the speed of light,
v is the relative velocity of the inertial systems *S*, *S'*.

One can transform the above equations in such a way that *x, y, z* and t are on the left:

(4.4)
$$x = \gamma(x' + vt')$$
$$y = y'$$
$$z = z'$$

(4.5)
$$t = \gamma(t' + vx'/c^2)$$

and γ remains unchanged:

$$\gamma = 1/\sqrt{1-v^2/c^2}$$

This transformation had a meaning: (4.4) and (4.5) inversely describe the transition from S' to S. In this the signs before v have reversed.

The concrete meaning of the Lorentz transformations is easy to see. If the coordinates of a point *P* in the inertial system *S* are given, one can calculate the coordinates of *P* in the inertial system *S'*, which moves with the velocity *v* relative to *S*, with the help of the Lorentz transformations, i.e. from $P(x, y, z, t)$ results in $P(x', y', z', t')$. (The prerequisite for this is that the clocks in *S, S'* show the times *t, t'*, see below chapter 4.3.)

Here are some numerical examples: a) If x, y, z, t is zero, the same applies to x', y', z', t' , i.e. the origin of S' moves past the origin of *S* at the time $t = t' = 0$.

b) Assuming that all clocks in *S* always show the same time. Let *x* = 300000 km, *y*, *z* arbitrary, $t = 0$, $v = 0.5c = 0.5 \times 300000$km/s, then $\gamma = 1.15$ and for x' as well as t' results:

$$x' = \gamma(x - v \times 0)$$
$$= 1.15 \times 300000 \text{ km}$$
$$t' = \gamma(0 - 0.5 \times 300000/\text{c})$$
$$= 1.15 \times (0 - 0.5)$$
$$= -0.575 \text{ s}$$

Accordingly, if

$$x = -300\,000 \text{ km}$$

then
$$x' = -1.15 \times 300\,000 \text{ km}$$
$$t' = +0.575 \text{ s}$$

y' as well as z' always have the same values as y and z. One can see that although, $t = 0$ t' is not equal to zero and different for different x' . This means in general: For inertial systems there is a "relativity of simultaneity". If all clocks in an inertial system S show the same time, the clocks in S' show different times for different x'. The $y'$and z'coordinates have no influence on the values of t', as the formula for t' shows. For further information see below.

In order to better understand the Lorentz transformations, they are to be compared with the classical transformation formulae for the transition between inertial systems, the Galilean transformations:

$$x' = x - vt$$
$$(4.6) \qquad y' = y$$
$$z' = z$$
$$t' = t$$

You can see immediately that there is no relativity of simultaneity here. All clocks show the same time in all inertial systems at all locations, because the formula $t = t'$ applies.

Numerical example: With the above values for x, v we get with (4.6)

$$x' = -300000 \text{ km}, \qquad \text{resp.}$$
$$x' = +300000 \text{ km}.$$

For $t = 0$ is also $x' = x$. The greater length $x' = 1.15 \times 300000$ km does not appear in S contracted to $x = 300000$ km. (The multiplication sign "×" will be omitted in the following).

Why the Lorentz transformations describe physical reality better than the Galileo transformations remains to be justified. As a first step, the aim here was to clarify by way of example what Lorentz and Galileo transformations say and what distinguishes them.

4.3. The relativity of simultaneity and the Einstein convention of clock synchronisation

While the Galilean transformation does not pose any difficulties when clock times are to be measured - all clocks are brought to one place, set to the same time and then distributed to all places in the different inertial systems, after which the same time $t' = t$ prevails everywhere - the numerical examples for the Lorentz transformations have shown a different situation. If all clocks in one inertial system are synchronous and show the time t, the clocks in other inertial systems, depending on x, show different times t'. (Relativity of simultaneity).

Why can't clocks be synchronised uniformly in the same way in all inertial frames? The catchword is: moving clocks go slower, this can be justified on the basis of the Lorentz transformations (see below) and means: as soon as you have synchronised all the clocks in an inertial system at one place, they have to be moved to other places. They are moved at a certain speed for different lengths and distances, during which time they go slower than the clocks at rest, the synchronisation is lost. There is no known method - and because of the extensive attempts to find one, one may say there is none - to take these effects into account mathematically or to make them arbitrarily small. If the clocks are already running synchronously for some reason, there is no method to confirm this.

The Einstein convention of clock synchronisation uses light signals to "solve" the synchronisation problem. If the speed of light were infinite, it would suffice to emit a flash of light in all directions from one point and set all clocks, as they are hit by it, to zero. Since the speed of light has the finite value c, one must take the distance into account. To travel the distance *s*, the light signal needs the time s/c . If the light signal is emitted at the time $t = 0$, the clock in *S* clock must be set to the time $t = s/c$ when the signal arrives. In this way, all clocks in *S* can be brought to a defined synchronisation state without having to move them. Proceed in the same way for *S'*.

The prerequisite for this procedure is the unusual assumption that the speed of light has the same value c in every inertial frame, regardless of how fast they are moving relative to each other. This requirement is called the 'principle of constancy of the speed of light'. It can be derived from the Lorentz transformations:

Be (4.7) $x = ct$ or $x - ct = 0$

Multiply formula (4.5) by c and subtract from (4.4):

$$x - ct = \gamma(x' + vt') - \gamma(ct' + vx'/c)$$

After simple interim calculation:

$$0 = x' - ct'$$

or (4.8) $x' = ct'$

i.e. for a light signal emitted in *S* from $x = 0$ in the positive x-direction, formula (4.7) applies and formula (4.8) follows in *S'*.

Thus, the Einstein convention for clock synchronisation is permissible as long as the Lorentz transformations apply. However, the applicability of the Einstein convention allows two different interpretations:

a) Clocks set according to the Einstein convention are "in reality" synchronous (and it follows that the statement "clocks are synchronous" is only valid relative to the respective inertial frame).

b) According to the Einstein convention, clocks are only purposefully synchronised. Based solely on this expedient definition, the Lorentz transformations have the same form for all inertial systems. Other synchronisations would be permitted in principle, especially if this would simplify the Lorentz transformations.

Viewpoint a) is called the classical or relativistic interpretation of the Einstein convention (or: *Einstein interpretation (EI)*) and is standard in (almost) all textbooks. Viewpoint b) characterises the *Lorentz interpretation (LI)*. The Lorentz transformations and thus the same mathematical formulae apply to both. Without knowing the experimental confirmations of the Lorentz transformations, it is too early to evaluate the points of view.

Both points of view involve a different interpretation of the principle of the constancy of the speed of light:

a) Light "in reality" has the value c in all directions.

b) The speed of light is measured at c in all directions because the clocks have been synchronised accordingly; no statements can be made about its "real" values without additional considerations.

Note for later: By Einstein's convention, the so-called one-way speed of light becomes c, which is more than has been experimentally proven. Only the constancy of the so-called two-way speed of light is experimentally proven. [371].

4.4. Conclusions from the Lorentz transformations

In this chapter we repeat the previous, elementary conclusions from the Lorentz transformations and supplement them with the further, important ones of length contraction, time dilation, and the addition theorem of the velocities, which will be needed in later considerations.

1. the maximum velocity of inertial systems

In the Lorentz transformations, the factor γ becomes infinitely large when v approaches c, because the factor $\sqrt{1 - v^2/c^2}$ becomes very small. This mathematical fact, interpreted physically, means that there must not be any inertial systems in reality that move relative to each other with c or even greater speed, because no transformation formulae exist for them. The assumption of particle velocities greater than or equal to c is not contradictory. The assumption of $x = wt,\ w \geq c$ does not lead to undefined values within the Lorentz transformations and in particular the value $w = c$ is fulfilled for light quanta, only for the relative velocity of inertial systems it is not fulfilled. Nevertheless, particle velocities $w > c$ do not exist in reality, as the discussion of the law of conservation of energy will show, since infinitely large amounts of energy would be required for normal particles, e.g. electrons, to accelerate them to the speed of light.

Numerical example: If $v = 0.99c$ then $\gamma = 7.089$
if $v = 0.999c$ then $\gamma = 22{,}37$

2. relativity of simultaneity

As already discussed, formula (4.2) of the Lorentz transformations shows different times t' in S' for constant t and variable x in S. The reverse is true for t at constant t' and variable x', see formula (4.5). This mathematical fact translated into the language of physics means: If clocks in an inertial system show the same time t an observer in S' notices that the clocks in S do not run synchronously for him, they only show the same time if they are compared in S' at different times t'.

It is not possible to prefer one of the two observers, the Lorentz transformations treat both as equal. Each of the above statements is real in the same way, there is no absolute simultaneity or weaker, it is not realisable. In no case can absolute simultaneity be experienced, observers in S and in S' experience the simultaneity of their clocks in the same way, and just as they experience their own simultaneity, they experience the foreign non-simultaneity - both are experience.

The disregard for the “relativity of simultaneity” is the cause of various paradoxes, as will be shown, chapter 10.

3. constancy of the speed of light

The speed of light is measured with c in all inertial frames, which already results from the fact that the clocks in inertial frames have to be synchronised with light signals. But we have also seen that because of the Lorentz transformations from

$$x = ct$$

$$x' = ct'$$

entails. (s. formula (4.10))

As already discussed, this may be the consequence of a merely expedient definition of simultaneity. The Lorentz transformations only apply in inertial systems when their clocks run synchronously according to this definition.

4. length contraction

One sets

$$\Delta x' = x_2' - x_1'$$

$$\Delta x = x_2 - x_1$$

$\Delta x'$ is the length of a rod at rest in S', x'_2 and x'_1 are the location coordinates of its starting and end point in S'. What is the length Δx of this rod in S at a certain time t_0?

About

$$x_1' = \gamma(x_1 - ct_0)$$

$$x_2' = \gamma(x_2 - ct_0)$$

follows

$$x_1' - x_2' = \gamma(x_1 - x_2) \text{ and}$$

(4.11)

$$\Delta x' / \gamma = \Delta x$$

Since $v>0$, γ is always greater than one, so the length $\Delta x'$ of the rod in S appears as a smaller, contracted length Δx. Moving lengths are contracted.

Note: A rod length could be more conveniently abbreviated with l. Δx expresses that the rod length is the difference between x-coordinates. Instead of Δx it would be:

$$l'_{rod}/\gamma = l_{rod}$$

A rod of length l' at rest in S' has a shorter length l in S.

Numerical example: $v = 0.5c,\ l' = 1$ m

$l = 1$ m $/\gamma = 0.866$ m

(l' is measured 13.4 % shorter in S).

The essential point about length contraction is that it is determined in S at a certain point in time t_0 , this is not a uniform point in time in S'. Conversely, a rod at rest in S appears to be moving in S' and therefore shortened there. One must always see clearly: The measurements of moving rods yield measurement results that have smaller values than the measurement results of rods at rest. There is nothing contradictory about this. The measurement results describe real experiences: If I hold a moving rod at t_0 at x_1 and x_2 , I feel the beginning and end of the rod in my hands.

5. time dilation

Moving clocks go slower, their time indications are dilated compared to clocks at rest.

Let $\Delta t' = t_2' - t_1'$ be the time interval of a clock at rest in S'. In S, the clock moves with the speed v in the positive x- direction. Let it pass a clock at rest in S at the point x_1 at the time t_1 and then another clock at rest at the point x_2 at the time t_2 . Then applies:

$$t_2' = \gamma(t_2 - vx_2 / c^2)$$

$$t_1' = \gamma(t_1 - vx_1 / c^2)$$

$$x_2 - x_1 = v(t_2 - t_1)$$

$$\begin{aligned} t_2' - t_1' &= \gamma(t_2 - t_1 - v(x_2 - x_1)/c^2) \\ &= \gamma(t_2 - t_1 - v^2(t_2 - t_1)/c^2) \\ &= \gamma(t_2 - t_1)(1 - v^2/c^2) \\ &= (t_2 - t_1)\sqrt{1 - v^2/c^2} \end{aligned}$$

(4.12)
$$\Delta t' = \Delta t\sqrt{1 - v^2/c^2}$$

$$\Delta t' = \Delta t / \gamma$$

Now the factor $1/\gamma$ is on the right If the time interval Δt is passed in S, the moving clock shows the smaller time interval $\Delta t'$.

Numerical example: If $v = 0.5c$ then $\Delta t' = \Delta t\ 0.866$. If 60 minutes have passed for the clock at rest, 52 minutes have passed for the clock in motion.

Muons are certain elementary particles like electrons, but they only have an average lifetime $\Delta t'$of 2.2 10^{-6} s, after which they decay. Muons moving with $v = 0.99$ c have a range of

$$s' = vt' = 0.99\ c\ 2.2\ 10^{-6}\ s = 653\ m$$

Due to time dilation, however, moving muons live longer, so they lay back the path

$$s = 0.99\ c\ 2.2\ 10^{-6}\ s/\sqrt{(1 - 0.99^2)} = 4629\ m$$

This path has been measured and agrees with the calculations.

How does the muon experience its dilated lifetime? It says my life span is 2.2 10^{-6} s, I will not live longer, but in *S* I cover the distance *s* because *s* is moving towards me and is therefore shortened by a factor of $1/\gamma$, 2.2 10^{-6} s are enough for me to reach the end point of *s*.

The experience and measurement results of two people moving against each other contradict each other. Each person measures and experiences a longer life in the other. Nevertheless, the following applies: The measured values in the inertial systems are at the same time the quantities that are experienced, therefore they describe something real. One cannot expect a moving muon to live longer on the one hand, but on the other hand, if one stops it, to have lived a shorter time. For the EI, this is self-evident: in relation to measured values and to one's own experience, all inertial systems are equivalent. For the LI, length and time measurements are something other than space and time; experienced simultaneity something other than presence..

6. addition theorem of the velocities

We have seen that inertial systems move relative to each other with velocities lower than c. S' moves relative to S with $v = 0.9$ c, S'' relative to S' also with $v' = 0.9$ c. What is the velocity w of S'' relative to S? Certainly not 0.9 c + 0.9 c > c, because the Lorentz transformations do not exist for $v > c$.

We adopt the addition theorem for velocities from the literature [sp1], since the derivation is mathematically simple and offers nothing fundamentally new:

(4.13) $$w = \frac{w' + v}{1 + w'v/c^2}$$

Or dissolved after w' : $$w' = \frac{w - v}{1 - wv/c^2}$$

S' has velocity v relative to S
S'' has velocity w relative to S
S'' has relative to S' the velocity w'

w, w' can also be regarded as a particle velocity, namely of all particles at rest in S''.

Numerical example: A particle is at rest in S'' and has the velocity in S' $w' = 0.9$ c, S' has the same velocity relative to S, v = 0.9 c, then $w = 0.994$ c, with this velocity the particle is moving in S.

4.5. Conservation of momentum, mass and energy in the special theory of relativity

Part of understanding special relativity is understanding how statements about the conservation of momentum, mass and energy can be derived from the Lorentz transformations, although at first glance they have nothing to do with coordinate transformations. Certainly, this is not possible with the Lorentz transformations alone. It must be added as a further prerequisite that conservation laws similar to those of classical physics also apply in special relativity theory, although it is not clear exactly what they are and what form they take.

The first thing to be shown is that if the three conservation laws above are to continue to apply in special relativity, mass must be a function of velocity. Initially, we are only concerned with this qualitative proof, not with the quantitative connection.

Consider in S' the ideal inelastic collision of two particles of the same mass m and opposite velocities v and $-v$

Conservation of momentum in classical physics (see Fig. 4.2):

(4.15.) $$m\,v + m\,(-v) = M\;0 = 0$$

Conservation of Mass

(4.16) $$m + m = M$$

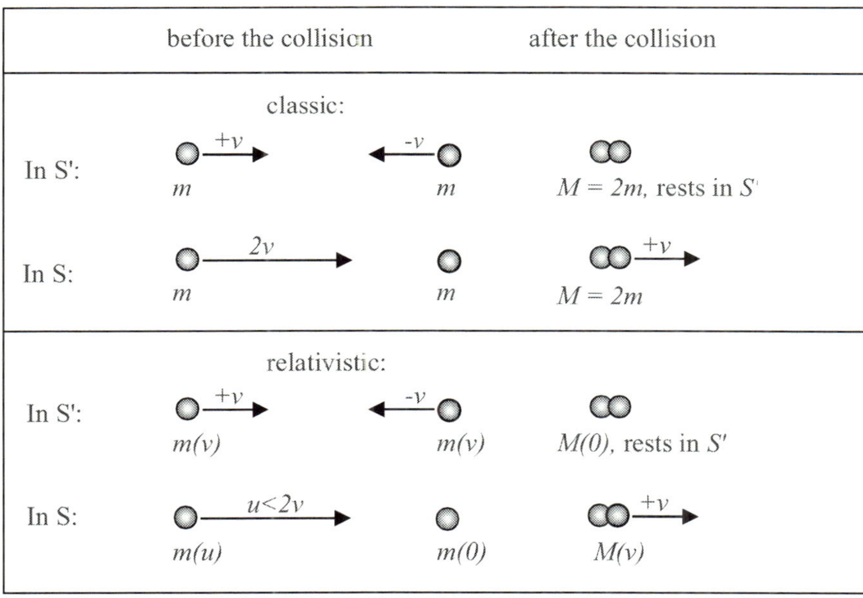

Fig. 4.2 The ideal inelastic collision of two identical particles of opposite velocity v

From S, but using the relativistic addition theorem for velocities, holds:

(4.17) $$m\frac{2v}{1+v^2/c^2} + m0 = Mv$$

$$m + m = M$$

In S', the sum of the momentum before and after the collision is zero; in other words, the total momentum is always zero and is thus conserved in S'. In S, the particle that moves in S' with $-v$ is at rest, its momentum in S is zero. Because of the conservation of momentum, the momentum of the other particle must be equal to the momentum of the particle with mass M created by the collision. Equation (4.17) would read classically:

(4.18) $$m\ 2v = (m + m)v = Mv$$

Equation (4.17) can only be valid if m has a greater mass in S than in S'. The only thing that distinguishes m relative to S' from m relative to S is its velocity. Hence the qualitative and, compared to classical physics, revolutionary consequence: mass is a function of velocity. The reason is the relativistic addition theorem, a consequence of the Lorentz transformations.

Now we are concerned with the quantitative connection and the exact form of the relativistic conservation laws. They are found using a procedure analogous to classical physics. There, one examines special cases – certain variants of elastic and inelastic collisions, but there are also completely different approaches - and thus tries to discover the laws.

We start with the least radical assumption: mass depends on velocity, it is a function of velocity, i.e.

$$m = m(v)$$

Incidentally, the law of conservation of momentum and mass applies to these masses $m(v)$, i.e. to the dynamic masses, but not to the so-called rest masses of bodies, these are the masses of bodies when they are at rest, i.e. when they have the value indicated by $m(0)$. The inelastic impact now has a somewhat more complicated form:

(4.19) $$m(v)v + m(v)(-v) = M(0)\ 0 = 0$$

(4.20) $$m(v) + m(v) = M(0)$$

In S:

(4.21) $$m(u)u = M(v)v$$

(4.22) $$u = \frac{2v}{1+\frac{v^2}{c^2}}$$

(4.23) $$m(u) + m(0) = M(v)$$

$M(0)$ is not equal to $2m(0)$, as (4.20) shows, if one observes that $m(v) > m(0)$ which follows from (4.18).

For the following calculation, it is advantageous to state the result in advance:

$$m(v) = \gamma\, m(0)$$
$$= \frac{m(0)}{\sqrt{1-v^2/c^2}}$$

γ is the same factor as in the Lorentz transformations.

One obtains from (4.22) and (4.23):

(4.24) $$M(0) = 2m(v)$$

(4.25) $$m(u)u = M(v)v$$

(4.26) $$m(u) + m(0) = M(v)$$

Eliminating $M(v)$ from (4.25) and (4.26)

(4.27) $$m(u)u = (m(u) + m(0))v$$

and substituting formula (4.22) for u, we obtain

$$m(u)\frac{2v}{1+v^2/c^2} = (m(u)+m(0))\,v$$

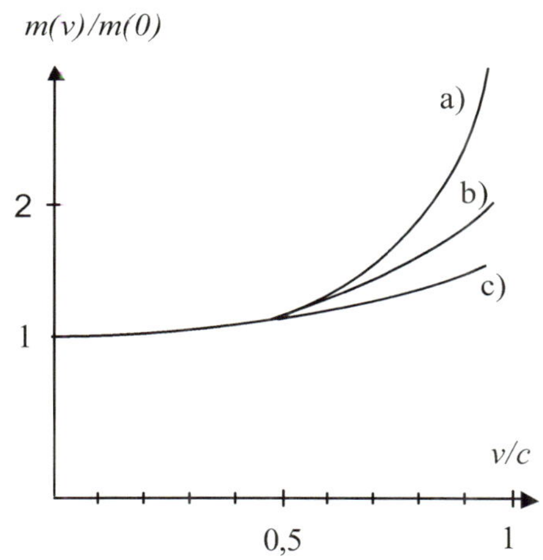

Fig. 4.3 The function $m(v)$ and approximation functions

a) $m(v)$, b) series development, formula (4.32), with 3 members and c) with 2 members. Only for high velocities, $v > 0.4c$, deviations are recognisable in the drawing.

After some interim calculations using

$$\frac{1-v^2/c^2}{1+v^2/c^2} = \sqrt{1-\left(\frac{2v/c}{1+v^2/c^2}\right)^2}$$

one obtains the result after back-transforming u:

$$m(u) = \frac{m(0)}{\sqrt{1-u^2/c^2}}$$

Using the factor γ from the Lorentz transformations, one generally has

(4.28) $$m(v) = \gamma m(0)$$

To confirm this result, one can still determine $M(v)$: At first because of (4.20) and (4.28) one gets

(4.29) $$M(0) = 2\gamma m(0)$$

Eliminating $m(u)$ from (4.21) and (4.23) and replacing $m(0)$ using (4.29) leads to:

$$M(v) = \frac{M(0)u}{2\gamma(u-v)} = \frac{M(0)}{\sqrt{1-u^2/c^2}}$$

if one sets u and γ accordingly.

One sees, assuming

(4.30) $$m(v) = \gamma m(0)$$

the law of conservation of mass and momentum applies, formulae (4.21) and (4.23) in S and formulae (4.19) and (4.20) in S', in the same form as in classical physics. One only has to replace the rest mass $m(0)$ by the dynamic mass $m(v)$.

So what is the law of conservation of energy? If one anticipates the famous result,

(4.31) $$E = mc^2$$

then the law of conservation of energy is the law of conservation of mass multiplied by the constant c^2. How to see this? Develop $m(v)$ into a series:

$$m(v) = \frac{m(0)}{\sqrt{1-v^2/c^2}} = m(0)\left(1+\frac{1}{2}\frac{v^2}{c^2}+\frac{1}{2}\frac{3}{4}\left(\frac{v^2}{c^2}\right)^2+\ldots\right)$$

This series development can be found in mathematical formulae [35], the simplest confirmation can be achieved by drawing both curves on a PC, see Fig. 4.3.

Except for members of a higher order, the following applies:

$$m(v) = m(0) + 1/2\, m(0)\, v^2/c^2$$

$1/2\, mv^2$ is the kinetic energy of a particle of mass m and velocity v in classical physics and therefore the thesis:

$$E = mc^2$$

It is plausible because, except for the constant $m(0)c^2$, it is the kinetic energy. One therefore says, $m(0)c^2$ is the rest energy and $m(v)c^2$ the total energy of a moving particle (the relativistic formula for the kinetic energy is their difference).

It is experience that shows that these formulae - derived with many hypotheses, but nevertheless not arbitrarily - really do represent the conservation laws sought in special relativity. *There is no exact derivation.*

Let's look back again: The Lorentz transformations give rise to the addition theorem (4.13) for velocities, from which the qualitative dependence of mass on velocity is compelling if one also demands the law of conservation of momentum. The quantitative correlations and finally $E = mc^2$ can be developed from special impact processes. One may therefore say: If one has derived the Lorentz transformations - from whatever principles - one has the entire special theory of relativity.

This simplifies our task of assessing the philosophical significance of relativity; it is sufficient to assess the principles from which the Lorentz transformations can be derived. As we shall see, this succeeds from two very different positions. Also the discussion of the paradoxes requires - apart from possibly minor details - only the knowledge of the Lorentz transformations, so that the above, more extensive remarks on the conservation laws need not be fully understood.

4.6. The different shape of natural laws in special relativity and classical physics

What remains right about classical physics compared to special relativity and what becomes wrong - wrong in the sense of only approximately valid?

The idea of formulating laws of nature relative to inertial systems remains correct, and the idea that laws of nature have the same form relative to inertial systems remains correct. Thus, momentum is classically defined as $m\,v$ relative to inertial systems and, accordingly, the law of conservation of momentum applies to classical physics in all inertial systems.

Wrong are the transition formulae from one inertial system to another. In classical physics, they are described by the Galilean transformations; the correct way is to use the Lorentz transformations. That is why the conservation laws of classical physics are also wrong; they have been formulated in such a way that they retain the same shape for Galileo transformations and not for Lorentz transformations and therefore only apply approximately.

The situation is different for classical electricity theory. As is well known, it is based on Maxwell's equations. The principle of relativity also applies to them, but they do not have to be rewritten, because they retain their classical form for Lorentz transformations and do not retain their form for Galileo transformations. If this were not so, Maxwell's equations would also have had to be changed (and EINSTEIN or another physicist would have become even more famous).

"Theories come and go, facts remain".
R. W. Pohl

5. The Experimental Evidence of Special Relativity

5.1. Introduction

The experimental proofs of special relativity are countless. As diverse as they are, they do not all have the same fundamental significance. We are concerned here with the experimental validation of the Lorentz transformations. Therefore, as will be shown, the proof of time dilation and length contraction and proofs of the principle of the constancy of the speed of light (which has not been experimentally proven in strict form) are of greater importance, because the Lorentz transformations are the basis of the relativistic space-time theories and the hang-up for the much-discussed paradoxes.

Of lesser importance in this context are the experimental proofs of the relativistic conservation laws, regardless of their outstanding physical significance. However, to round off our discussion of the conservation laws in Chapter 4.5, the experimental confirmation of the dependence of mass on velocity will be included.

All the experiments on the special theory of relativity have been repeated in many variants with ever-increasing precision, among them some with an intense desire to get the theory of relativity into trouble, so that one can only ask one question: How completely and without gaps has the experimental proof of special relativity succeeded? Are there essential experiments that have not yet been carried out?

What is to be regarded as essential depends on the respective theoretical standpoint, ours lies in possible variants to the space-time theory. The investigation will show that the synchronisation rules for clocks are not experimentally proven (and probably cannot be proven), which is of considerable importance for the space-time concepts. The length contraction and time dilation of the Lorentz transformations are confirmed with sufficient accuracy.

5.2. Experimental proof of time dilation

Of all the theoretical predictions of special relativity, that of time dilation is the most impressive and the one that has provoked the most opposition. Experimental proof has since been provided in a variety of ways.

a) Through the experiments of HÄFELE and KEATING [142], [143], highly sensitive clocks were flown in an aircraft around the earth from east to west and vice versa and the time dilation was measured directly. Fig. 5.1 describes the experiment. A more precise measurement by [2] confirmed the time dilation to be 1 %.

b) A series of elementary particles have only a finite lifetime, after which they decay. Such elementary particles are clocks that indicate a certain time interval. Their velocity dependence has been measured in various ways. This is also the case for pions, elementary particles with a half-life of $t_H = 1.8\ 10^{-8}$ s, i.e. out of 10,000 particles, on average 5,000 have decayed after this time. If pions move at 92% of the speed of light, they live by the factor

$$\gamma = 1/\sqrt{1-0.92^2} = 2.55$$

longer. Only after a time of $t_H \gamma$ half of them will have decayed. The other half travels a path greater than

$$s = 0.92ct_H \cdot \gamma = 12.7 \text{ m}$$

back. Without time dilation, it would only be

$$s = 0.92ct_H = 4.97 \text{ m}$$

The experiment confirms the prediction with an accuracy of 1 %. [305], [116].

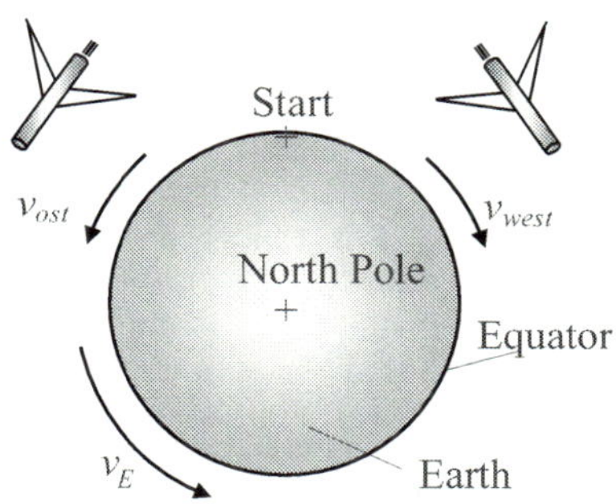

Fig. 5.1 Aircraft around the Earth.

Relative to the center of the earth, clocks in aircraft and on the surface of the earth move at different speeds. Moving clocks run slower, the slowest running clock moves in the direction of the east, $v_E + v_{Ost}$, the fastest running clock is in the direction of the west, $v_E - v_{West}$. The speed of the clock on the Earth's surface is in between. (Gravitational forces are neglected, the center of the earth is an inertial system for periods of a few days). Measurements with atomic clocks confirm the predictions.

c) The first proof of time dilation was provided by H. E. IVES and G. R. STILWELL 1938 with the help of channel beams, moving particles emitting light [169]. In this experiment, the frequency shift between stationary and moving light sources was measured. Light emitted by a moving light source has a frequency lower by a factor of $1/\gamma$ because it represents a moving clock that beats at light frequency. (This consideration applies to observations perpendicular to the direction of movement, in the direction of movement or against the direction of movement a normal Doppler effect is added).

All experiments confirm the time dilation of formula (4.12) that follows from the Lorentz transformations. Accuracies of 0.05% were achieved [136].

5.3. The dependence of mass on velocity

At the turn of the century, there were electromagnetic theories which, like the theory of relativity, predicted an increase in the mass of electrons with increasing speed. The experiments therefore carried out by KAUFMANN [181] were only later recognised as confirmation of the special theory of relativity.

In this experiment, electrons are accelerated to a certain speed in electric fields and then their dynamic mass $m(v)$ is determined by deflection in magnetic fields. If the magnetic fields are perpendicular to the direction of flight of the electrons, they are deflected onto circular paths and their radius depends on the mass because of the centrifugal forces. The quantitative relationships are not difficult to comprehend [115], we are interested in the result: In Fig. 5.2, the relativistic statement, formula (4.28), is drawn in the form

$$\frac{m(v)}{m(0)} = \frac{1}{\sqrt{1-v^2/c^2}} \tag{5.2}$$

and the determined measuring points are marked by crosses. They show good agreement. Corresponding experiments have also been carried out for other elementary particles. An experiment for protons with seven times their rest mass agrees with the theory to within 0.05% [136]. Today's particle accelerators are a visible proof of the theory's correctness.

Fig. 5.2 also clearly shows the character of c as a limiting speed. The closer v approaches the value c, the steeper this curve becomes.

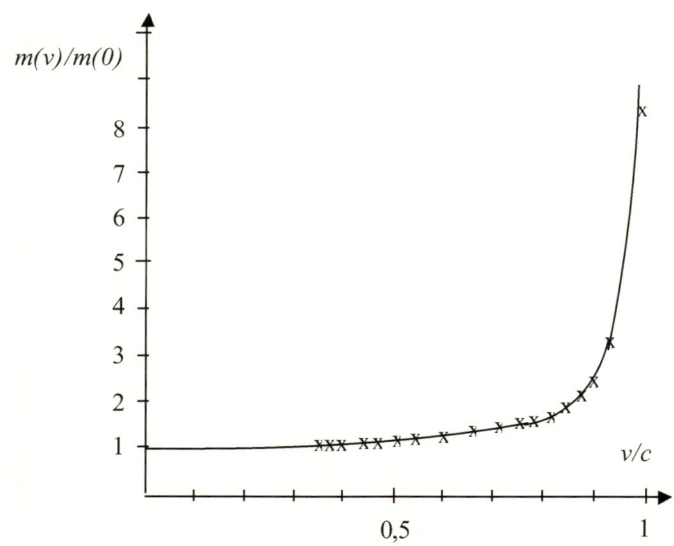

Fig. 5.2 The relativistic dependence of the mass m on the velocity v (black line) with measuring points for electrons and protons (crosses).

5.4. On the experimental proof of the constancy of the speed of light

A particularly simple arrangement for measuring the speed of light is to emit a light signal, as in Fig. 5.3, along the path $\Delta x'$, reflect it at mirror *A* and measure the time $\Delta t'$ it took for it to travel there and back along $\Delta x'$. If the measuring apparatus is moved at different speeds and $\Delta t'$ does not change, the constancy of the speed of light is proven. This is feasible, but it is not very accurate and, on the other hand, does more than one would like. The aim is only to prove the constancy of the speed of light and not to measure its magnitude.

Therefore, it is better to split the same light signal with mirrors into two directions perpendicular to each other and check whether the light signals need different times to travel back and forth along $\Delta x'$and $\Delta y'$ if you move the apparatus at different speeds *v*.

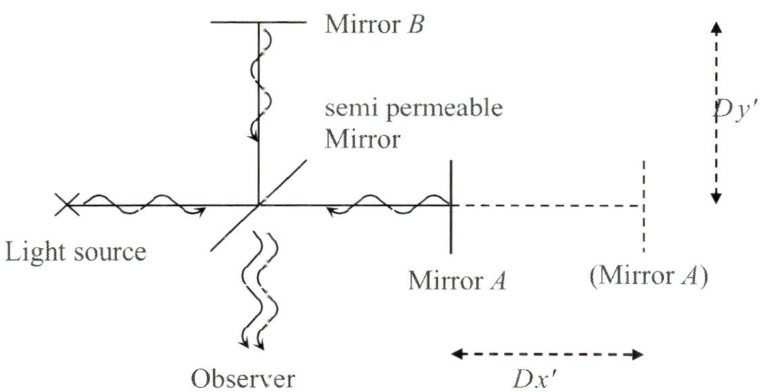

Fig. 5.3 Interferometer-experiment by MICHELSON-MORLEY and KENNEDY-THORNDIKE.

Light of a certain wavelength is split into two beams at a semi-transparent mirror, reflected at mirror *A* and *B* and then deflected again at the semi-transparent mirror towards the observer. Both light beams remain in phase when the measuring apparatus is rotated by 90°, because the interference fringes do not change. KENNEDY-THORNDIKE's experiment has unequal arms; for the sake of simplicity, the arm in the *x*-direction has twice the length $2\Delta x$ (dashed: longer arm).

Differences in the transit times of light signals can be measured very precisely. In Fig. 5.3, the wave fronts of the split light flash on return are drawn. Wave crests and troughs are in the same place, they are in phase if they return at the same time, they are not phase, wave crests and troughs are shifted against each other if outward and return times along $\Delta x'$ and $\Delta y'$ are different. If light signals have wave crests (and wave troughs) at the same point, they amplify each other, if they are shifted, light signals cancel each other out completely or partially. This can be easily observed (and is called superposition or interference).

Performing the well-known experiments of MICHELSON-MORLEY and KENNEDY-THORNDIKE, [228] and [184], using corresponding measuring equipment, which had different speeds and speed directions due to the rotation of the Earth and the movement of the Earth around the Sun, no differences in transit time could be detected and confirmed that the speed of light signals travelling back and forth is constant. The Earth's orbital velocity is about 30 km/s, and with today's equipment, changes could be detected for velocities as little as 3 cm/s.

What these experiments cannot prove is the constancy of the so-called "one-way velocity of light". To do this, you have to set up a clock at the start and end points of the route $\Delta x'$ and measure the travel times of the light signals in a single direction. However, this is only possible if the clocks have been synchronised beforehand - the special theory of relativity uses light signals for this purpose (see chapter 6.3) and thus sets the one-way light speed to c if it is not already so and therefore cannot determine the one-way velocity of light experimentally. This will be dealt with in more detail in ch. 6, where it will be discussed what consequences can be derived from this.

The real significance of the experiments of MICHELSON-MORLEY and KENNEDY-THORNDIKE lie in more than just proving the constancy of the two-way velocity of light. If one examines the experiments quantitatively in arbitrary inertial frames moving relative to the measurement set-up, one finds that they are a proof of the Lorentz contraction, the constancy of the two-way velocity of light is only a necessary consequence of this. This investigation is very important for our objectives and shall now be carried out.

5.5. The Michelson-Morley and Kennedy-Thorndike experiments to prove relativistic length contraction.

The experiments by MICHELSON-MORLEY [228] and KENNEDY-THORNDIKE [184] were intended to test how the speed of light changes when the apparatus measuring it moves. The movement was thought to be relative to the medium of propagation of the light waves. The null result of the experiments allows two interpretations. In addition to that of EINSTEIN in 1905 [96] also that of LORENTZ in 1896 and 1904 [205], [204]. They are:

Thesis I. The speed of light is independent of the propagation medium, i.e. there is *no* propagation medium (EINSTEIN, Einstein interpretation of the MICHELSON-MORLEY and KENNEDY-THORNDIKE experiments).

Thesis II. Because of the principle of relativity *it is not possible to prove the existence of* a propagation medium. Specifically, in this experiment, it is the occurring length contraction and time dilation that force the zero result (LORENTZ, Lorentz interpretation of the experiments by MICHELSON-MORLEY and KENNEDY-THORNDIKE).

It is compatible with both theories to assume a so-called excellent or distinguished inertial frame. An excellent or distinguished inertial frame is an inertial frame in which the speed of light has the same value c in all directions. In the Einsteinian interpretation, all inertial systems are excellent, in the Lorentzian interpretation only one. Whether a real propagation medium must therefore be assumed in this interpretation or whether there are other explanations for the existence of only one excellent inertial system is irrelevant here - see Chapter 6.

The experimental apparatuses are so-called interferometer devices that detect the interference of light beams of the same wavelength, see Fig. 5.3. In principle, interferometers are two perpendicular arms of equal (MICHELSON-MORLEY [228]) or unequal (Kennedy-Thorndike [184]) length at whose ends light signals from a light source are reflected by mirrors and then caused to interfere. If the light signals along the two arms have different transit times, the interference fringes would have to change when the apparatus is rotated, as explained in Chapter 5.4.

No effect is seen. Without assuming length contraction and time dilation, an effect should be detectable. The result of the quantitative analysis will be: The MICHELSON-MORLEY EXPERIMENT alone proves a length contraction, but it could also take place in the y- and z-directions. Together with the KENNEDY-THORNDIKE EXPERIMENT and the IVES-STILWELL EXPERIMENT on time dilation, it can be shown:

(5.3)
$$\Delta x' = \gamma \Delta x \\ \Delta y' = \Delta y \\ \Delta z' = \Delta z$$

Fig. 5.3 shows: Light of a certain wavelength is split into two beams at a semi-transparent mirror, reflected at mirror A and B and then deflected again at the semi-transparent mirror towards the observer. Both light beams are in phase. When the apparatus is rotated 90°, the interference fringes do not change. KENNEDY-THORNDIKE's experiment has arms of unequal length; for simplicity's sake, the arm in the x' direction has twice the length $2\Delta x'$.

The excellent inertial system is denoted by $S = S(x, y, z, t)$ and the inertial system in which the measuring apparatus rests by $S' = S'(x', y', z', t')$. Relative to S'applies:

(5.4)
$$\Delta y' = \Delta x'$$

$\Delta t'$ is the time interval for light signals back and forth along $\Delta x'$ or $\Delta y'$. The light travel time $\Delta t'$ is the same in all directions, since no interference occurs during rotation (more precisely: the *measured* values $\Delta t'$ of the light travel times are the same).

(5.5)
$$\Delta t = \Delta x \left(\frac{1}{c-v} + \frac{1}{c+v} \right)$$

(5.6)
$$\Delta t = \Delta x \frac{2\gamma^2}{c}$$

(5.7)
$$\Delta t = \frac{2\Delta y}{\sqrt{c^2 - v^2}}$$

Δt is the time interval for light signals back and forth along Δx, Δy. It is the same for both cases because no interference occurs.

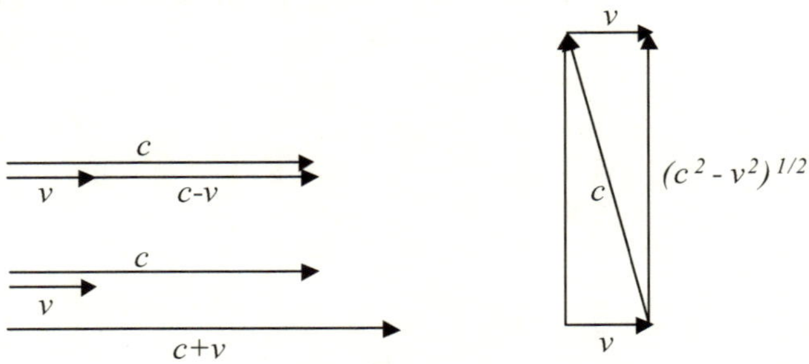

Fig. 5.4 The vectorial velocity addition for light signals in the *x*-direction, against it and perpendicular to it.

(c, *v* are measured in the same inertial frame, if c and *v* are measured in different inertial frames, formula (4.13) applies and the sum is c).

If the light moves in the positive *x*-direction, it has the velocity direction of the measuring apparatus and the magnitude $c - v$, after reflection at mirror *A* it has the opposite direction and the magnitude $c + v$. In the y' direction the resultant of both velocities is perpendicular to *v* and has the magnitude$(c^2 - v^2)^{1/2}$, see Fig. 5.4.

To explain the relationship $c^2 - v^2$. If a swimmer with speed c wants to cross a river flowing at *v* vertically, he must swim slightly upstream. Per second, he must swim as many meters upstream as he is driven downstream by the river. Its total velocity has a direction perpendicular to the bank, its magnitude is less than c, as shown in the diagram in FIG. 5.4, and because of Pythagoras' theorem it has the value

$$(c^2 - v^2)^{1/2}.$$

From (5.6) and (5.7) it follows by simple calculation:

(5.8) $$\gamma\, \Delta x = \Delta y$$

Δx is smaller than Δy , but both can be smaller than $\Delta y'$.

From the MICHELSON-MORLEY experiment, it follows because of $\Delta y' = \Delta x'$ and the same $\Delta t'$ in all directions:

$$\frac{2\Delta x'}{\Delta t'} = \frac{2\Delta y'}{\Delta t'} = \text{konstant}$$

However, the constant could be a function of the velocity *v* of the inertial system *S'* relative to *S*. The KENNEDY-THORNDIKE experiment removes this uncertainty. It proves that the constant has the same value for all velocities *v* and is therefore equal to the speed of light c:

(5.9) $$\frac{2\Delta l'}{\Delta t'} = \frac{2\Delta x'}{\Delta t'} = c$$

$\Delta l' = \Delta x'$ is valid because the arm of the interferometer in the x-direction was chosen in this way. If one assumes

$$\frac{2\Delta x'}{\Delta t'}$$

changes with the velocity v, interference would have to be observed in the experiment because, due to the unequal arm length, this change would no longer be compensated by an equally large change in the y-direction as in the MICHELSON-MORLEY experiment. The wave crests in Fig. 5.3 would have to shift against each other because the difference in travel time changes. Different velocities and velocity directions resulted in the KENNEDY-THORNDIKE experiment from measurements over longer periods of time already due to the rotation of the Earth and the rotation of the Earth around the Sun.

From the previous derivations it follows

(5.10) $$\Delta x' = \frac{c\Delta t'}{2} \quad \text{because of (5.9)}$$

(5.11) $$\Delta x' = \frac{c\Delta t}{2\gamma} \quad \text{because of (4.12), time dilation,}$$

(5.12) $$\Delta x' = \gamma \Delta x \quad \text{because of (5.6)}$$

From (5.4) and (5.12) follows

(5.13) $$\Delta y' = \gamma \Delta x$$

(5.14) $$\Delta y' = \Delta y \quad \text{because of (5.8)}$$

Formulas (5.12) and (5.14.) agree with (5.3) and we see: The experiments of MICHELSON-MORLEY, KENNEDY-THORNDIKE and IVES-STILWELL together are the experimental proof of the Lorentz contraction, as H. A. Lorentz had correctly assumed from the beginning.

What is the statement of the MICHELSON-MORLEY experiment on the principle of the constancy of the speed of light? For the outward and return velocity, the value c is measured in S′ when it has the value c in S: Viewed from S, the speed of light in S' is $c - v$, $c + v$, $(c^2 - v^2)^{1/2}$ depending on the direction - i.e. always unequal to c. However, it is measured in S':

(5.15) $$\begin{aligned} v_{Licht} &= \frac{2\Delta x'}{\Delta t'} \\ &= \frac{2\gamma\Delta x}{\Delta t'} && \text{because of (5.3)} \\ &= \frac{2\gamma^2\Delta x}{\Delta t} && \text{because of (4.12)} \\ &= c && \text{because of (5.6)} \end{aligned}$$

Correspondingly in y' direction.

What value would be measured for the one-way speed of light measured in S' if the clocks are synchronised relative to the excellent inertial frame S?

(5.16) $$\Delta t_+ = \frac{\Delta x}{v_{Licht}} = \frac{\Delta x}{c-v} = \frac{\Delta x'}{\gamma(c-v)}$$

Δt_+ is the time for a light signal to reach mirror A in S, $\Delta t'_+$ is the corresponding time in S'. Because of $\Delta t_+ = \gamma \Delta t'_+$

(5.17) $$\frac{\Delta x'}{\Delta t'_+} = (c-v)\gamma^2$$

This would be the one-way speed of light to mirror A measured in S'.

The following applies to the corresponding values $\Delta t'$ and $\Delta t'_y$ for light signals back from mirror A and along $\Delta y'$ in one of the two directions, respectively:

(5.18) $$\frac{\Delta x'}{\Delta t'} = (c+v)\gamma^2$$

(5.19) $$\frac{\Delta y'}{\Delta t'_y} = \sqrt{c^2 - v^2}\,\gamma = c$$

If the clocks in S' are synchronised relative to the excellent system S, the value c is only measured for light in the y' direction. The formulas (5.16) - (5.19) have not been experimentally verified; this would be a test of the one-way speed of light (see chapter 7.3).

Summary: The Lorentz contraction (5.3) and the time dilation (4.12) are proven by the experiments of MICHELSON-MORLEY, KENNEDY-THORNDIKE and IVES-STILWELL. They can be explained by both the Einstein and Lorentz interpretations. Because of their importance, all these measurements were carried out with special care, repeated many times and varied, so that doubts about the results do not seem justified.

6. Derivation of the Lorentz transformations

6.1. Introduction

There are a number of different, correct ways of deriving the Lorentz transformations from general assumptions [96], [326], [he2]. Elegant as they are, they are not physically compelling because there is no complete experimental evidence of such general assumptions. Therefore, we first take a different route and ask: *Can the Lorentz transformations be derived from experimentally proven assumptions without additional assumptions, or are there other, albeit improbable, coordinate transformations that are consistent with experience?*

If S is the excellent inertial system and thus light has the speed c in all directions (precondition I), the moving rod shortens (precondition II), the moving clock runs slower (precondition III). If clocks are also synchronised in S' with light, either the inverse applies (Einstein interpretation (EI),) or the inverse is only *measured* as a result of synchronisation effects (Lorentz interpretation (LI), POINCARÉ , LORENTZ and successors). Both lead to the same formulae: the Lorentz transformations.

(The Lorentz interpretation has great philosophical significance but is not treated in usual textbooks).

Answering the above question first has the following advantage: One can save oneself the trouble of checking whether the theory of relativity could get into difficulties if the derivation of the Lorentz transformations turns out to be incomplete because the assumptions made are not correct. *The principles from which the Lorentz transformations can be derived may be doubtful, but the Lorentz transformations themselves must be accepted because of the experimental confirmations, as will be shown now.*

We consider the following facts to be experimentally confirmed:

a.) moving lengths are shortened in the direction of movement,
b.) moving clocks go slower,

relative to each inertial frame when considering the *measured values*, i.e. we leave open the question of what influence the co-changing scales can have. The experimental evidence for this are the experiments of MICHELSON-MORLEY, KENNEDY-THORNDIKE and IVES-STILWELL, as examined in ch. 5, as well as their later clarifications.

6.2. Heuristic derivation of the Lorentz transformations from length contraction and time dilation.

In the system *S*:

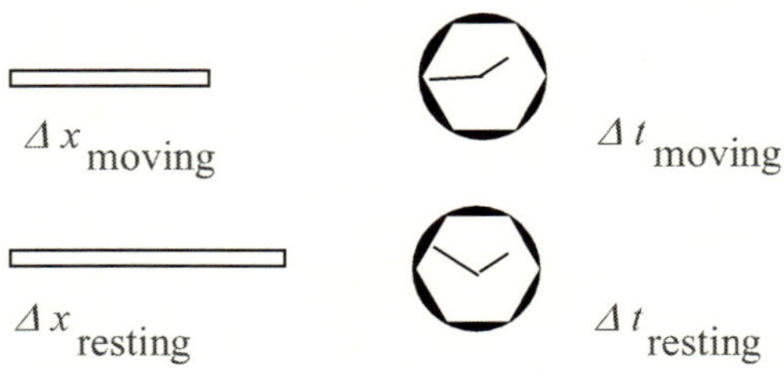

Fig. 6.1 On the derivation of the Lorentz transformations.

In accordance with [he1], we assume (Fig. 6.1):

I. $S = S(x, y, z, t)$ is an excellent inertial system, i.e. an inertial system in which the speed of light has the constant value c in all directions.

At least one such inertial system must exist. Any other assumption would mean that light has a preferred direction in space, which contradicts experience.

II. Bodies moving in $S(x, y, z, t)$ with v are shortened in the direction of movement by the factor $1/\gamma$:

(6.1)
$$\Delta x' = \gamma \Delta x$$
$$\Delta y' = \Delta y$$
$$\Delta z' = \Delta z$$

$\Delta x', \Delta y', \Delta z'$ are the lengths of the bodies when they are at rest. Their numerical value agrees with the value obtained when moving lengths are measured with moving scales (because then both are shortened in the same way. Hence the designation $\Delta x'$ instead of Δx_{ruhend}). $\Delta x, \Delta y, \Delta z$ are the real, shortened lengths of the moving bodies in $S = S(x, y, z, t)$. Shorter: The length Δx of a rod measured in *S* corresponds to the length $\Delta x'$ measured in *S'*.

III. Clocks moving with v in $S(x, y, z, t)$ run slower by a factor of γ than clocks at rest in $S(x, y, z, t)$.

(6.2)
$$\Delta t' = \Delta t/\gamma$$

$\Delta t' = \Delta t_{bewegt}$ is the proper time or the proper time interval of a clock, which is the time or time span displayed by the moving clock. Δt is the real time span measured in $S(x, y, z, t)$ with clocks at rest.

Assumptions (6.1) and (6.2) are experimentally confirmed, as shown in chapter 5.

IV. Clocks are synchronised in moving inertial frames according to the Einstein convention. Thus, clocks are synchronised "in reality" (Einstein interpretation) or only "expediently" (Lorentz interpretation). The synchronisation rule is not (completely) proven or demonstrable experimentally, see ch. 5.

To derive the Lorentz transformations, we introduce an inertial system $S' = S'(x', y', z', t')$ that moves relative to S in the positive x-direction with velocity v. The following applies to rods at rest in S':

(6.3)
$$\Delta x' = x' - x_0'$$
$$\Delta y' = y' - y_0'$$
$$\Delta z' = z' - z_0'$$

if they lie on the x'-, y'-, or z'-axis, x_0', y_0', z_0' are zero if the rod beginnings lie at the origin of S'.

If S and S' coincide at the time $t = t' = 0$, then also applies:

(6.4)
$$\Delta x = x - x_0$$
$$\Delta y = y - y_0$$
$$\Delta z = z - z_0$$

Now let x_0, y_0, z_0 be the position of the origin of S' in S. For it holds

(6.5)
$$x_0 = vt$$
$$y_0 = y_0' = 0$$
$$z_0 = z_0' = 0$$

$x_0 = vt$ expresses that the coordinate origin of S' in S moves with v in x-direction.

From (6.1) follows with (6.3), (6.4) and (6.5)

(6.6)
$$1/\gamma\left(x' - x_0'\right) = x - x_0$$
$$x' = \gamma\left(x - x_0\right)$$
$$x' = \gamma\left(x - vt\right)$$

(6.7)
$$y' - y_0' = y - y_0$$
$$y = y'$$

(6.8)
$$z' - z'_0 = z - z_0$$
$$z = z'$$

(6.6), (6.7), (6.8) are the first part of the Lorentz transformations, the time context is still missing.

The obvious idea is, since *S* represents an excellent inertial frame, to synchronise the clocks in *S′* relative to *S*. Because of (6.2) then holds:

$$t' - t'_0 = \gamma(t - t_0)$$

or $$t' = \gamma t \quad \text{if } t_0 = t'_0 = 0 \text{ is set.}$$

However, this is practically not feasible, since *S* exists according to the prerequisite, but is otherwise not known. In order to get defined states, a synchronisation rule for clocks is needed that is realisable independently of *S*. This applies to the well-known Einstein definition of simultaneity IV with the help of light signals. One can clearly see its importance here in the derivation of the Lorentz transformations: length contraction and time dilation already exist; the synchronisation rule is only an expedient ingredient. (See also the further discussion in chapter 7.)

Expressed in formulas, the Einstein definition of simultaneity is called:

(6.9) From $x = ct$ $x' = ct'$ should follow.

Because of (6.2) t' must depend on *t* and because of (6.9) it must also depend on *x*. The simplest assignment that does this is the linear relationship

(6.10) $$t' = at + bx$$

Now *a* and *b* must be determined. The most elegant way is to differentiate formula (6.10). One obtains:

(6.11) $$\frac{dt'}{dt} = a + b\frac{dx}{dt} \qquad x' = \text{const}$$

Because of (6.2) $$\frac{dt'}{dt} = \frac{\Delta t'}{\Delta t} = \frac{1}{\gamma}$$

and because of $x' =$ const one gets

$$\frac{dx}{dt} = v$$

This gives the intermediate result

(6.12) $$1/\gamma = a + bv$$

Without differentiating, one obtains (6.12) as follows:

Consider a clock at rest in *S′* at the location $x' = 0$, which shows the time $t' = 0$ at the time $t = 0$. Then

$$\frac{t' - t'_0}{t - t_0} = \frac{\Delta t'}{\Delta t}$$

$$= \frac{1}{\gamma} \qquad \text{because of (6.2)}$$

$$t'/t = \frac{1}{\gamma} \qquad \text{because of } t'_0 = t_0 = 0$$

Furthermore $x = vt$

The clock at rest at the location $x' = 0$ moves like the coordinate origin of S' with the speed v.

From (6.10) follows

$$t' = t(a + bx/t)$$
$$t'/t = (a + bx/t)$$

and thus again, but without differentiating

(6.12) $$1/\gamma = a + bv$$

Replacing x and x' in (6.6) and (6.10) by ct and ct', respectively, which is allowed because of (6.9), and eliminating t', we get

(6.13) $$\gamma(1 - v/c) = a + bc$$

Together with (6.12), a and b can be determined:

(6.14) $$a = v$$

(6.15) $$b = -\gamma v/c^2$$

(6.14) and (6.15) can be checked by substitution. Thus, one has because of (6.10)

(6.16) $$t' = \gamma(t - vx/c^2)$$

the time dependence we are looking for.

Formulas (6.6), (6.7), (6.8) and (6.16) are the Lorentz transformations we wanted to get.

The differences in the Einstein and Lorentz interpretations lie in the interpretation of the relationship (6.9).

6.3. Alternative derivation of the Lorentz transformations with the help of the reciprocity principle.

In addition to the Einstein synchronisation rule (6.9), there is another that applies: the reciprocity principle, [24]. It says: Synchronise the clocks in S' in such a way that not only for $x' = 0$ $x = vt$ is valid, but also the other way round: for $x = 0$ $x' = -vt'$ is valid, too. This aligns S' with S, now the coordinate origin of S in S' has the velocity $-v$, if S' in S has the velocity $+v$. It is astonishing that this actually insignificant assumption proves to be equivalent to Einstein's synchronisation rule.

$x = 0$ and $x' = -vt'$ used in (6.6), 2nd formula, and in (6.10) lead to:

$$t' = \gamma t$$
$$t' = at$$

so $$a = \gamma$$

On the other hand, for:

$$x' = 0$$
$$x = vt$$

and after (6.2)

$$\Delta t = \gamma \Delta t'$$
$$t - t_0 = \gamma(t' - t'_0)$$
$$t = \gamma t'$$

Inserted in (6.10):

$$t/\gamma = \gamma t + bvt$$
$$b = (^1/_\gamma - \gamma)/v$$
$$b = -\gamma v/c^2$$

With this, one knows the constants *a*, *b* of the formula (6. 10) and the Lorentz transformations are derived. In the Einstein interpretation of the reciprocity principle, it is rightly pointed out that the coordinate origins must move relative to each other with the same *v* just like any bodies and that this corresponds to experience, the Lorentz interpretation can object that velocities measured with moving and therefore slower moving clocks yield different measurement results for *v*.

Because of the so-called group property of the Lorentz transformations, the Lorentz transformations apply between any inertial systems S' and S'' because they apply between *S* and S' as well as *S* and S'' and thus *S* mathematically loses its preferential position. We will spare ourselves this calculation.

(The inertial or SELLERI transformations use an absolute simultaneity and do not have a group property [328]. Lorentz transformations in arbitrary spatial directions and not only in *x*- or *x'*-direction: see [328]).

Let us summarise: The Lorentz transformations can be derived from the experimentally secured assumptions of length contraction and time dilation relative to an excellent inertial frame, if one synchronises clocks in the other inertial frames appropriately. For this purpose, one uses the Einstein definition of simultaneity or the reciprocity principle. Both can be interpreted in the sense of LORENTZ and in the sense of EINSTEIN.

6.4. The Einstein and Lorentz Interpretation of the Lorentz Transformations

The Lorentz transformations could be derived from the two different premises:

a) All inertial systems are "in reality" equivalent. This is the Einstein interpretation of the Lorentz transformations and the viewpoint of special relativity as it is the subject of textbooks.

b) All inertial systems "appear" as equivalent. Even if one assumes absolute changes, the *measurement results* of lengths and times are - as chapter 6.2 has shown - only dependent on the relative velocities (Lorentz interpretation).

Which of the interpretations is the simpler one can be left open, since the simpler one need not also be the correct one. But we still have to generalise the two principles of relativity to all physical effects, which, however, is not very difficult:

a) All laws of nature have the same form in all inertial systems (EINSTEIN'S principle of relativity). This cannot be otherwise, because then certain inertial systems would be different from others because, for example, some laws are easier to formulate in them.

b) All laws of nature have the same form in all inertial frames, as far as they describe measurable properties. One must add: Other properties – e .g. the absolute change in length of a moving rod - are initially of no physical interest because they cannot be verified experimentally.

The Lorentzian principle of relativity is an extension of the Galilean principle of relativity. Although one moves absolutely in space, as GALILEO [328, p. 85] describes it, e. g. on a ship's voyage, the laws of fall apply here as in a state of rest. Absolute motion exists, as proven by Newton's bucket experiment. This is the Galilean principle of relativity and is deepened: Despite one's own absolute motion, one experiences the course of one's own clock and the length of one's own scales as unchanged and the relatively moving clocks and scales as slowed down or shortened. In addition to the Newtonian bucket experiment, there are other absolute relativistic effects for rotating clocks and scales, as explained in Chapter 9. In contrast to Einstein's principle of relativity, there are non-measurable properties here, so-called hidden parameters. A simple example is the speed of light. It is measured at c in all inertial systems, but of course in a system moving absolutely at v it has all values between $c + v$ and $c - v$ depending on the direction of propagation (see Fig. 5.4). (Philosophically, this interpretation should fit well with protophysics, JANICH [174], but it also works relativistically, SCHONEFELD [322]).

At this stage of our discussion, both interpretations are equally valid. We shall see: Experimentally, it is impossible to decide between them (see next chapter), but it remains open whether philosophical-scientific considerations, thought experiments or the solution of paradoxes favours one of the two. If you want to know the result already, read chapter 12.

As was shown in Chapter 4, with the derivation of the Lorentz transformations, apart from the unproblematic addition of the conservation laws, (almost) the entire special theory of relativity has been derived. This has been achieved in two ways, and in this sense, it is permissible to speak of the Einstein and Lorentz interpretations of special relativity. Should one prefer to imagine two different theories, it should be borne in mind that both should bear EINSTEIN'S name and therefore already become similar in the naming, for the physical meaning of the special theory of relativity is expressed in its formulae and it was EINSTEIN [96] who was the first to derive them completely and link them to observable physical quantities. Therefore, it seems more appropriate to leave it at the term 'interpretation'. Y. PIERSEAUX [268] goes the other way in naming it. (Lorentz theory).

6.5. Derivation from the principle of relativity and the principle of constancy of the speed of light for both interpretations

The following derivation of the Lorentz transformations follows the explanations of U. E. SCHRÖDER [326] and is intended to show how the theoretical assumptions required for this apply in an analogous manner to both interpretations of the theory of relativity and thus, in the end, hardly any differences can be seen in the elegance of their derivation. The principle of relativity and the principle of the constancy of the speed of light are applied in the respective variant. **The Lorentz interpretation is indicated in angle brackets** where it differs from the Einstein interpretation; the formal differences are minor.

1) For small velocities, the transformation formulae sought are to be transformed into the Galilean transformation

$$\begin{aligned} x' &= x - vt \\ y' &= y \\ z' &= z \end{aligned}$$

which describe the transition from S to S'. Therefore the approach

$$\begin{aligned} x' &= \gamma(v)(x - vt) \\ y' &= n(v)y \\ z' &= n(v)z \end{aligned} \tag{6.20}$$

is tried.

2) $\gamma(v)$ is an even function, i.e. $\gamma(v) = \gamma(-v)$, as can be shown: Because of the homogeneity and isotropy of the space the transformation formulae must not change if both the direction of the x-axis and the direction of the x'-axis are reversed and also the value of v becomes $-v$.

$$\begin{aligned} -x' &= \gamma(-v)(-x - (-v)\,t) \\ y' &= n(-v)y \\ z' &= n(-v)z \end{aligned} \tag{6.21}$$

Let us take one and the same point $P(x, y, z)$. It initially has the coordinates x, y, z and x', y', z'. After reversing the axes, the same point has the coordinates $-x, y, z$ and $-x', y', z'$ and v becomes $-v$, since the direction in which S' moves remains the same. Or to put it a little differently, if there is a coordinate system with (6.20), there must also be one with (6.21) because the space behaves the same in all directions.

From (6.20) and (6.21) follows the assertion:

$$\begin{aligned} \gamma(v) &= \gamma(-v) \\ n(v) &= n(-v) \end{aligned}$$

3) The inertial systems S and S' are **<or appear to be>** equal, i.e. one may exchange x and x', y and y', z and z' in (6.20), replace the value of v by $-v$ and the transformation formula must remain unchanged. Otherwise, the principle of relativity would be violated.

(6.22)
$$x = \gamma(-v)\,(x' + vt')$$
$$y = n(-v)y'$$
$$z = n(-v)z'$$

It follows:

$$yy' = n(v)^2 yy'$$
$$n(v) = 1$$

As $n(v)$ must not be negative for reasons of symmetry.

4) Now the principle of the constancy of the speed of light is applied. It says c is constant in all S **< or c is measured like this in all S>**. In formulas: From

$$x = ct$$
$$x' = ct' \text{ should follow.}$$

The two equations thus follow from (6.20) and (6.22):

(6.23)
$$x' = \gamma(v)x(1 - v/c)$$

(6.24)
$$x = \gamma(v)x'(1 + v/c)$$

If both formulas are multiplied together, the factors before xx' must be equal:

$$xx' = \gamma(v)^2(1 - v^2/c^2)xx'$$
$$\gamma(v) = \frac{1}{\sqrt{1 - v^2/c^2}}$$

5) With this one has

(6.25)
$$x' = \gamma(x - vt)$$
$$y' = y$$
$$z' = z$$

and the missing transformation for time is obtained by substituting

$$t' = x'/c$$

into the relation (6.25) for x'

$$t' = \gamma(x/c - vct/c^2)$$

and because of $x/c = t$ and $x = ct$ it follows

(6.26)
$$t' = \gamma(t - vx/c^2)$$

The formulae (6.25) and (6.26) are the Lorentz transformations, they are derived for both interpretations of the theory of relativity, the different interpretation of the coordinate values has not entered into the mathematical evaluation.

As far as the derivation - e. g. of (6.26) - should appear unsatisfactory, it is so for both interpretations. Ultimately, it is justified by its agreement with experience. By

U. E. SCHRÖDER one finds in [326] a further, more detailed derivation of the Lorentz transformations but without application of the principle of the constancy of the speed of light, which - appropriately varied - is also valid for the Lorentz interpretation and thus proves that in mathematical respect almost identical argumentations are allowed in both interpretations, both deliver the same formulae and therefore neither of the two is to be preferred to the other

6.6. Definition of EI and LI - somewhat more abstract

EI:

1) In inertial systems, all laws of nature have the same form. (Principle of relativity)

2) In inertial systems, the speed of light has the value c. (Principle of constancy of the speed of light).

The reason for the validity of the two principles is that there is neither space nor time, but only a four-dimensional space-time continuum, the so-called Minkowski space. This leads to the Lorentz invariance of the laws of nature. The two principles presuppose that all laws of nature are measurable. Only the reverse is experimentally verifiable: what is measurable from the laws of nature fulfils both postulates.

LI:

1) What can be measured by natural laws can be described in every inertial system by the same formulae. (Relativity principle of LI).

2) The clocks of an inertial system can be synchronised in such a way that the measurement of the speed of light always gives the value c. (Principle of constancy of the speed of light of LI).

It follows from the 2nd principle that the speed of light is independent of the speed of the source. The 1st principle is experimentally well confirmed and applies to inertial systems whose clocks are synchronised in accordance with the 2nd principle (Einstein clock synchronisation). The reason for the changed principles is that motion changes the measurement objects as well as the measurement instruments, but both effects together can be described in the Minkowski space - as is done for the EI. This leads to the Lorentz invariance of the laws of nature, as far as they can be measured.

So, the same formulas for EI and LI, but other physical explanations, the meaning of which will become clearer later. There are other definitions of EI and LI, there are open questions (what is movement within LI?). The next chapters lay the foundations and bring application cases that provide further clarity here. As it turns out (and has already been shown), the LI must additionally demand the

3) existence of an excellent inertial system.

7. Thought experiments to differentiate between the Einstein and Lorentz interpretations

7.1. Introduction

If one accepts a different interpretability of the Lorentz transformations, the search for an experimental separation is obvious. To say it in advance, there are no special relativistic experiments with which to distinguish between the two interpretations. (Fig. 24.6 would be a counterexample worth discussing.) Nevertheless, such considerations are useful because they make the differences between the two interpretations clear.

One cannot experimentally distinguish between them because a principle of relativity applies to both interpretations and so the same measurement results are predicted for all natural processes, although in the Lorentz interpretation only the measurement values relative to the distinguished inertial system describe the "real" value of a physical quantity. The following thought experiments refer to this difference. They therefore have an immediately recognisable null effect for the classical view of relativity theory; the same applies to the Lorentz interpretation, but it is more difficult to see.

7.2. Thought experiment on particle decay

Let elementary particles of the mean lifetime t_m fly with the same speed w' in S' in all directions. Then they decay on average after the same period of time and have travelled the path r' :

$$r' = \frac{t_m w'}{\sqrt{1 - w'^2 / c^2}} \tag{7.1}$$

whereby time dilation must be taken into account for t_m. All particles decay on average on a circle with the radius r', when they start from a point.

Due to the principle of relativity, this is true for all inertial systems and not only for an excellent one and is predicted in the same way in both interpretations. But in this particular case, the principle of relativity can be verified. For the Einstein interpretation this is trivial, formula (7.1) is valid for any inertial system S', everywhere the same experimental result is expected. For the Lorentz interpretation, this is more complicated, formula (7.1) initially only applies to the distinguished inertial frame, but S' should be any inertial frame. If the elementary particles in S' have the velocity w', then they have different velocities in the different directions in the excellent inertial system S - in the direction of motion of S' $v + w$, against $v - w$, perpendicular to it $\sqrt{(v^2 + w^2)}$.

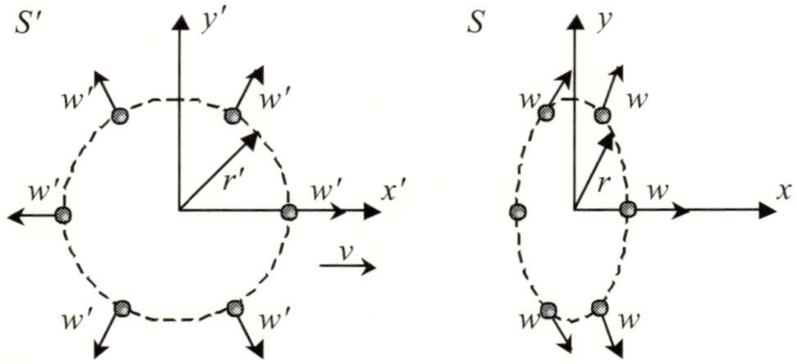

Fig. 7.1 Thought experiment on the Einstein and Lorentz interpretations.

If elementary particles,● , in S' have the velocity w' in all directions, they have different velocities and different lifetimes in S. Viewed from S, it is not readily apparent for the Lorentz interpretation whether the elementary particles in S' decay on average on a circle with the radius r'.
($v = w' = 0.866c, \gamma = 2$, the positions and velocities of the elementary particles at the time of decay are shown).

The idea of distinguishing between the two theories is, Fig. 7.1: Since relative to S the elementary particles have different lifetimes because of their different velocities, and since this is their real size, they must move differently in S' where they have the same velocity. On average, they will not decay on a circle of radius r'. This would then be a contradiction to experience.

The exact calculation shows that this idea does not work. The reason is: The higher lifetime in S even includes the higher velocity in S, but S' also moves in S, the elementary particles with the high velocity run behind the inertial system S' and vice versa. Let us calculate for particles in positive and negative x-direction: S' is an inertial system and the relativistic addition theorem applies, formula (4.13), because in S' clocks are synchronised according to Einstein's convention:

(7.2) $$w_r = \frac{v + w'}{1 + vw'/c^2}$$

(7.3) $$w_l = \frac{v - w'}{1 - vw'/c^2}$$

The special case examined here corresponds with other designations to the movement of clocks in Fig. 10.8. Particles flying to the right in the positive x-direction have the velocity in S denoted by w_r , they cover the path s_r. Correspondingly in S' w'_r and s'_r and to the left w_l, w_l', s_l, s_l'. For the time dilation applies:

$$t_m(w_r) = \frac{t_m}{\sqrt{1 - w_r{}^2 / c^2}}$$

$$t_m(w_l) = \frac{t_m}{\sqrt{1 - w_l^2 / c^2}}$$

Distance travelled in *S*:

$$s_r = w_r t_m(w_r)$$
$$s_l = w_l t_m(w_l)$$

S' moves with *v* in *S*, it covers the paths:

$$s(S'_r) = v t_m(w_r)$$
$$s(S'_l) = v t_m(w_l)$$

Viewed from *S*, the elementary particles in *S'* cover the paths:

$$s_r - s(S'_r) = (w_r - v) t_m(w_r)$$
$$s_l - s(S'_l) = (w_l - v) t_m(w_l)$$

In *S'* this is a longer path by the factor γ

$$s'_r = \gamma(s_r - s(S'_r))$$
$$s'_l = \gamma(s_l - s(S'_l))$$

s'_r, s'_l must be equal to r', which confirms the calculations:

(7.12) $$s'_r = \frac{\gamma(w_r - v) t_m}{\sqrt{1 - w_r^2 / c^2}}$$

(7.13) $$s'_l = \frac{\gamma(w_l - v) t_m}{\sqrt{1 - w_l^2 / c^2}}$$

With the formulas

(7.14) $$w_r - v = w' \frac{1 - v^2 / c^2}{1 + vw' / c^2}$$

and (7.15) $$\sqrt{1 - w_r^2 / c^2} = \frac{\left(1 - v^2 / c^2\right)\left(1 - w'^2 / c^2\right)}{1 + vw' / c^2}$$

and analogously for w_l the assertion results when inserted in s'_r and s'_l, formula (7.12), (7.13). The result is not different for the y'direction.

(Formulas (7.14) and (7.15) are mathematical identities which can be confirmed by elimination of w_r using formula (7.2)).

(In general: Conversely, one can use any natural law to synchronise clocks, e. g. with two-particle decays of resting (elementary) particles. The result is the same as for light signals).

7.3. A thought experiment on measuring the one-way speed of light using lasers.

According to the Einstein interpretation, the speed of light is constant c in all inertial frames. According to the Lorentz interpretation, it is only constant in the excellent inertial frame *S.* In inertial frames moving relative to *S*, however, it is measured as c if the clocks are synchronised in the usual way. In itself, for example in the direction of motion one has $c - v$ and $c + v$ in the opposite direction. This gives rise to the following thought experiment with two lasers to distinguish between the two interpretations.

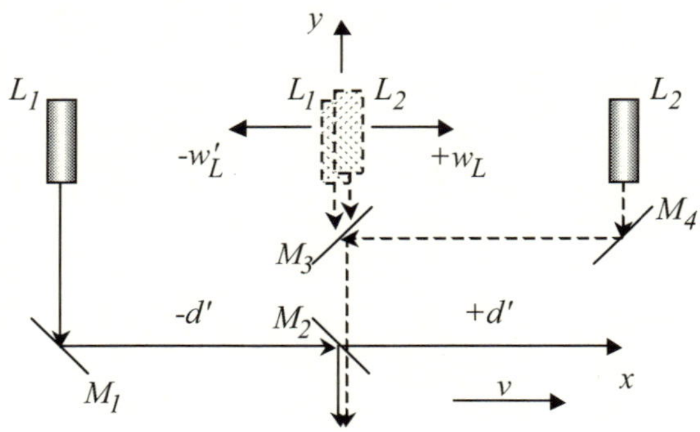

Fig. 7.2 Test of the second postulate (principle of constancy of the speed of light).
M_1 - M_4 : Mirror, L_1 , L_2 : Laser

If we look at the experimental arrangement in Fig. 7.2, light waves emitted by two lasers L_1 and L_2 via the mirrors M_1 and M_4 at a distance of $2d'$are not in phase when they meet at the mirrors M_2 and M_3, if they were in phase at M_1 and M_4, because the speed of light relative to the excellent system $S(x, y, z, t)$is $c - v$ and $c + v$. But one does not know whether the light waves or the lasers L_1 and L_2 are in phase at M_1 and M_4. Therefore, the lasers L_1 and L_2 are first moved to M_2, M_3 , brought into phase and then moved at constant speed w' to M_1 and M_4.. During this transport, the lasers change their phase in a defined way and the phase of L_1 and L_2 can now be calculated in advance at this point and any phase difference can also be determined experimentally. According to the Einstein interpretation, there is no phase difference, the exact calculation carried out in the following shows the same for the Lorentz interpretation.

(This experiment examined here has been proposed in variants in the literature [54], [387], [102] variously and wrongly for the measurement of the one-way speed of light [371]. The author himself had succumbed to this error and submitted the proposal of Fig. 7.2 for publication in 1974. It was accepted, and the author was able to withdraw his proposal just in time).

Since the result is clear, the calculation has been compressed, but is therefore somewhat more complicated.

The phases of the light waves of the lasers L_1 and L_2 are designated $\Phi_{L1}(x,t)$ and $\Phi_{L2}(x,t)$. At the time $t' = t = 0$ L_1, L_2 are at the position $x' = x = 0$ and the phases are $\Phi_{L1}(0,0) = \Phi_{L2}(0,0) = \Phi_0$. The velocity w'_L of L_1, L_2 is measured in S' whose clocks are synchronised according to the Einstein convention. Therefore w_L can be determined according to the relativistic addition theorem:

(7.20) $$w_L = \frac{v + w'_L}{1 + w'_L v / c^2}$$

The time taken to move L_1. L_2 in S' to $x' = d'$ is $\Delta t' = d'/w_L$. In the excellent system $S(x,y,z,t)$ the corresponding time intervals are Δt_{L1} and Δt_{L2}. They are different from each other, for L_2 one gets

$$w_L \Delta t_{L2} - v \Delta t_{L2} = d'/\gamma$$

$$\Delta t_{L2} = \frac{d'}{\gamma(w_L - v)}$$

$$\Delta t_{L2} = \gamma d'(1 + w'_L v/c^2)/w'_L$$

Correspondingly for L_1:

$$\Delta t_{L1} = \gamma d'(1 - w'_L v/c^2)/w'_L$$

During the shift of the lasers, they have the frequency f_{L1}, f_{L2} :

$$f_{L1} = f/\gamma(w_{L1})$$

$$f_{L2} = f/\gamma(w_{L2})$$

where f is the frequency of the lasers at rest in S. Only relative to S' do the lasers have the same speed w'_L

With these formulae, the phase of the two light waves with the wavelengths l_+ and l_- at the point $x' = 0$ can be given as a function of time:

$$\Phi_{L2}(x' = 0, t > \Delta t_{L2}) = \Phi_0 + \Delta t_{L2} 2\pi f_{L2} + \frac{d' 2\pi}{\gamma l_-} + \frac{2\pi f(t - \Delta t_{L2})}{\gamma}$$

$$\Phi_{L1}(x' = 0, t > \Delta t_{L1}) = \Phi_0 + \Delta t_{L1} 2\pi f_{L1} + \frac{d' 2\pi}{\gamma l_+} + \frac{2\pi f(t - \Delta t_{L1})}{\gamma} - \frac{(\Delta t_{L2} - \Delta t_{L1}) 2\pi f}{\gamma}$$

Because $$\Delta t_{L2} 2\pi f_{L2} = \Delta t_{L1} 2\pi f_{L1}$$

you get

(7.30) $$\Phi_{L1}(x' = 0, t) - \Phi_{L2}(x' = 0, t) = \frac{2\pi d'}{\gamma}\left(\frac{1}{l_+} - \frac{1}{l_-}\right) - \frac{d' 2 v 2\pi f}{c^2}$$

The formulas for the wavelengths are:

$$l_+ = \frac{(c-v)\gamma}{f}$$

$$l_- = \frac{(c+v)\gamma}{f}$$

Putting them into the phase difference (7.30) results in zero, as predicted.

The two thought experiments above illustrate why it is not possible to distinguish experimentally between the Einstein and Lorentz interpretations. They also show that the weaker assumption that length contraction and time dilation only exist relative to an excellent inertial frame is sufficient to explain relativistic experiments.

7.4. Statements on the Lorentz Interpretation - MITTELSTAEDT, TREDER, SEXL

As convincing as the considerations on the Lorentz interpretation may be, it will surprise some readers why, on the one hand, so many scientists plead for it, [he1], on the other hand, there is practically nothing to be found about it in the usual (and more numerous) textbooks. As examples, we will therefore refer to the exceptionally thorough works of P. MITTELSTAEDT [232], H.-J. TREDER [368], and R. SEXL [215ff]. But before, a reference to H. A. LORENTZ, NOBEL NobeL Prize winner like EINSTEIN: BERENDS [19.1] *"Hendrik Antoon Lorentz: his role in physics and society"*.

P. MITTELSTAEDT surprises in his widely distributed work "Der Zeitbegriff in der Physik" [232] by not even mentioning the Lorentz interpretation, though he is aware of it, as can be seen from the cited literature. For him, there is no reason to do so because, as experience shows, all inertial systems are equal, none is distinguished. If, however, one tries to distinguish an inertial system and defines the relative simultaneity that exists here as absolute for all other inertial systems, then - as is correctly shown if one considers only the formulae - the principle of relativity is lost, which is therefore not justified.

MITTELSTAEDT'S point of view can be quoted with EINSTEIN [90]: "Why should I distinguish the system K, to which the systems K' are physically completely equivalent, in theory from the latter by the assumption that [there is an excellent inertial system]? Such an asymmetry of the theoretical edifice, to which no asymmetry of the system of experiences corresponds, is intolerable to the theorist. It seems to me that the physical equivalence of K and K' with the assumption that [there is an excellent inertial system], though not downright incorrect from the logical point of view, is nevertheless unacceptable."

Logically "not downright incorrect", Mittelstaedt concedes that to the Lorentz interpretation - (probably) too little to even mention it.

H.-J. TREDER, whose point of view also does not deviate far from Einstein's on the other hand, examines alternative interpretations in his book "Philosophical Problems of Physical Space" and in his work "Active and Passive Generalisations of the

Lorentz-Poincaré Transformations and the Light- and Relativity Principle of EINSTEIN" [369] on the example of the theses of Holst [159] and Janossy [175]. He says there and justifies it somewhat more abstractly than can be reproduced here: "If, on the other hand, there were, according to the conception of LORENTZ (1895), a privileged (absolute) reference system ... a distinction could be made between the increase and the decrease [of an absolute velocity]." and from this the relativistic formulae could be derived alternatively. For H.-J. TREDER the LI is logically flawless.

As H.-J. TREDER - in agreement with EINSTEIN [90] - notes, the concept of an ether is also legitimate in relativity theory. By aether is meant that which physically distinguishes a certain inertial system from the others and can, for example, cause the Lorentz contraction, (s. ch. 8.7). "However, [in the classical theory of relativity] ... any kinematic reference to the aether drops out of the dynamics of particles and fields. ... This can of course also be interpreted in such a way that the interaction between "ether and matter" is regulated precisely by the Lorentz contraction, the Einstein dilation etc. in such a way that all effects dependent on the velocity relative [to the privileged inertial frame] are compensated. - This 'causal' interpretation of [special relativity], which goes back to LORENTZ, was adopted by HOLST ... and has recently been systematically performed by JANOSSY." This is nothing other than our Lorentz interpretation and for it applies: "BORN'S criticism hits the core of the problem: kinematic effects do not need a 'cause'." [369]

How is BORN's criticism to be understood? Kinematic effects are the consequences (effects) of a force-free movement. Length contraction, time dilation, mass increase $m(v)$ etc. are solely dependent on the movement, the relative velocity, and therefore cannot be associated with forces. These effects have no cause in the true sense of the word, they are explained by the new space-time structure. For BORN and TREDER an ether is superfluous.

Thus, the reply is clear: In curved space, as formed by a spherical surface, a circular motion is purely kinematic; in normal space it is the effect of a (centripetal) force. The principle of relativity has different meanings for different spaces. And the question arises: Is necessarily that space-time concept our reality for which the principle of relativity can be formulated particularly elegantly?

H.-J. Treder concedes the Lorentz interpretation, as said, logical consistency. This point of view can be well quoted with V. LAUE [380]: "An actual experimental decision between the extended Lorentzian and the theory of relativity, on the other hand, is probably not to be made at all, and if the former has nevertheless receded into the background, then this is mainly due to the fact that, as close as it comes to the theory of relativity, it nevertheless lacks the great, simple, general principle, the possession of which gives the theory of relativity something imposing even in its present (note:

Roman U. Sexl

A Test Theory of
Special Relativity

Relativity, Groups,
Particles

White dwarfs -
Black holes

Space - Time - Relativity

Gravitation and cosmology

Fig. 7.3 Roman U. SEXL, Vienna, 1939 - 1986.

Until 1986: Professor and Head of Theoretical Physics, University of Vienna
Member of the international committees for physics education as well as for general relativity and gravitation

1972-1976: Head of Department at the Institute for Space Research
of the Austrian Academy of Sciences

Known for his publications: A Test Theory of Special Relativity; White Dwarfs - Black Holes; Gravitation and Cosmology; Space - Time - Relativity; Relativity - Groups - Particles, [215], [338], [334], [335].

<table>
<tr><td colspan="2">The validity of the Lorentz transformations has been experimentally proven.</td></tr>
<tr><td colspan="2">They can be derived from:</td></tr>
<tr><td>Einstein's principle of relativity</td><td>effects relative to an excellent inertial system.</td></tr>
<tr><td colspan="2">With the important results:</td></tr>
<tr><td>Relativity of simultaneity, space and time exist as Minkowski space</td><td>Absolute simultaneity, Newtonian space and absolute time</td></tr>
<tr><td colspan="2">How should we distinguish between the two interpretations?</td></tr>
<tr><td colspan="2">Not experimentally ([215], ch. 7. (Contrary to common views.)</td></tr>
<tr><td colspan="2">Not with the help of the paradoxes (because of the same formulas, [335], ch. 10).</td></tr>
<tr><td colspan="2">The general theory of relativity can be interpreted twice in the same way ([338], [334], ch. 16)</td></tr>
<tr><td colspan="2">But: Einstein's principle of relativity is philosophically-scientifically superior ([335], ch. 11)</td></tr>
</table>

Fig. 7.4 On the space-time theory of Roman U. SEXL, Vienna, 1939-1986.

SEXL's space-time theory proves both interpretations to be experimentally equivalent - within special as well as general relativity. Therein lies the paramount importance of this theory. The Einstein interpretation is ultimately preferred not for experimental but for philosophical-scientific reasons.

1911) form, which is still very much in need of further development". - by which EINSTEIN'S formulation of the principle of relativity s meant. A ontribution to this development: TREDER, H.-J., Philosophische Probleme des physikalischen Raumes [368].

The clearest agreement of generally accepted scientists is found in the much-cited work of R. MANSOURI, R. U. SEXL "A test theory of special relativity" [215], [216], [217]. There we find our considerations on the Lorentz interpretation and it says: "In this way we arrive at the remarkable result that a theory with an absolute simultaneity is equivalent to the special theory of relativity" and further: "Thus the much discussed question can be answered approvingly whether the special theory of relativity and an ether theory which incorporates time dilation and length contraction but preserves absolute simultaneity, are empirically equivalent."

Clear words from MANSOURI and SEXL which a commentary can only diminish. However, empirically equivalent does not in itself mean equivalent. The Lorentz interpretation bears the burden of clarifying the ether concept, the Einstein interpretation the burden of paradoxes and philosophical objections, as further considerations will show.

On SEXL's mindset - of particular weight, since it is held by a widely known and recognised scientist - some further, partly anticipatory references, see Figs. 7.3 + 7.4:

The empirical equivalence of both alternatives is emphasised by SEXL several times: "... one cannot distinguish between the two theories by measuring space and time intervals" [335]. "We see, then, that here the ... Theory of EINSTEIN demands exactly the same as the ... Theory of LORENTZ. ... And here it should be noted in general right away: quite in principle there is no experimentum crucis (note: distinguishing experiment) between these two theories." (SEXL quoting P. EHRENFEST from the year 1913 in [335])

SEXL denies a possible advantage of the alternative Lorentzian theory in clarifying the paradoxes: "However, one easily convinces oneself that all observable consequences such as "clock paradox" etc. result in this version of the theory just as in the standard version. Both versions differ, to emphasise this again, only by a convention about synchronisation of clocks."

The detailed investigation of this is postponed to ch. 10.8 Twin paradox. SEXL essentially contents himself with pointing out the asymmetrical situation of both clocks and says: "The error in this argument lies in the fact that the ... coordinate system (note: see Fig. 10.11) ... is curvilinear. This corresponds to the fact that the system moving with the clock 2 (note: with the clock M_r) is not an inertial system. The use of such a system is of course ... permissible, ... only one must transform all the formulae suitably ..." and the predictions become the same for both clocks. [335]

SEXL leaves open the extent to which a modified principle of relativity, as in ch. 6.5, makes both alternatives equivalent in terms of scientific theory. But with regard to the difficulties of the Lorentzian alternative, he says: "The negative outcome of all these experiments (note: futile proof of an absolute motion through space) could be interpreted without contradiction in the ether theory by suitable modification of the basic equations, for example by incorporating the Lorentz contraction. However, every positive result of these experiments could also have found an ... explanation by the ether theory." [333]. Only Einstein's theory clearly predicts the results and is therefore considered superior. This affects the explanations in chapters 10 and 11.

Also of great importance are the considerations of SEXL (together with URBANTKE) on the different interpretations of the curved space of general relativity. In [334] they lead to (almost) an alternative theory of relativity and are vividly justified in [338]. This alternative cannot be reasonably disputed either, as will be shown in Chapter 13.

The outstanding performance of R. U. SEXL (together with R. MANSOURI and H. K. URBANTKE) lies in the proof that the Einstein and Lorentz interpretations of special and general relativity are experimentally equivalent. As chapters 6.3 and 6.5 have shown, the Lorentz interpretation of special relativity can also be derived from principles that are equivalent to the Einstein interpretation. The same applies to general relativity, as will be shown. SEXL'S considerations on general relativity also match those of KIP S. THORNE, Nobel Prize 2017, see end of ch. 23.

With the knowledge acquired or updated up to this point, the following contribution by CZERNIAWSKI is easier to understand. An important result of the contribution is the flawless proof, satisfying philosophical criteria, that the LT can be derived alternatively.

8. JAN CZERNIAWSKI – On rods, clocks, and relativity

There is a long-lasting controversy between the advocates of two different interpretations of the relativity theory, the orthodox one formulated by EINSTEIN and MINKOWSKI, and the alternative formulated by LORENTZ. It is not always clear what the main points of controversy are. The polemic often concentrates on inessential point, or results from misunderstandings with terminological origin. The arguments of both sides are often inconsistent with generally accepted views. Even the respective parties are far from agreement and unite representatives of views which are mutually contradictory in essential points. Consequently, it is not easy to clarify the matter to the extent that would enable to settle the controversy. Nevertheless, we will try.

8.1. Physical meaning of kinematical transformations

As we are going to discuss two interpretations of a theory, it is essential to make clear what is to be discussed. To simplify our task, we initially restrict our considerations to the special theory of relativity. In principle, it reduces to the Lorentz kinematics and its consequences. Consequently, one must know what it means that in the world of special relativity the Lorentz kinematics "holds". It is worth noting that such a world should not be identified with our real world, in which the special theory holds only approximately, in contrast to an idealized model.

It is usual to explain the meaning of kinematical transformations on the basis of their derivation. As a rule, such a derivation starts with stating the famous EINSTEIN's postulates, which are often, not quite adequately, regarded as the only "assumptions" of the theory. Then, a general form of linear transformations is introduced, containing some free parameters, to be defined later. Their linear form is usually regarded as obvious; only few authors are aware of the necessity to adopt some additional assumptions to justify the requirement of the linear form. Next, the restrictions resulting from the postulates are imposed on the transformation formulae, which results in defining the parameters. Finally, they are interpreted physically.

In general, almost until the end of the derivation, formal transformations concern formally introduced quantities, the physical meaning of which remains unclear. Such a way of clarifying their meaning is not very convincing. The derivation would be more instructive if all the introduced quantities were specified just from the beginning. Moreover, it is not certain that all the assumptions should be stated at once. Maybe, it is better to introduce them one by one, in correspondence with the progress of the derivation.

We start with the following assumption:

A1. The relative space of a refence system S is three-dimensional and Euclidean, i.e., isotropic, uniform, and flat.

“Isotropic” and “uniform” means here: If any local experiment is performed, its result does not depend on the spatial orientation and place of the apparatus as well as of the examined physical system. “Flat” means here roughly the same as “affine”, i.e., the notion of rectilinear motion (translation) is applicable.

Then, next assumptions follow:

A2. The relative time in S is uniform.

This is equivalent to the assumption that physical phenomena, as they are observed in S, are in a definite sense repeatable, i.e., in principle, the same phenomenon can appear at different times.

A3. There are rigid bodies and reliable clocks.

“Rigid” means here that the objects in question retain their linear sizes (in the sense of the metric of space) after change of their orientation and/or place. “Reliable” clock is a clock that runs uniformly, i.e., ascribes the same duration to equal phenomena appearing at different times. By means of them, an observer operating in S can introduce a Cartesian system of coordinates in his surrounding space, and a time coordinate.

Whereas the procedure of introducing spatial coordinates is relatively simple, it is not so in the case of the time coordinate. The problem lies in the notion of the same coordinate at different places. It can roughly be reduced to readings of identical clocks that run synchronously and rest at respective places. Consequently, a synchronization procedure is needed.

In usual presentations, the clocks are synchronized by light signals. This has the disadvantage of leaving kinematics depending on such a special phenomenon as light propagation. Moreover, this leads to apparent circularity, which results in some speculations about alleged conventional components of relativity theory. The result of such a procedure depends on the assumption of direction independence of light propagation, which in turn depends on the simultaneity defined by the procedure. As it will be shown later, actually there is no real circularity here, but it is not easy to see this, so it is better to avoid the whole problem by choosing another procedure, that does not depend on such a problematic assumption.

As an alternative procedure, it seems natural to choose synchronization by clock transport. A clock is moving uniformly and rectilinearly in such a way that it passes in sequence both clocks to be synchronized. When passing the first of them, it is synchronized with this clock. Next, when it passes the second of them, the latter is synchronized with it.

However, a problem is associated with this method. One cannot exclude a priori any influence of motion on clock rate. Of course, such an influence would affect the outcome of the procedure. To avoid this, it is advisable to adopt so-called "adiabatic" clock transport, which is equivalent to the limit of the above procedure for the velocity of transport approaching zero.

Another problem on this stage of our derivation relates to the notion of uniformity of the clock transport. One might object that it does not make sense to speak about the uniformity of motion before having defined the time coordinate. However, this is not a serious problem, since the notion in question can be defined easily by comparison of distances travelled by the clock to its indications. (cf. [8-8])

After the spacetime coordinates have been introduced in *S*, the notion of uniform and rectilinear motion is well-defined. Now we can make the next assumption:

A4. In *S* any free body either rests, or moves uniformly and rectilinearly.

This means that in *S* the principle of inertia holds, and thus, *S* can be legitimately regarded as an inertial reference system (inertial frame). In consequence of this assumption, not only an observer resting in *S* can introduce spatial coordinates and a temporal coordinate in his frame, but also any observer who rests in any reference system *S'* that moves uniformly and rectilinearly with velocity *v* relative to *S*. It is then interesting to find out how the coordinates introduced in *S* relate to coordinates introduced on the basis of the same procedure in a freely moving system.

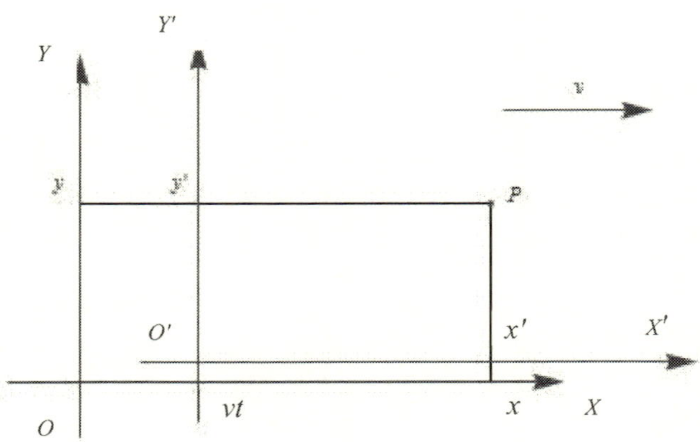

Fig. 8.1 The kinematic transformation for a uniform and rectilinear motion of the reference system

Without any significant loss of generality, we assume that the direction of motion of S' coincides with the OX axis and that for $t = 0$, the origins and the directions of all axes of both coordinate systems coincide – Fig. 8.1. If, in such a case, there was no influence of motion onto the dimensions of rigid bodies, then the spatial coordinates in S' would be related to the spacetime coordinates in S by the well-known formulae of Galilean transformations. Yet, one must consider the possibility that there is some such influence. From the symmetry conditions, it follows that it will express itself by two factors ε and δ, that depend on the absolute value of the velocity with which the body moves. This leads to the following transformation formulae:

$$\begin{aligned} x' &= \frac{1}{\varepsilon(v)}(x - vt) \\ y' &= \frac{1}{\delta(v)}y \\ z' &= \frac{1}{\delta(v)}z \end{aligned} \tag{8.1}$$

The formula for the time coordinate cannot be obtained so easily. Apart from possible influence of motion on the readings of a clock resting in S', expressed by the factor γ, the influence of motion on the outcomes of the synchronization procedure applied in S' must be taken into account – Fig. 8.2. Let us consider two clocks resting in S', and the third clock for carrying out the synchronization procedure. Let the reading of the first clock at the beginning of the procedure be t'_0. If the time interval for carrying out the procedure is T, then the first clock will read the time $t'_0 + \frac{T}{\gamma(v)}$, whereas the second and the third clock will read the time $t'_0 + \frac{T}{\gamma(u)}$, where u is the velocity of the third clock in S.

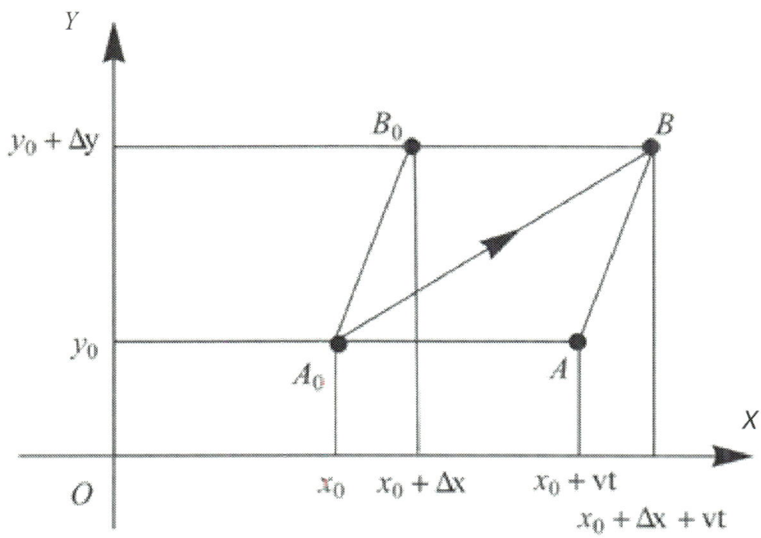

Fig. 8.2 Synchronization of clocks in the moving reference frame, observed from the resting frame

We call the limiting case of the "adiabatic" clock transport procedure ideal synchronization. Such limiting case corresponds to the situation in which u tends to v, and T to infinity, so the difference of readings of ideally synchronized clocks is:

(8.2) $$\Delta t' = \lim_{u \to v} \left(\frac{T}{\gamma(u)} - \frac{T}{\gamma(v)} \right) .$$

With $$\propto = 1/T :$$

(8.3) $$\Delta t' = \lim_{\propto \to 0} \frac{1}{\propto} \left(\frac{1}{\gamma(u)} - \frac{1}{\gamma(v)} \right) ,$$

thus:

(8.4) $$\Delta t' = \frac{d(1/\gamma)}{d\propto} (\propto = 0) = \frac{d(1/\gamma)}{d\propto} (0)$$

(the derivative of $\frac{1}{\gamma}$ with respect to $\propto$ at the point $\propto = 0$ analogous later for $u = v$), and finally:

(8.5) $$\Delta t' = -\frac{1}{\gamma^2(v)} \frac{d(1/\gamma)}{du} (v) \frac{du}{d\propto} (0) .$$

It is easy to see that $u = \sqrt{(v + w_x)^2 + {w_y}^2 + {w_z}^2}$, where $w_x = \alpha \Delta x$, $w_y = \alpha \Delta y$, $w_z = \alpha \Delta z$ and Δx, Δy, Δz – differences of spatial coordinates of the positions of synchronized clocks simultaneous in S. As a result:

(8.6) $$\frac{du}{d\propto} (0) = \Delta x ,$$

so:

(8.7) $$\Delta t' = -\frac{1}{\gamma^2(v)} \frac{d(1/\gamma)}{du} (v) \Delta x .$$

Now, let for $t = 0$, the first clock be at the origin of the coordinates system and shows the time $t' = 0$. In such a case, this clock has at any time t the spatial coordinates $(vt, 0,0)$ and shows time ${t'}_0 = \frac{1}{\gamma(v)} t$, whereas the second clock at the point with coordinates (x, y, z) shows time $t' = t'_0 + \Delta t'$. It is easy to see that, in such a case, $\Delta x = x - vt$. From this, together with (8.1) and (8.7), we obtain:

(8.8) $$\begin{aligned} x' &= \frac{1}{\varepsilon(v)} (x - vt) \\ y' &= \frac{1}{\delta(v)} y \\ z' &= \frac{1}{\delta(v)} z \\ t' &= \frac{1}{\gamma(v)} t - \frac{1}{\gamma^2(v)} \frac{d(1/\gamma)}{du} (v)(x - vt) . \end{aligned}$$

Based on what has been said up to this point, this is the most general form of kinematical transformations. For $\varepsilon(v) = \delta(v) = \gamma(v) = 1$, one obtains, as a special case, the Galilean transformations. This describes the situation when motion has no influence on dimensions of rigid bodies nor on readings of clocks. On the other hand, for $\delta(v) = 1$ and $1/\varepsilon(v) = \gamma(v) = 1/\sqrt{1 - v^2/c^2}$, one gets the Lorentz transformations. It is noted that the possible influence of motion onto dimensions of rigid bodies is the only ground for the deviations of the formulae for spatial coordinates from the Galilean transformations; consequently, it has nothing to do with simultaneity, nor with the problem of infinitely fast propagation of interactions between distant objects. Similarly, the possible influence of motion onto the rate of moving clocks is the only ground for the deviation of the general formula for the time coordinate from the simple Galilean formula. In particular, it is also the necessary condition for the effect of relativity of simultaneity, which consists in the circumstance that the clocks synchronized in a moving frame are not synchronized in a resting one (and vice versa). Nevertheless, it is not a sufficient condition for this effect, since it does not exclude defining the absolute simultaneity on the basis of some synchronization procedure different from the one applied above.

Another important property of the obtained above general form of transformations is their linearity. This is a consequence of assumptions A1 and A2, and it implies, when also A4 is taken into account, that, apart from *S*, any reference system *S'* that moves relative to *S* uniformly and rectilinearly is an inertial system, i.e., in any such system the principle of inertia holds. Then, we come to the following assumption:

A5. In any reference system that moves relative to S uniformly and rectilinearly, any law of nature has the same form as in *S*.

This means that the special (or restricted) principle of relativity holds. Since the transformation formulae contain statements about the influence of motion onto rigid bodies and clocks, they can be seen as a kind of laws of nature. This means that they must have the same form in all inertial systems. The only way to guarantee satisfying this condition is to demand that the kinematical transformations form a group [8-49]. In particular, this means that the inverse transformation for (8.8) must have the same form as the transformation itself. This follows also from A5, since (8.8) can be seen as a law of nature. Let us then apply this condition to (8.8). The formulae of the inverse transformation can be obtained easily by resolving the system of equalities (8.8) with respect to x, y, z, t, and this gives:

(8.9)
$$\begin{aligned}
x &= \varepsilon(v)\left(1 + \frac{1}{\gamma(v)}\frac{d\gamma}{du}(v)v\right)x' + \gamma(v)vt' \\
y &= \delta(v)y' \\
z &= \delta(v)z' \\
t &= \gamma(v)t' + \varepsilon(v)\frac{1}{\gamma(v)}\frac{d\gamma}{du}(v)x' .
\end{aligned}$$

On the other hand, as a consequence of the above condition, (8.9) must be equivalent to:

(8.10)
$$
\begin{aligned}
x &= \frac{1}{\varepsilon(v\prime)}(x' - v't') \\
y &= \frac{1}{\delta(v\prime)}y' \\
z &= \frac{1}{\delta(v\prime)}z' \\
t &= \frac{1}{\gamma(v\prime)}t' - \frac{1}{\gamma^2(v')}\frac{d(1/\gamma)}{du}(v')(x' - v't') .
\end{aligned}
$$

By comparison of the corresponding factors in both forms of the formulae, we obtain the following conditions:

(8.11a)
$$\frac{1}{\delta(v\prime)} = \delta(v)$$

(8.11b)
$$-\frac{1}{\varepsilon(v')}v' = \gamma(v)v$$

(8.11c)
$$\frac{1}{\gamma(v\prime)}\left(1 + \frac{1}{\gamma(v')}\frac{d(1/\gamma)}{du}(v')v'\right) = \gamma(v)$$

(8.11d)
$$\frac{1}{\varepsilon(v\prime)} = \varepsilon(v)\left(1 + \frac{1}{\gamma(v)}\frac{d\gamma}{du}(v)v\right)$$

(8.11e)
$$-\frac{1}{\gamma^2(v')}\frac{d(1/\gamma)}{du}(v') = \varepsilon(v)\frac{1}{\gamma(v)}\frac{d\gamma}{du}(v) .$$

Moreover, v' must be the same function of v as v of v', and for $v = 0$ holds also $v' = 0$.

If the fact is taken into account that ε, γ and δ are by definition positive and depend at most on the value of v (i.e., they are even functions of v), then it is easy to see that $v' = -v$ is the only function that satisfies (8.11b). If it is so, then (8.11a) and (8.11b) give:

(8.12a)
$$\delta(v) = 1$$

(8.12b)
$$\varepsilon(v) = 1/\gamma(v) .$$

With (8.12a) and (8.12b), one obtains from (8.8) the following simplified transformation formulae:

(8.13)
$$
\begin{aligned}
x' &= \gamma(v)(x - vt) \\
y' &= y
\end{aligned}
$$

$$z' = z$$
$$t' = \frac{1}{\gamma(v)}t - \frac{1}{\gamma^2(v)}\frac{d(1/\gamma)}{du}(v)(x - vt)\ .$$

As it is easy to see, what remains to be determined is only the form of the factor $\gamma(v)$ as a function of velocity.

Because of (8.12b) and the condition $v' = -v$, (8.11e) becomes an identity, and both equations (8.11c) and (8.11d) turn out to be a differential equation:

(8.14)
$$\frac{1}{\gamma(v)}\frac{1}{\gamma^2(v)-1}\frac{d\gamma(v)}{dv} = \frac{1}{v}\ ,$$

that can be integrated easily. The most general solution can be expressed in the form:

(8.15)
$$\gamma(v) = 1/\sqrt{(1-\lambda v^2)}\ ,$$

wherein λ can be any real number. Now the transformation has the form:

(8.16)
$$x' = \gamma(v)(x - vt)$$
$$y' = y$$
$$z' = z$$
$$t' = \gamma(v)(t - \lambda v x)\ .$$

Now the formulae look very similar to the Lorentz transformation. What remains to be determined is the constant λ. However, let us first consider some special cases of our solution. For $\lambda = 0$, we obtain $\gamma(v) = 1$, and the transformation turns into the Galilean transformation:

(8 17)
$$x' = x - vt$$
$$y' = y$$
$$z' = z$$
$$t' = t\ .$$

For negative λ, we can introduce a new parameter ϕ with $\tan\phi = v\sqrt{(-\lambda)}$. Then we get:

(8.18)
$$x' = x\cos\phi - \left(t/\sqrt{(-\lambda)}\right)\sin\phi$$
$$y' = y$$
$$z' = z$$

$$t' = \left(t/\sqrt{(-\lambda)}\right)\cos\phi + x\sin\phi \,.$$

Thus, for a suitable choice of the time unit, we obtain the form of a spatiotemporal rotation in the plane XOT by the angle ϕ smaller than 90 degrees. Whereas the solution (8.17) is to be expected, (8.18) may be seen as an unpleasant surprise, since nothing physical seems to correspond to it. Fortunately, the principle of relativity enables to impose a further restriction onto the form of kinematical transformations. If they must form a group, then any composition of two transformations must have the same form. Yet the composition of two rotation by angles smaller than 90 degrees is not necessarily a rotation by less than 90 degrees. Thus, the transformations of the form (8.18) are incompatible with the principle of relativity. The only transformations compatible with this principle express the Galilean transformations (8.17) and the transformations (8.16) with positive λ. The remaining indeterminacy can be eliminated in different ways. One of them is the following assumption:

A6. Any light ray propagates (in empty space) in *S* with the same constant velocity c, independently of the motion of its source.

As we see, this is the second of the famous EINSTEIN's postulates. Together with A5, this enables to derive that the light velocity has the same constant value in all inertial reference systems. It is commonly known that this theorem, which is often confused with the above second postulate, is incompatible with the Galilean kinematics. Now let us consider the transformation law for velocities, as it derives from (8.16). With a few transfigurations, one has:

(8.19)
$$u'_x = \frac{u_x - v}{1-\lambda v u_x}$$
$$u'_y = \frac{u_y}{\gamma(v)(1-\lambda v u_x)}$$
$$u'_z = \frac{u_z}{\gamma(v)(1-\lambda v u_x)} \,.$$

In the special case $u_y = u_z = 0$, the above formulae reduce to:

(8.20)
$$u' = \frac{u-v}{1-\lambda v u} \,.$$

As it is easy to see, for $u = c$ holds also $u' = c$ if and only if $\lambda = \frac{1}{c^2}$. Thus, the only kinematics compatible with A5 and A6 must go out with the transformations (8.16) with $\lambda = \frac{1}{c^2}$, i.e.:

(8.21)
$$x' = \gamma(v)(x - vt)$$
$$y' = y$$

$$z' = z$$
$$t' = \gamma(v)(t - vx/c^2)\ .$$

Of course, this is the Lorentzian transformation (Lorentz transformation).

As we have just seen, the often-overestimated role of the second postulate reduces to defining the constant c. This role can be taken over by other assumption, for instance:

A6'. Any body that moves with constant velocity v relative to S undergoes a contraction in the direction of motion by the Lorentz factor $\gamma(v) = 1/\sqrt{(1 - v^2/c^2)}$, where c is a universal constant.

This is the Lorentz-FitzGerald contraction hypothesis, the compatibility of which with the principle of relativity is sometimes wrongly denied.

A more significant role is played by the first postulate. Nevertheless, the assumptions A1-A4 are also indispensable. In particular, the linear form of the transformations cannot be justified without A1 and A2. This means that the authors who apparently dispense with them, actually take them implicitly for granted.

The principle of relativity has a further consequence. If it did not hold, then it would be possible to distinguish between inertial systems based on physical measurements. Consequently, there would be some freedom in the choice of the form of kinematical transformations, connected with the possibility of applying different procedures for introducing coordinates in different reference systems. The differences could be justified by different velocities of motions of the systems relative to the distinguished reference system. For instance, one could prefer, instead of the Lorentz transformations, the Galilean transformations, or the so-called generalized Galilean transformations (Tangherlini transformations) [8-59]:

(8.22)
$$x' = \gamma(v)(x - vt)$$
$$y' = y$$
$$z' = z$$
$$t' = \frac{1}{\gamma(v)}t\ .$$

Although the latter do not form a group, it would be no significant flaw, since there would be no justification for the requirement that they must so. Yet, if the principle of relativity holds, then the Lorentz transformations are in an unequivocal sense "truer" than all other transformations, as the ones that connect the coordinates introduced in different reference systems with the same simple procedure. In still another sense, "truer" are the Galilean transformations – cf. Sec. 8.5.

Now, let us compare the final form of transformations (8.21) with their form (8.8), before EINSTEIN's postulates have been applied. Taking solely (8.21), one might get the impression that they establish a special connection between the time coordinate

t and the spatial coordinate x, since in the latter formula the same factor γ appears, and the formulae resemble each other very much. One may make them still more similar, in which a new parameter φ is introduced, with $\tanh\varphi = \frac{v}{c}$. Now the transformations acquire the form:

$$\begin{aligned} x' &= x\cosh\varphi - ct\sinh\varphi \\ y' &= y \\ z' &= z \\ t' &= ct\cosh\varphi - x\sinh\varphi\ , \end{aligned} \tag{8.23}$$

very similar to (8.18). With a suitable choice of time unit, they turn into spatiotemporal pseudo-rotations in the plane XOT. Hence the opinion that the Lorentz kinematics introduces a deep and indissoluble connection of space and time, in sharp contrast to the Galilean kinematics. Some authors interpret this connection as "spatialization of time", others as "temporalization of space" [8-1]. After comparison with (8.8), one recognizes that the similarity of the formulae in question results from restrictions imposed by the principle of relativity onto the most general form of kinematical transformations. Thus, it is not the case that in both formulae the same factor appears, but only, that the specific values of factors in the formulae are interconnected.

The alleged "temporalization of space" might also be understood as some fundamental dependence of spatial measurements on time measurements. However, in our procedure of introducing coordinates, spatial coordinates are defined independently of time. Of course, one can apply another procedure, that binds spatial measurements with time measurements, for instance the "radar method" of distance measurements, but the resulting "temporalization of space" would be an artifact of the procedure. The same holds also for the alleged "spatialization of time", understood as some dependence of time measurements on spatial measurements, since in our procedure the time coordinate is defined solely by clocks. It seems that this time the opposite opinion is reached, because too high significance is ascribed to EINSTEIN's procedure of synchronization by light signals, which at least in some versions depends on having already defined the light velocity, and this, in turn, requires spatial measurements, apart from temporal ones.

Another fact from which speculations on the alleged indissoluble connection of space and time are drawn is the presence of the time coordinate t in the formula for the x'- coordinate, and of the x – coordinate in the formula for t'. The presence of the time coordinate in the formula for x' is indispensable in any form of kinematical transformation, in particular in Galilean transformations. On the other hand, no kinematical transformation produces any link between space and time coordinates in one and the same reference system. Their only function is "converting" of the description of physical phenomena in one reference system into their description in another.

8.2. Reality of relativistic effects

Despite the relative simplicity of special relativity, the debates on its interpretations are still current. In particular, the problem of reality of relativistic effects (time dilation, length contraction and mass increase) is taken up repeatedly. The situation is in this particular controversy strikingly confused. On the other hand, its solution is of fundamental significance for understanding the physical meaning of relativity theory, so it would be better if it had a generally accepted solution.

Unfortunately, the textbooks either do not mention this problem, or represent mutually contrary views. Sometimes this results in grotesque situations, like the one described by J. BELL. He carried out a Gedankenexperiment, in which two spaceships are set one after the other and connected with a fragile thread. Then, they begin to accelerate at the same time and in the same way. It was clear to him that the thread must break at one time, as a result of the Lorentz contraction. Yet one of his colleagues refused to agree. They appealed to the CERN theory division for arbitration and had received an unequivocal but false verdict that the thread will not break. This illustrates the significance of this problem. Thus, it must be taken up again.

At first glance, there seems to be a big difference between time dilation and mass increase. Whereas the former is mostly regarded as real, the latter is considered to be a mere product of a terminological convention. As a rule, in the discussion, even the very problem of mass increase is denied. The reason is simple: whether or not the mass increase appears, depends on the way in which the dynamical laws are formulated. It may be expressed without any use of the velocity-dependent relativistic mass. Its role may be taken over by the invariant rest mass, if the form of physical laws is adjusted accordingly [8-48]. As we will see, such a notational convention is not quite innocent, so the very problem remains.

The length contraction takes an intermediate position. The reason for its non-reality seems not as persuasive as in the case of mass increase. On the other hand, there is no comparatively convincing argument for its reality as in the case of time dilation. The latter is associated with the so-called clock, or twin, paradox. Some authors are convinced that, whereas the reciprocal effect of time dilation, as inertial observers attribute it mutually to their clocks, is synchronization dependent and thus unreal, the effect observed in the clock paradox is synchronization independent and therefore real. There is no comparable synchronization independent effect for length contraction.

What is so peculiar about the synchronization independent effects, in view of what the length contraction could not be real? There is no general agreement about this among the advocates of the non-reality thesis. Some regard the synchronization as conventional, the others stress only its relativity, i.e., dependence on the reference system. Below we begin with the former, stronger thesis.

The non-reality thesis rests on the recognition that the ratio of lengths of bodies moving relative to each other depends on the synchronization definition, since the length of a moving body is defined as the distance between simultaneous positions of its endpoints. Then, the length contraction is seen as a consequence of adoption by the observer of EINSTEIN's synchronization procedure, whereby one obtains a relative, i.e., frame dependent result. If one adopts other simultaneity defining procedure, the result of lengths comparison is different. In particular, the lengths of the moving relative to each other objects can be made equal by an appropriate synchronization.

This argument is relevant only if simultaneity is conventional in some deeper sense. Actually, it is not – cf. [8-47]. If the principle of relativity holds, there is no reasonable alternative to the standard simultaneity, as defined by EINSTEIN's procedure of synchronization by light signals. For example, the same is obtained as a result of adoption of a very natural procedure that defines simultaneity by adiabatic clock transport (see the previous section). Any other, non-equivalent simultaneity definition breaks at least one of well confirmed symmetries of space, without empirical justification. Moreover, the standard synchronization is the only definable by the sole causal structure of spacetime – see [8-43].

Furthermore: Even if the simultaneity were conventional, it would not imply the non-reality thesis. At most, it would mean that the relativistic effects contain conventional components. On the other hand: If they contained no real components, then it would be possible to eliminate them altogether for any velocity of objects by suitable choice of synchronization definition, which is not the case.

The same argument is valid, if instead of alleged conventionality of relativistic effects, their pure relativity is considered. In a sense, they are obviously relative, since they always depend on standards of length, time and mass units in a given reference system. Yet it does not cause in any way they non-reality. If they were mere results of the perspective of observer, it would be possible to find a reference system in which the relativistic effects would disappear for all velocities of objects. It can be shown that this would entail the validity of Newtonian Mechanics, in place of relativity theory – cf. [8-58]. Therefore, it is not true that reality of these effects means the end of the theory of relativity. On the contrary, the relativity theory can only be right if they are real.

Let us concentrate on the length contraction. H. REICHENBACH considered two interpretations of this effect, the Lorentz and the Einstein contraction. The Lorentz contraction was described by him as the difference between the lengths of a moving rigid body and its classical, not contracted counterpart. The Einstein contraction can be understood as the difference between the length of a resting rod as measured in a reference system moving relative to this rod and its rest length – see Fig. 8.3 and 8.4.

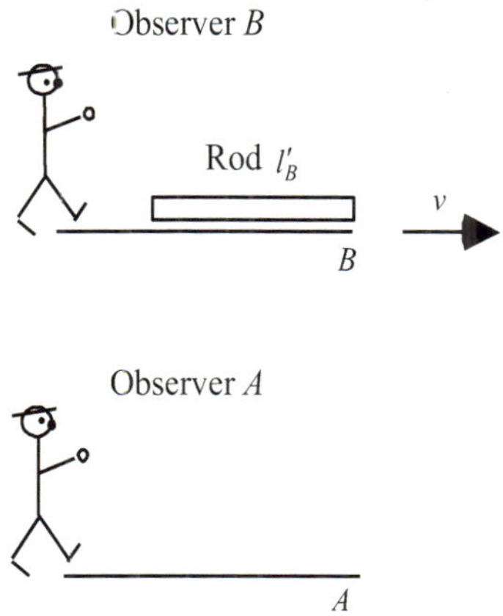

Fig. 8.3 Lorentz and Einstein contraction
Observer A: The rod l'_B is moving, therefore it is shorter than at rest (the Lorentz contraction, it is real).
Observer B: The rod l'_B is resting, so it is not contracted. Observer A regards it as shortened, because the clocks in the frame A go non-synchronously and thus the estimation of the length of the rod is wrong (the Einstein contraction, it is apparent).

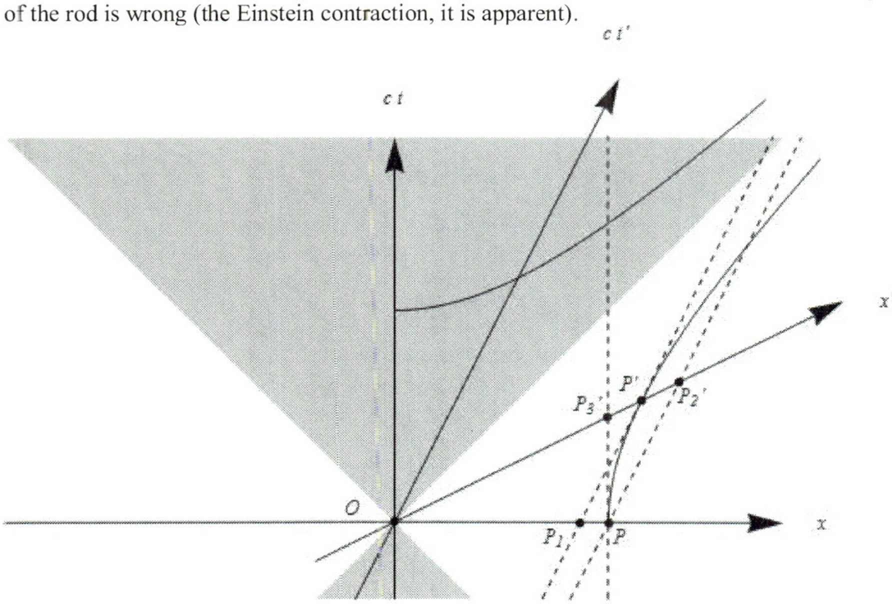

Fig. 8.4 The length contraction in Minkowski diagram
The ratio $OP_1{:}OP$ and $OP'{:}OP'_2$ represent the Lorentz contraction, $OP'_3{:}OP'$ the Einstein contraction.

Is it meaningful to consider, inside the conceptual building of relativity, the differences between a body and its (non-existing) classical counterpart? Moreover, can such a difference between two different things be reasonably called "contraction"? It is an essential property of classical body that its length is always equal to its rest length, independently of its motion. This means that also the Lorentz contraction can be understood as the difference between the length of a moving body and its rest length. If this is so, then the only difference between the contractions is the circumstance that for the Lorentz contraction, the reference system is regarded as resting and body as moving, whereas for the Einstein contraction, the opposite is true. Is this difference significant?

Indeed, it is. Although the viewpoints of observers resting in different reference systems are in a definite sense equivalent, this cannot be so understood that it is irrelevant, which of them we decide to adopt. If two bodies move relative to each other, we can regard as resting any of them, but not both at the same time. Some authors seem to have overlooked this simple fact.

The above observation sheds some light on the problem of interpretation of the rest length. Usually, it is understood as the length of the body as measured in its rest frame. This interpretation is fully correct as long as the viewpoint of the observer is adopted which rests relative to the body. However, once we adopt the point of view of an observer that moves relative to the body, such an interpretation would mean that two different viewpoints have been adopted at the same time. In addition, this would mean that the Lorentz contraction as well as the Einstein contraction should be understood as differences between the lengths of a body as measured in two different reference systems. Once again, there would be no reason to call this "contraction". If one regards a body as moving, then its rest length must be somehow interpreted as a property of the body related to the reference system that has been chosen as resting.

It is easy to find such an interpretation. We may namely see the rest length of a moving body as the length it would have if it were at rest. For a resting body, its rest length coincides with its length in the reference system regarded as resting. Therefore, what must be ascribed to the Lorentz contraction is just the difference between the length of a moving body and its rest length.

In regard to the Einstein contraction, the circumstances are more complex. Here the whole situation must be considered from the point of view of the observer that rests relative to the body in question. The length of the body is then identical with its rest length. On the other hand, its length as measured by a moving observer cannot be regarded as its true length, since all results of his measurements are faked by two causes: the Lorentz contraction of his measuring instruments and the results of the simultaneity defining procedure as applied in his frame. This is why the Einstein contraction can be legitimately regarded as apparent. It has a purely metrogenic (or metrokinematic) character.

The latter circumstance is largely blurred by misunderstanding the role of simultaneity. Some people maintain that the Einstein contraction is a consequence of relativity of simultaneity. Actually, it is not the only cause of this effect. The relativity of simultaneity causes an apparent contraction, but it is not identical with the Einstein contraction. Whereas the latter is given by the Lorentz factor $1/\sqrt{(1-v^2/c^2)}$, the former by its square $1/(1-v^2/c^2)$. The factor for the Einstein contraction is reduced, since as a result of the Lorentz contraction of moving standards of length unit, an apparent elongation enters. The Einstein contraction has a more complex origin as is usually attributed to it. It is a composition of two different, purely apparent effects, one of which is a result of the Lorentz contraction and the other a result of the simultaneity defining procedure. If there were no Lorentz contraction, the resulting effect would be stronger.

Let us imagine a body A resting in some inertial system, and another body B resting in another inertial system, that moves relative to A. From the point of view of an observer that rests relative to A (A-observer), the body B undergoes the Lorentz contraction, and the body A is not contracted, whereas, from the perspective of A-observer, an observer that rests relative to B (B-observer) observes the (apparent) Einstein contraction of the body A, and sees the body B as (apparently) non-contracted. Yet, from the viewpoint of B-observer, the body A is Lorentz-contracted, and simultaneously the body B must appear to A-observer Einstein-contracted. Thus, because of the exchange of points of view, the Lorentz contraction and the (apparent) Einstein contraction exchange their roles. Does this allow concluding that the Lorentz contraction is apparent as well?

One can get such an impression only if the points of view are mixed up. Actually, the Lorentz contraction and the Einstein contraction turn out to be just two interpretations of the same physical phenomenon. For example, the same length contraction of the body B will be interpreted as the Lorentz contraction if considered by A-observer, and as the Einstein contraction if the observation by A-observer is assessed from the point of view of B-observer. It is apparent from the point of view of B-observer, but it does not mean that it is also apparent from the point of view of A-observer. On the contrary, in order to regard the Einstein contraction as apparent, it is essential to look at the Lorentz contraction as real (see above).

It seems that we have successfully identified the main source of confusion in the debate on the reality of the relativistic length contraction. Some interpret it as the Lorentz contraction and assert its reality, others interpret it as the Einstein contraction and proclaim its non-reality. Still worse, the points of view are often mixed up, and the authors shift unconsciously from one to another. It is a fact that the length contraction, as experienced by the observer that moves relative to a body, is real for this observer, and apparent for the one that rests relative to it. For other observers it is a mixture of (real) Lorentz contraction and (apparent) Einstein contraction, thus being partly real and partly apparent. Therefore, the reality of length contraction is in a certain sense relative, i.e., frame dependent. The same is true about time dilation and mass increase.

Our solution of the problem of reality of relativist effects may make the impression that we see all the effects as an observer ascertains them as real from his point of view, and (at least partly) not real from the points of view of other observers. This would be a self-deception. There are yet effects that are apparent also from the point of view of the observer that experiences them. If, for example, a moving body approaches the observer, he sees it as elongated, whereas if the body moves away, he sees it as stronger contracted than results from the Lorentz contraction; let us call it (relativistic) Terrell effect [8-60]. An analogous effect refers to the frequency of a moving light source, and it is called relativistic Doppler effect.

Strictly speaking, the above-mentioned two effects are not fully apparent from the above adopted point of view. Both are composed of two ingredients. One of them is real, namely the Lorentz contraction and the (real) time dilation, respectively. The other is apparent and results from finiteness of the velocity of light as an information carrier. In the case of the relativistic Doppler effect, the apparent component is the well-known non-relativistic Doppler effect for the case when the observer rests and the source moves.

Once again, from the point of view of the observer that rests relative to the body, the visual effect is fully apparent, since it consists of two apparent effects: the Einstein contraction and another effect, that results from finiteness of light velocity. In the case of the relativistic Doppler effect, from the point of view of comoving observer, the effect is also fully apparent, as composed of two apparent effects. One of them is the non-relativistic Doppler effect for resting source and moving observer; the other is the apparent frequency increase as the consequence of the influence of time dilation on the moving frequency standard. It is worth noting that without the consequences of (real) length contraction and time dilation, the effects as seen by the moving observer for resting objects would be different from the ones seen by the resting observer for moving objects. The real relativistic effects restore the symmetry (reciprocity), which is indispensable if the principle of relativity is to hold.

A commonly known argument for reality of the Lorentz contraction is that it can be – at least in principle – photographed. This argument rests on a false assumption. Actually, the length contraction is not photographable, since due to finite light velocity, an additional effect enters. Only after some reformulation the argument will be right. The length contraction is real not because it can be photographed, but because what can be photographed depends on it. Were there no contraction, the visual effect would look to the observer otherwise.

The same holds for the visual effects as seen by any other observer that moves relative to him. So, the observer must take into account the real contraction to understand what he actually sees, and what see other observers at the same time. The same holds for the Doppler effect. Were there no real time dilation, we would observe the non-relativistic Doppler effect, and it would be possible to prove the absolute motion of reference systems. Thus, although the effects of length contraction and time dilation are real only from the point of view of the observer who can observe them directly, they are not only really observable, as the above-mentioned visual effects, but real in a deeper sense.

For clarifying the visual effects, only relative reality of relativistic effects is needed, i.e., their reality from the point of view of the observer. This can be understood so that one may conclude that they are lacking any absolute reality. It is not so simple. This is so only in the case when there is no privileged inertial system. The presence of such a reference system changes the situation completely. If it exists, then, for example, the length contraction of a moving body as observed from this distinguished reference system can be regarded as real in a more fundamental sense, for instance as absolutely real. On the other hand, the contractions as observed from other reference systems for bodies that rest relative to the distinguished reference system are in a definite sense absolutely apparent. Other contractions, observed from these systems, are in an absolute sense partly real and partly apparent. The same holds for other relativistic effects.

Irrespective of the question about the existence of such a privileged reference system, some effects are indeed absolutely real. One of them is the above-mentioned twin paradox. It is sometimes misunderstood as a support for the thesis that a privileged reference system exists. It is really a fact that the existence of such a reference system implies the presence of absolutely real effects, but not vice versa. We find nothing unusual in the fact that an interplay of relative quantities results in invariant (i.e., independent of the reference system), thus in a sense absolute quantities.

What is the main motivation of the non-reality thesis? It seems to be the invariance of the rest mass, the rest length and proper time. Some authors maintain that, e.g., the rest length is the actual length of a body, whereas its length, which is measured by an observer moving relative to it, is only apparent. The length contraction is then regarded as the difference between its apparent length in some reference system and its actual, velocity-independent length. Thus the view is taken that it refers only to the apparent length, and not to the actual one.

But are there reasonable grounds for regarding the rest length of a body as its actual length? In the non-relativistic physics, the length was invariant, thus there was no need for distinguishing between the rest length and the length in a reference system. In the relativity theory, the question arises what the relativistic counterpart of the non-relativistic notion of length is.

At first glance, the invariant rest length seems to be a better candidate. Yet, after some thoughts, it proves unsuitable. For a moving body, the measurement of the rest length is secondary in comparison with its length as the distance between simultaneous positions of its endpoints, which is common for the relativistic and non-relativistic physics. Therefore, there are better grounds for regarding the latter as the actual length of the body. The rest length should be seen not as the actual length, but as the length it would have if it were at rest, as we have already observed earlier.

The proper time can be treated alike. It is not appropriate to maintain that the moving clock measures the actual time between two events that mark the beginning and the end of a process in a comoving material system. Certainly, the measured proper time coincides with the time that passes between similar events if the material system is at rest. The same holds for the rest mass of a moving body, that shall be regarded as the mass a body would have if it were at rest. Otherwise, the relativistic explanation of the behavior of high-energy particles at acceleration would lose its intuitiveness: the key for it is the growing inertia, and mass is the measure of inertia. Moreover, the mass-energy equivalence would lose its generality, and in such a case, it would be restricted to the rest energy.

But how about the natural intuition that, apart from relative lengths (masses) in various reference systems, a body must have something like a true length (mass)? And that, apart from relative time intervals in various reference systems, there must be something like true time interval between two given events? Is this not a good reason for treating the proper time, rest length and rest mass in the way we have just denied?

For the reasons mentioned above, this is impossible. If there is no privileged reference frame, as it is generally assumed, then there is no "true" time (length, mass), but only relative times (lengths, masses) in various reference frames. This means that we must regard the notions in question as, by their very essence, relative, i.e., frame dependent. Therefore, we must give up the above-mentioned intuition altogether. Yet, if there is a privileged frame, then this intuition may be retained. In such a case, the roles of "true" time, length and mass may be taken over by the (relative) time, length, and mass in the privileged frame. This possibility is fulfilled in the framework of the Lorentz interpretation of the relativity theory. Cf. e.g. [8-3, 8-7, 8-10, 8-15, 8-25, 8-27, 8-29, 8-33, 8-34, 8-39, 8-40, 8-41, 8-45, 8-53, 8-57, 8-58, 8-62].

Sometimes it is held that the reality thesis could be true only if a dynamical (Lorentz-) interpretation of relativistic effects were allowed. Surprisingly, exactly this is the case. However, their dynamical interpretation does not contradict their standard kinematical explanations, but it should rather be regarded as their supplement.

The problem of causal explanation of relativistic effects has nothing to do with the problem of absolutely real effects. Not only the absolutely real effects need such an explanation, but the relative real ones as well – cf. [8-3, 8-19]. The only difference consists in the fact that, in the case of the latter, their explanation shares their relativity, e.g., they must be referred to the reference system in which they are observed. Therefore, for them, no assumption about the existence of a privileged frame, or the ether, is needed, as it is sometimes maintained. Only for the explanations of absolutely real effects, as the absolute Lorentz contraction, the existence of the ether rest frame is indispensable, namely for both the explanation and the very existence of the effects.

8.3. Clarifications of terminology

Up to now, we have restricted ourselves to the considerations neutral with respect to the confrontation of the competing interpretations of relativity. We hope to have successfully eliminated some widespread misunderstandings about its contents. Now we will try to do this with some others, this time of terminological origin.

First of all: The advocates of the non-orthodox Lorentz interpretation are not always fully aware what they actually defend. They put too often too much emphasis onto their opposition against the orthodox Einsteinian interpretation and confuse the Lorentz interpretation as a theory alternative to the relativity theory. This facilitates the orthodox relativists to portray it as one more kind of "crank" anti-relativism.

In particular, they dislike the concept of spacetime too much. At the same time, it is not at all clear what of this concept would be false. It has proved to be a useful mathematical and heuristic tool. One should only not to attribute to it too much ontological significance. About the statement that the physical spacetime has (in the local approximation) the structure of a four-dimensional pseudo-Euclidean space, there is nothing at all that a Lorentzian cannot accept under any circumstances. The only thing he cannot accept under any circumstances is the inseparable union of space and time, which is portrayed as a consequence of this statement. At the same time, it is not logically excluded that the spacetime has some deeper structure, from which an unequivocal decomposition of spacetime into space and time is derivable

The negative attitude of some Lorentzians against the principle of relativity is equally unjustified. This well-established physical principle, as stating that the laws of nature have the same form in all inertial systems, is very reasonable, in contradistinction to the extreme version of the philosophical principle of relativity of motion, according to which it is meaningless to speak about the motion of a single body, thus the motion is only relative to another body. The latter is, in contradistinction to the principle of relativity, no consequence of the relativity theory. Still more, it is even incompatible with the latter, because the theory knows at least an absolute difference between accelerated and unaccelerated motions. On the other hand, the principle of relativity by no means excludes the existence of a privileged reference frame. Anyway, it is excluded, to detect the motion relative to such a frame by means of local experiments. Even the philosophical principle of relativity of motion, properly understood, does not exclude the existence of a privileged reference frame. On the contrary, the right way of freeing oneself of absolute motion is to substitute it with the motion relative to the ether, which is a kind of a material object, even if not of ordinary matter.

Next, it is not clear what false would be about EINSTEIN's second postulate. Taken literally, it says yet only that the velocity of a light wave is independent from the motion of its source (ILV) [8-2, 8-56]. Such an assumption the relativity theory has in common with all ether theories, and is incompatible only with ballistic theories, e.g., of RITZ. Unfortunately, it is usually confused with the hypothesis of constancy of light velocity (CLV) [8-56], according to which the value of light velocity is the same in all inertial systems.

On the other hand, the constancy of light velocity is by no means incompatible with the existence of a privileged reference system. If in such a system, the value of light velocity is direction independent and equal to the universal constant c and the

Lorentz kinematics is satisfied, then the value of light velocity is equal to c in all inertial systems. Thus, the constancy of light velocity is also satisfied.

Some Lorentzians insist on making the difference between one-way velocity of light and the one along closed path. Then, they accept the constancy of light velocity along closed paths and deny it with respect to the one-way velocity of light. Yet this attitude is irreconcilable with the thesis that the Lorentz kinematics is satisfied, which, as it was demonstrated in section 8.1, results from the hypotheses of length contraction and time dilation, which are denied by no Lorentzians.

Moreover, such a standpoint is a result of an inconsistent attitude with respect to space and time measurements. Its advocates take it as admissible to extend the simultaneity in the privileged frame onto all inertial frame, but they refuse to extend in analogous way the spatial metric and time scale. As a result, they obtain a kinematics based on the generalized Galilean transformations of section 8.1. A more consistent attitude would lead to the Galilean kinematics, which results in the rejection of the constancy not only of the one-way light velocity, but also for closed paths. Both above-mentioned kinematics have no physical meaning due to the principle of relativity, since this presupposes to demonstrate which inertial system is the privileged one.

If the principle of relativity and the Lorentz kinematics hold, then any physical synchronization procedure gives relative, i.e., frame dependent outcomes. Such a procedure defines a relation between events. Some Lorentzians agree to call it simultaneity only if EINSTEIN's standard procedure, or some equivalent one, is applied in the privileged frame, whereas they hesitate with the agreement if the same procedure is applied in a frame moving relative to it. There is no necessity of such reservations. Instead, it is advisable to distinguish between simultaneity and absolute simultaneity, the latter coinciding with the simultaneity in the privileged frame. This would enable to agree with Einsteinians that simultaneity is in a certain sense relative, whereas refuse to agree also that there is no absolute simultaneity.

It is equally advisable to distinguish between the velocity of a physical object in a reference system and its velocity relative to another object. The latter can be defined as the result of the subtraction of velocities of the objects in question. The so-called relativistic law of addition of velocities is actually the law of transformation of them (see section 8.1) and by no means contradicts the above ascertainment. It allows also agree, on the one hand, that the velocity of light is in all inertial systems direction independent and equal to c, but on the other hand to maintain, that the velocity of light relative to a given body is in general direction dependent and equal to c only in the rest frame of the body. This conclusion can be to the same extent accepted both by Einsteinians and Lorentzians.

The difference of views becomes more evident, when the Lorentzians introduce the notion of the absolute velocity of an object as identical with the one in the privileged frame, as Einsteinians deny the existence of any absolute velocity in this sense. Nonetheless, sometimes it is claimed that the velocity of light is absolute. In a certain sense this is justified, since, if there is an absolute velocity, then the value of the velocity of light rays in any reference frame is the same as of their absolute velocity. Yet, since this leads to confusion if one does not believe in any absolute velocities, it is advisable to speak about invariance of light velocity, instead of its absoluteness, and to restrict the qualification "absolute" to the quantities defined in the privileged frame. On the other hand, it would be better for the Lorentzians to speak not about the alleged anisotropy of the velocity of light in moving frames, but rather about the anisotropy of absolute velocity of light relative to absolutely moving frames. The latter formulation is more complicated, but less misleading, its complication being the price of accuracy. Of course, one may use the former formulation as some abbreviation, but only after a necessary terminological clarification.

A similar distinction can be made in relation to the notions of length, time and mass. For example, a body has an invariant rest length and a spectrum of different lengths in different reference frames. If there is a privileged reference frame, it has an absolute length equal to the one in this frame. Therefore, the absolute length is, in general, different from its rest length and the length in a given reference frame. It is equal to the rest length only if the body is absolutely resting, and to the length in a reference frame only if the latter coincides with the privileged frame. However, if there is no privileged frame, then there is also no absolute length, although it is legitimate to speak on the rest length as absolute, since it is invariant. The same holds for the absolute time compared to the proper time, and for the absolute mass compared to the rest mass.

It might seem as though the absolute quantities were here defined as identical with the quantities corresponding to them in the privileged frame. However, there is no necessity to proceed this way. It can be assumed instead that a quantity can be absolute, i.e., frame independent, in yet another two ways: either it is invariant, i.e., the same in all reference frames, or it stands in no relation to any reference frame. Due to this, the privileged frame can be defined as the one in which all the absolute but not invariant quantities coincide with their relative counterparts. Thus, the notion of absolute quantity can be regarded as primitive, and the notion of the privileged frame as derivative. This enables the Lorentzians to abstract from their assumption of existence of an ether, while retaining the kernel of their view. Naturally, Einsteinians will deny the existence of such non-invariant absolute quantities, and say, e.g., that length is purely relative and does not have any absolute meaning, while the rest length is absolute.

Once the above distinctions are met, the Lorentzians may distinguish, e.g., between length contraction as the ratio of the length of a body in a given inertial frame to its rest length, and absolute length contraction as the ratio of the absolute length of the body to its rest length. Then, they may say that the absolute length contraction is absolutely real, whereas the length contraction is only relatively real (see the previous section). The Einsteinians would agree that the length contraction is relatively real, but add that this effect has no absolute meaning, and thus no absolute reality. The same holds for other relativistic effects.

Once the privileged frame is defined, other important notions can be defined: first, absolute rest, as identical with the rest in the privileged frame; second, absolute simultaneity, as the simultaneity in the privileged frame; and finally, absolute motion, as the motion in this frame. The Einsteinians deny the existence of absolute rest, absolute simultaneity and absolute motion, but they must admit that motion is not purely relative, since one may define invariant quantities that characterize the acceleration of a motion, or the difference of two non-accelerated motions, though those quantities do not coincide with the absolute acceleration, and the absolute velocity of the objects relative to each other. Thus, the Einsteinians must admit, that some aspects of motion have absolute meaning.

It must be added that, although absolute rest in the above sense can be introduced, this is not recommendable. Next, this possibility is restricted to the system of the special theory of relativity. The identification of absolute rest with the rest in the ether frame is, as a rule, no longer possible in the dynamical spacetimes of general relativity – see the next section. We will see (Sec. 8.5) that the same holds for identification of the absolute quantities with their counterparts in the ether frame.

8.4. The decisive argument

All preparations have been made. Now it's time to make a choice between the rival interpretations of relativity theory, at which we shall apply only such arguments that are both relevant and non-circular. This means that some arguments reported in this study are now pointless. Some others have already been mentioned somehow in the previous sections.

It should be clear that we are not going to question the theory of relativity as a whole. So any supposedly justified rejections of the principle of relativity, the Lorentz kinematics or the constancy of light (see the previous section) are of no significance for us. This does not mean that we do not relate to them, but only, that we make no use of them.

Currently, one such supposedly successful rejection attracts the interest of various authors. They maintain that the so-called Sagnac effect contradicts directly the constancy of light velocity, and indirectly the Lorentz kinematics. What they propose in place of it, is a kinematics based on generalized Galilean transformations (see the first section).

The Sagnac effect arises as a result of the phase shift of coherent light beams that propagate along the same path on a rotating platform in opposite directions (clockwise and counterclockwise). The phase shift can be observed by means of an interferometer and derived theoretically for the non-rotating frame from the assumptions that in inertial frames the principle of constancy of light velocity holds and furthermore a moving light source undergoes a frequency change as a result of the time dilation. The derivation of this effect in the co-rotating frame is not as unproblematic. As a result, it is sometimes maintained that on the basis of the theory of relativity, the absence of this effect must be predicted, since in any inertial frame tangent to the rotating platform the principle of constancy of light velocity holds, and the routes of both beams are equal. In this way, the special theory of relativity would be inconsistent in relation to the principle of constancy of light velocity, and incompatible with the experimental results.

It can be shown that such a simple inference would be premature. Since the rotating frame is not an inertial frame, the temporal relations are not as simple as in an inertial frame. This means, in particular, that, although the light velocity is direction independent, the time that a light beam takes along a closed path can be direction dependent.

The authors who are conscious of this fact do not insist on this apparent empirical rejection, but they stress the contradiction in basing the time in a rotating reference frame on the principle of constancy of light velocity. They propose to base the time description on such a definition of simultaneity that ensures the consistency of this description. The most natural way of doing this consists in transferring to the rotating reference frame the simultaneity from an inertial frame relative to which the motion is a pure rotation. In other words, a so-called external synchronization (see [8-44]) should be applied. In this way, transformation formulae will be obtained which are quite similar to the generalized Galilean transformations and can be regarded as their generalization to rotating frames. Finally, it will be shown that in the same way the generalized Galilean can be further generalized on any non-inertial frame, and that there is no such generalization for the Lorentz transformations.

No doubt, the possibility of such a generalization is a certain advantage. On the other hand, there are also some disadvantages. The generalized Galilean transformations do not form a group, with the consequence that a composition of two such transformations, as a rule, does not belong to them. What is more important, a transformation from one frame to another depends not only on their relative motion, but also on their motions (or at least one of them) relative to the inertial frame from which the simultaneity has been transferred to them.

One might argue that the role of such special frame should be taken by an absolutely resting frame, but such a postulate presupposes the possibility of finding a frame of this kind, which is excluded by the principle of relativity. If this is true, the "privileged" frame must be chosen arbitrarily. Much of this arbitrariness is taken over by the transformations themselves (in contradistinction to the Lorentz transformations), and by the spacetime coordinates. To introduce the latter, any observer must establish his motion relative to an arbitrarily chosen frame. This makes his task more complicated than the one of his special-relativistic counterpart. Moreover, this introduces to his description of physical phenomena an arbitrary anisotropy, that is lacking in their special-relativistic description.

Perhaps there would be a certain justification for preferring the generalized Galilean transformations to the Lorentz transformations, if there were some phenomena that would break the principle of relativity, i.e., if this principle would not hold universally. Such phenomena would enable to choose non-arbitrarily one of the otherwise equivalent frames. Up to now, no such phenomenon is known. On the other hand, even if there were such a phenomenon, it would still be possible to prefer the Lorentz transformations. It is not proven that the description of this anomalous phenomenon with the application of the generalized Galilean transformations would be simpler than with the Lorentz transformations, whereas it is certain that just the opposite holds for all other phenomena. The Sagnac effect is no exception, since even its allegedly non-relativistic description in the co-rotating frame utilizes its relativistic description in a non-rotating frame.

Another phenomenon that is often mentioned by the advocates of the Lorentz interpretation is the cosmic background radiation. It is a fact that there is only one reference frame in which this radiation is isotropic. Is this not the privileged, absolutely resting frame?

In connection with the theory of relativity there is no ground for such a conclusion, since one would have to show that the background radiation could not "prefer" other inertial frame. It is not a special phenomenon, but a special physical object, so this "preference" does not mean breaking the principle of relativity. Yet this desirable conclusion gets some support from the cosmology based on the general theory of relativity. If the Big Bang hypothesis is assumed, then there is a class of so-called fundamental frames, which sets out at any place a local state of "absolute rest". Cf. e.g. [8-11, 8-53].

It goes without saying that it is not the global absolute rest of the Lorentz interpretation, since the fundamental frames recede from each other due to the Big Bang. Nevertheless, there remains something of the spirit of this interpretation, as there is in such models the cosmic time, that is defined by the spacetime geometry and thus, in a way, absolutely – e.g. [8-61]. The cosmic time is global, and it contains a global relation of simultaneity, that coincides locally with the simultaneities in the fundamental frames. This is why the background radiation is significant above all as a confirmation of the Big Bang hypothesis.

We remember that there was such an absolute time in the non-relativistic physics, and then it disappeared in the usual Einstein-Minkowski interpretation of the special theory of relativity. Now it seems to come back in the general-relativistic cosmology. Such a development looks peculiar. There seems to be two distinct discontinuities in it. How distinct they are, will be clear soon.

In general, there is a very fundamental understanding of the nature of time. According to it, the world extends only spatially – cf. [8-45]. What extends temporally is not the world itself, but its history. The events the history consists of do not exist eternally, but they become, or come into being. This phenomenon is often addressed as the so-called time flow.

The non-relativistic approach retains this insight. But the standard interpretation of the special theory of relativity dares to deny it, since the becoming requires a global absolute simultaneity, that coincides with the relation of co-becoming. As a special-relativistic result, the block universe model of the world was invented, in which there is no time flow, and the future, as well as the past events have absolutely the same existential status as the present, i.e., they exist in a certain sense timelessly, and time is only the fourth dimension of the world, on a par with its spatial dimensions.

It must be stressed that the relativistic kinematics itself does not result in the elimination of time flow (see the first section). It is always possible to assume that the absolute simultaneity coincides with the (relative) simultaneity in a certain, though unknown, inertial frame. Also the general theory of relativity does not have such a consequence. Although there are solutions of the Einstein equation which do not allows for the foliation into hypersurfaces of absolute simultaneity (as e.g. Gödel's solution), one does not have to take them really seriously. On the contrary, the lack of such a possibility can be regarded as a ground for excluding such solutions as non-physical.

However, it is also not the case that the relativistic kinematics does not relate in any way to this question. In the non-relativistic kinematics, the simultaneity is absolute, i.e. the same in all reference frames, so the elimination of the privileged frame does not result in the elimination of the absolute simultaneity and of time flow. In contradistinction to this, the Lorentz transformations introduce a close relationship between the existence of a privileged frame and the presence of time flow. If there is time flow, then there is also a privileged frame, and vice versa. It is impossible to deny the existence of such a frame and to assume time flow at the same time – cf. [8-54].

We see that the choice between the (two) interpretations of the theory of relativity is dramatic. Either one accepts the orthodox interpretation as well as the block universe picture of the World, or one assumes time flow. Both possibilities are in agreement with the results of physical observations. Nevertheless, as long as one does not restrict the notion of empirical evidence to observations of this kind, there is a clear evidence for the second possibility. It is our experience of time flow itself. Although the advocates of the standard interpretation are trying to explain it away as a mere appearance, they do not have in their picture of the world any possibility to explain this alleged appearance.

Another argument for the Lorentz interpretation can be based on a methodological principle, namely on the correspondence principle. According to it, a new theory should correspond with the old one, e.g., the old one should be a special, or limiting case, or at least an approximation of the new one. Yet in Minkowski's spacetime there is no element that would correspond to the absolute time of non-relativistic physics, or to the cosmic time of relativistic cosmological models. Thus, this spacetime is neither special nor limiting case, nor an approximation of such a model. Of course, it is a solution of the Einstein equation, just as Gödel's model, but nothing more.

The same holds for the spacetime of non-relativistic physics with respect to the Minkowski spacetime. However, the circumstances are different for the spacetime of the Lorentz interpretation. Although the so-called Galilean spacetime (without a privileged frame, i.e., without the absolutely resting frame) correspond neither to the Minkowski nor to the Lorentz spacetime, the spacetime that NEWTON preferred as a model of his mechanics, with the absolute rest as a non-empirical component, corresponds with the LORENTZ's quite well, since it is an approximation for small absolute velocities.

On the other hand, the Lorentz spacetime is a local approximation of all realistic cosmological models, where their fundamental frames coincide locally with the Lorentzian absolutely resting frame – and its absolute time with the cosmic time. The novelty is that now the notions of absolute rest and absolute time acquire a certain empirical meaning, as they relate to the observations of the background radiation and to some other astronomical observations.

8.5. Protophysics and spacetime geometry

LORENTZ's interpretation of relativity reintroduces some notions which are metaphysical in the sense that their application cannot be defined on the basis of observations. One of them is time flow. Another is the ether. However, there is some ambiguity in the latter case. Ether can be understood either as the substantival bearer of the fields of fundamental interactions (material ether), or as the distinguished reference frame, marked by a continuous timelike four-vector field (geometrical ether) – cf. [8-35]. Whereas the former notion is unquestionably metaphysical, the latter's status is not as clear. In some cosmological models it can even be a physical notion. In particular, the frame of fundamental observers, which is definable by the physical spacetime geometry, can be regarded as some geometrical ether.

The same is true about the notions of absolute simultaneity, absolute time scale and absolute space geometry. There is something peculiar about them, that arises some hesitation about qualifying them as metaphysical. In contradistinction to clear cases of metaphysical notions, their undefinability on the basis of measurements does not seem essential for them, but it is merely a consequence of the principle of relativity in its form holding in the relativity theory. The latter precludes the existence of material instruments unaffected by the relativistic effects of motion. However, if one could have, for instance, a perfect clock unaffected by the time dilation effect, then the absolute simultaneity, along with the absolute time scale, would be definable by measurements with such a clock. Thus, it is quasi-operational in a sense quite analogous to the operationality of the physical, relative simultaneity and is not simply operational only due to the contingent fact that in our world the principle of relativity holds in the form based on the Lorentz transformations.

The above observation shows the need for a different qualification of such notions, which clearly are not metaphysical in the usual sense. They are not physical, either, but seem something in between. Fortunately, there is such an alternative qualification of them. They can be namely regarded as protophysical.

Protophysics was intended by P. LORENZEN as a program for a new constructivist foundation for physics, alternative to A. EINSTEIN's relativistic program – cf. [8-42]. He regarded the latter as unsatisfactory, since violating the so-called principle of methodical order [8-32], according to which solving problems of a less fundamental theory, like electrodynamics or gravitation theory, cannot entail revisions of more fundamental theories, like kinematics and chronometry or geometry, respectively. In this respect, protophysics is some continuation of H. DINGLER's criticism of relativity theories, though it does not share his postulate of rejecting them.

The program of protophysics was rejected by the scientific community in the face of difficulties in its realization and doubts regarding the above principle. However, an additional factor which might have made mutual communication difficult was the aversion of protophysicists to the language of spacetime geometry, which is very common among the specialists in relativistic physics. Thus, one can believe that a translation of the program into that language may lead to its re-evaluation.

So far, such a translation has been attempted solely by its opponents, who adapted it to their own notions, which were unfavorable to its adequacy. Nevertheless, they noticed rightly that the program of protophysics entails ascribing to spacetime an absolute background geometry, apart from its physical geometry, defined by a metric satisfying some dynamical equations, as in so-called bimetric theories of gravitation [8-6]. What may arouse doubts is Minkowskian geometry, which they assumed as the background geometry.

The idea of absolute background geometry can be related to DINGLER's idea of the "zero-point" [8-20]. If physical, material standards of length and time units are influenced by gravitation, as the general theory of relativity implies, then their state must be conceivable in which they would be if this influence were absent. Equivalently, some ideal standards must be conceivable, insensitive to the effects of gravitation. Whereas measurements with physical standards define the physical spacetime geometry, satisfying some dynamical equations like the Einstein equation, or just the Einstein equation as a special case, which is somehow distorted by those effects, analogous measurements with ideal standards would define some "true", undistorted spacetime geometry.

To the bimetric theorists it seemed obvious that such a "true" geometry must be Minkowskian. However, one may object that the standards that define it are not sufficiently ideal. Although they are insensitive to gravitational effects, they still undergo relativistic effects of motion. In particular, ideal measuring rods of this kind undergo the length contraction and ideal clocks – time dilation of special relativity. Thus, they do not satisfy the idea of "zero-point". This idea can be satisfied only by some perfect standards, unaffected not only by the effects of gravitation, but by the effects of motion as well. Of course, there are no such standards in reality. Nevertheless, they are still conceivable. Moreover, the results of measurement with such standards can be reproduced by the measurements with physical standards corrected appropriately.

It is easy to see that perfect clocks in this sense would enable defining absolute simultaneity, absolute time scale, and thus absolute time. In particular, the absolute simultaneity can be defined by transport of such a clock (cf. [8-42]) in an entirely arbitrary way, since it is, by assumption, insensitive to effects of motion. Perfectly rigid rods, in turn, would enable defining some absolute space geometry, identical with the geometry of the hypersurfaces of constant absolute time. It is clear that it must be Riemannian, since the physical geometry is Riemannian. Then, it must be of constant curvature, since it seems reasonable to demand that space be strictly isotropic and homogenous. Alternatively, one might argue that only such a space allows for rigid motions. Finally, if one agrees that the only source of space curvature is physical interactions, then it must be flat. Another argument might be based on the intuition that it is meaningful to speak on the same orientation of vectors at different places. This is so only in flat spaces. As a result, it must be Euclidean. One can add that, among Riemannian spaces, only Euclidean space allows for similar figures, i.e., of the same shape, but different sizes.

Consequently, the absolute background geometry must contain absolute time and absolute Euclidean metric of space. It is clear that LORENTZ's interpretation of relativity requires at least as much. What more? The geometry that contains only those elements is the geometry of so-called Leibnizian spacetime [8-21, 8-22]. Many Lorentzians, including the present author at earlier stages of development of his view, assume above them the absolute rest, as identical with the rest relative to the ether. This is tantamount to enriching the background geometry to so-called Newtonian spacetime [8-21, 8-22]. However, as we saw in the previous section, such an identification is not recommendable in the framework of the special theory of relativity, and even impossible in the framework of the general theory. Thus, it is better to dispense with the notion of absolute rest altogether and satisfy oneself with the geometrical ether as its substitute, which is, however, dynamical, and consequently cannot belong to the absolute background geometry.

Another candidate for an element of the background geometry is a flat affine connection, which would enable defining motions non-accelerated in an absolute sense. This, in turn, would mean choosing, as the background geometry, so-called Galilean spacetime, which is the structure implicit in Newtonian mechanics, just as the structure of Minkowskian spacetime is implicit in the relativistic mechanics [8-35]. Both in Newtonian mechanics and in special relativity, a flat connection belongs to the physical geometry of spacetime and enables defining inertial frames. However, both in Newtonian gravitation theory and in general relativity, the connection is no longer flat, and it becomes dynamical, thus non-absolute.

In addition, the notion of inertia is clearly dynamical, and thus it is enough that inertial frames are defined in the physical geometry. This means that the dynamical connection belonging to the latter does not need any counterpart in the absolute background geometry. Therefore, the background geometry LORENTZ's interpretation needs is just Leibnizian geometry. In contradistinction to Minkowskian geometry, it does not contain any absolute velocity, which defined there the light cone structure. Moreover, in contradistinction to both Minkowskian and Galilean geometry, it does not contain any absolute standard of non-acceleration, nor any standard of vanishing of higher derivatives of velocity with respect to time. As a result, unlike them, it satisfies the principle of relativity of motion in full extent.

Introducing protophysical notions is a way of formulating the protophysical interpretation of relativity, which can be regarded as a version of the Lorentzian interpretation. It is worth noting that it is not necessary to introduce a geometrical ether just at the beginning. It is definable in the full spacetime geometry, including the physical and the background geometries, by the relation of orthogonality to the hypersurfaces of constant absolute time. Thus, introducing the Leibnizian geometry into relativistic spacetimes as the background geometry is sufficient for translating the protophysical interpretation into the language of spacetime geometry. From the point of view of physics, its introduction is always arbitrary and can be done in many different ways. Nevertheless, it does not mean that it is a matter of convention. One of those ways is the right one. Only the question which of them it is cannot be decided on the basis of observations.

The Leibnizian background geometry provides means for attributing absolute meaning to the relativistic effects of length contraction and time dilation. The absolute effects can be defined as the ratio of the length of an object and the duration of a process in it, respectively, as determined relative to perfect standards, to the proper length and proper duration, as determined relative to physical standards in their rest frame. In other words, as a measure of the effects of the ether wind can serve the difference between the physical and background geometries. In this way, the objective character of the relativistic effects is established. It is worth noting that the absolute effects do not necessarily coincide with their relative counterparts as observed in the ether frame. This is because neither the absolute time nor the absolute space metric necessarily coincide with the relative time and relative space metric in the ether frame. Assuming such a coincidence is natural only in the absence of gravitation.

Introduction of Leibnizian background geometry can also serve as a means to solve some well-known theoretical problems. Just as well as Minkowskian background geometry, it can help in solving the problem of the energy-momentum in general relativity (cf. [8-61]). To avoid the problem, one should calculate the energy not with respect to the physical geometry, but with respect to the absolute background geometry. However, in contradistinction to Minkowskian background geometry, it does not lead as easily to the conflict of causal structures (cf. for instance [8-50]), since it does not contain absolute light cone structure, but it enables to regard as causal any worldline which is time orientable with respect to the absolute time. It is a natural background for the very intuitive so-called "river model", in which gravitation is interpreted as the effect of the ether flow ([8-17, 8-28]; it is easy to see that HAMILTON & LISLE do not need Galilean, but just Leibnizian geometry, as the background geometry). Last but not least, it also enables avoiding the (in)famous problem of time in quantum gravity (cf. for instance [8-31]) thanks to the absolute time Leibnizian geometry contains.

Literature

[8-1] Arzelies H,, Relativistic Kinematics, Pergamon Press, Oxford 1966.
[8-2] Baierlein, R., Two myths about special relativity, Amer. J. Phys. 74, 193 (2006).
[8-3] Bell J.S., How to teach Special relativity, Progress in Scientific Culture 1, No 2 (1976); repr. in: Bell J.S., Speakable and Unspeakable in Quantum Mechanics, Cambridge Univ. Press, Cambridge 1987.
[8-4] Böhm D. & Hiley B.J., Nonlocality and the Einstein-Podolsky-Rosen experiment understood through quantum-potential approach, in: Selleri F. (ed.), Quantum Mechanics Versus Local Realism, Plenum Press, New York 1988.
[8-5] Bondi H., The space traveler's youth, Discovery 18, 505 (1957).
[8-6] Borzeszkowski, H.-H. von & Wahsner, R., Messung als Begründung oder Vermittlung? Ein Briefwechsel mit Paul Lorenzen über Protophysik und ein paar andere Dinge, Academia, Sankt Augustin 1995, p. 7.
[8-7] Brandes J., Die relativistischen Paradoxien und Thesen zu Raum und Zeit 2 Aufl., VRI, Karlsbad 1995.
[8-8] Bridgman P.W., A Sophisticate's Primer of Relativity, Routledge & Kegan Paul, London 1962 (pp. 26-27).
[8-9] Builder G., The resolution of the clock paradox, Austr. J. Phys. 10, 246 (1957).
[8-10] Builder G., Ether and relativity, Austr. J. Phys. 11, 279 (1958).
[8-11] Craig W.L., The elimination of Newton's absolute time by relativity theory, in: Physical Interpretations of Relativity Theory II, London 1990.
[8-12] Czerniawski J., Dwie interpretacje teorii względności (Two interpretations of Relativity), Zagadnienia Naukoznawstwa 26, 315 (1990), in Polish.
[8-13] Czerniawski J., Physical content of kinematic transformations, in: Physical Interpretations of Relativity Theory III, London 1992.
[8-14] Czerniawski J., Zrozumieć teorię względności (Understanding the Theory of Relativity), Tasso, Kraków 1993, in Polish.
[8-15] Czerniawski J., What is and what is not essential in Lorentz's relativity, in: Barone M. & Selleri F. (eds), Frontiers of Fundamental Physics, Plenum Press, New York 1994.
[8-16] Czerniawski J., Visual appearances and reality of the length contraction, in: Physical Interpretations of Relativity Theory IV, London 1994.
[8-17] Czerniawski, J., What is wrong with Schwarzschild's coordinates?, Concepts of Physics 3, 307 (2006).
[8-18] Czerniawski J., Ruch, przestrzeń, czas: Protofizyczne i metafizyczne aspekty podstaw fizyki relatywistycznej (Motion, Space, Time: The Protophysical and Metaphysical Aspects of Foundations of Relativistic Physics), Jagiellonian Univ. Press, Kraków 2009, in Polish.
[8-19] Dieks D., The 'reality' of the Lorentz contraction, Zschr. für Allg. Wissenschaftstheorie 15/2, 330 (1984).
[8-20] Dingler, H., Die Grundlagen der Physik: Synthetische Prinzipien der mathematischen Naturphilosophie, 2nd ed., Gruyter, Berlin 1923.
[8-21] Earman, J., World Enough and Space-Time, MIT Press, Cambridge (Massachusetts) 1989.
[8-22] Ehlers, J., The nature and structure of spacetime, in: Mehra, J. (ed.), The Physicist's Conception of Nature, D. Reidel, Dordrecht 1973.
[8-23] Einstein A., Zur Elektrodynamik Bewegter Körper, Ann. d. Phys. 17, 891 (1905).

[8-24] Gödel K., A remark about the relationship between relativity theory and idealistic philosophy, in: Schilpp A. (ed.), Albert Einstein: Philosopher-Scientist, Open Court, La Satte (Illinois) 1949.
[8-25] Grabińska T., Analiza operacjonistycznych założeń kinematyki H.E. Ivesa i kinematyki relatywistycznej (The analysis of operationistic foundations of the H.E. Ives' kinematics and relativistic kinematics), in: Z Zagadnien Filozofii Przyrodoznawstwa i Filozofii Przyrody 8, 7 (1986), in Polish.
Grabińska T., Relativity and space-time measurements, Astrophys. and Space Sci. 191, 23 (1992)
[8-26] Hafele J. C.& Keating R. E., Around-the-world atomic clocks, Science 177, 166 (1972).
[8-27] Hartmann E., Raum und Zeit im Lichte der neuesten physikalischen Theorien, Philosoph. Jb. der Görres-Gesellschaft 30, 1 (1917).
[8-28] Hamilton, A.J.S. & Lisle, J P., The river model of black holes, Am. J. Phys. 76, 519 (2008).
[8-29] Ives H.E., Derivations of the Lorentz transformations, Philosoph. Magazine 36, 392 (1945).
[8-30] Ives H.E. & Stilwell G.R., An experimental study of the rate of a moving atomic dock, J. Opt. Soc. Amer. 28, 215 (1938).
[8-31] Isham, C.J., Canonical quantum gravity and the problem of time, in: Integrable Systems, Quantum Groups, and Quantum Field Theories, Ibot, L.A. & Rodriguez, M.A. (eds.), Kluwer, Boston 1992.
[8-32] Janich, P., Die Protophysik der Zeit, Suhrkamp, Frankfurt/M 1980; Englisch Version: Protophysics of Time, D. Reidel, Dordrecht 1985.
[8-33] Janossy L., Theory of Relativity Based on Physical Reality, Akad. Kiado, Budapest 1971.
[8-34] Kajfosz J., On the Problem of Inertial Motion, Inst, of Nucl. Phys., Kraków 1982.
[8-35] Kopczyński, W. & Trautman, A., Czasoprzestrzeń i grawitacja, PWN, Warszawa 1981, in Polish; English version: Spacetime and Gravitation, Wiley & Sons, Chichester 1992.
[8-36] Kroes P., The dock paradox, or how to get rid of absolute time, Phil. Sei. 50, 159 (1983).
[8-37] Malament, D., Causal theories of time and the conventionality of simultaneity, Noûs 11, 293 (1977).
[8-38] Landau L.D. & Lifschitz E.M., Lehrbuch der theoretischen Physik Bd. II, Klassische Feldtheorie, Akademie-Verlag, Berlin 1967.
[8-39] Lenard P., Über Relativitätsprinzip, Äther, Gravitation, S. Hirzel, Leipzig 1921.
[8-40] Lorentz H.A., Die Relativitätstheorie für gleichförmige Translationen, Akad. Verlag, Leipig 1929; English version: The principle of relativity for uniform translations, in: Lorentz H.A., Lectures on Theoretical Physics, vol. 3, MacMillan, London 1931.
[8-41] Lodge 0., Beyond Physics, G. Allen & Unwin, London 1930.
[8-42] Lorenzen, P., Eine Revision der Einsteinschen Revision, Philosophia Naturalis 16, 383 (1977).
Lorenzen, P., Relativistische Mechanik mit klassischer Geometrie und Kinematik, Mathematische Zeitschrift 155, 1 (1977).
[8-43] Malament, D., Causal theories of time and the conventionality of simultaneity, Noûs 11, 293 (1977).
[8-44] Mansouri R. & Sexl R.U., A test theory of special relativity: I. Simultaneity and dock synchronization, Gen. Rel. Gravit. 8, 497 (1977).

[8-45] Maxwell N., Are probabilism and special relativity incompatible?, Phil. Sci. 52, 23 (1985).
[8-46] Minkowski H., Raum und Zeit, Phys. Z. 10, 104 (1909).
[8-47] Mittelstaedt P., Der Zeitbegriff in der Physik, Wissenschaftsverlag, Mannheim 1989.
[8-48] Okun L.B., The concept of mass, Physics Today 42 No 6, 31 (1989).
[8-49] Pars LA, The Lorentz transformation, Phil. Mag. 42, 249 (1921).
[8-50] Pitts, J.B. & Schieve, W.C., Null cones and Einstein's equations in Minkowski spacetime, Found. Phys. 34, 211 (2004).
[8-51] Poincare H., La mesure du temps, Rev. Mét. Mor. 6, 371 (1898); repr. in: Poincare H., La Valeur de la Science, Flammarion, Paris 1905.
[8-52] Post E.J., Sagnac effect, Rev. Mod. Phys. 39, 475 (1967).
[8-53] Prokhovnik S., Light in Einstein's Universe, D. Reidel, Dordrecht 1985.
[8-54] Putnam H., Time and physical geometry, J. Phil. 64, 240 (1967). Czerniawski J., Time and physical geometry: A formalization of Putnam's proof, Logic and Logical Philosophy 29, 97 (2020)
[8-55] Reichenbach H., Philosophie der Raum-Zeit-Lehre, W. de Gruyter & Co., Berlin 1928; English transl.: The Philosophy of Space and Time, Dover Publ. New York 1958.
[8-56] Schaffner, K.F., Einstein versus Lorentz: research programs and the logic of comparative theory evaluation, Brit. J. Phil. Sci. 25, 45 (1974).
[8-57] Selleri F., Noninvariant one-way velocity of time, Found. Phys. 26, 641 (1996).
[8-58] Sjödin T., Synchronization in special relativity and related theories, Nuovo Cim. 51B, 229 (1979).
[8-59] Tangherlini F.R., An introduction to the general theory of relativity, Nuovo Cim. Suppl. 20, 351 (1961).
[8-60] Terrell J., Invisibility of the Lorentz contraction, Phys. Rev. 116, 1041 (1959)
[8-61] Weinberg S., Gravitation and Cosmology, J. Wiley, New York 1972.
[8-62] Wiechert E., Relativitätsprinzip und Äther, Phys. Z. 12, 689 (1911).

9. The concepts of length and distance in special relativity

9.1. Lengths and distances interpreted relativistically

As will be shown, what is unsatisfactory about the Einstein interpretation relates to the questions: What is real about length contraction? Does length exist in itself? At first, these are philosophical questions, bearing in mind that for all kinds of length *measurements* there is agreement between theory and experiment, but as soon as absolute effects similar to the Newtonian bucket experiment arise, the situation changes. Then the incompleteness of EI becomes apparent. It always finds an inertial frame relative to which the effect can be explained, but there are others relative to which it does not work. The predictions of the LI are the same for all inertial frames.

EINSTEIN commented on the question: Are Lorentz contractions real? in a publication [92]: There he opposes the interpretation of the Lorentz contraction as only apparent. First of all, the view of being apparent is obvious. The Lorentz contraction is determined by comparing the starting and end points of a moving rod at a certain point in time with a stationary scale. Due to the relativity of simultaneity, this is not simultaneous in the rest system of the rod. In the system of rest, the rod has an invariable length at arbitrary times. So in reality, in real terms, the rod has not become shorter.

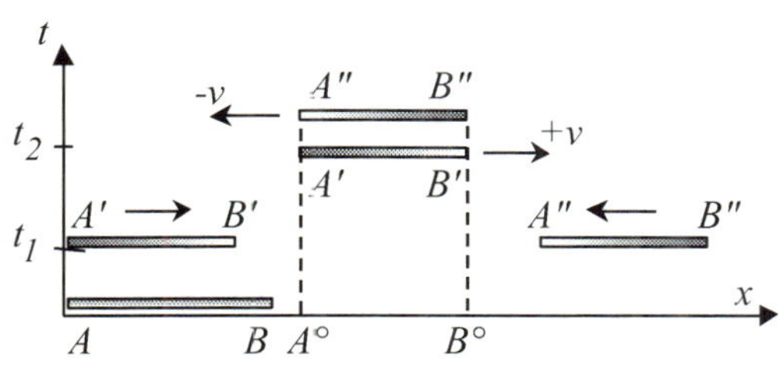

Fig. 9.1 EINSTEIN'S thought experiment on the Lorentz contraction

Rods of the same length approach each other with the same speed *v* from opposite directions. The Lorentz contraction is "not a subjective phenomenon" but real, because *A°B°* measured when *A'A''* and *B'B''* coincide, is smaller than *AB*, the rest length of the rods.

EINSTEIN objects to this (see Fig. 9.1):" ... Mr Varicak would perhaps admit this, i.e. in a certain sense retract his statement that the Lorentz contraction is a subjective phenomenon.... Let there be two rods of equal length (compared at rest) *A'B'* and *A''B''*, which can slide along the *x*-axis of an acceleration-free coordinate system . . ... *A'B'* and *A''B''* are supposed to slide past each other, whereby *A'B'* in the sense of the positive, *A''B''* in the sense of the negative *x*-axis is moved with a constant speed of any size.

In this case, the end points A' and A'' meet at a point $A°$ the end points B' and B'' meet at a point $B°$. The distance $A°B°$ is then smaller than the length of each of the rods $A'B'$ and $A''B''$ according to the theory of relativity, which can be stated with one of the rods by applying it to the distance $A°B°$ in the state of rest".

Length contraction is real, as EINSTEIN'S elegant argumentation shows. It is convincing that it can be proven without clock synchronisation. But on the other hand, it is now true that a body has any number of real lengths, because all lengths of the same body measured in the arbitrarily moved inertial systems are real in the same way. If inertial systems are moved perpendicular to the length of the rod, the length again remains unchanged.

As real as all these length measurements are, can't one of them be labelled? The first idea is to take the resting length of a rod or body. The resting length is characterised by being independent of a time measurement. It is measured with stationary scales and is valid for all times. The rods moving against each other in EINSTEIN's example above are not of equal length, their end points do not meet at the same time - they only behave like this in a certain inertial frame, in an arbitrary moving one they have different lengths and their end points meet one after the other. This does not apply to lengths at rest, they are excellent compared to other real lengths.

However, there is another length that is similarly time-independent as the rest length - circularly moving rings in the thought experiment, ch. 9.4. One can deform any resting rod in thought into a ring of the same rest length and then set it in rotation. Its actual length of rotation can be measured at any time with a scale at rest and does not change in any length of time. Attach a pencil to the periphery of the ring. Then the rotating ring draws a circle on a suitable support, the circumference and radius of which can be measured at rest. The length of rotation relative to all inertial frames is smaller than the length of the non-rotating ring. This is true regardless of the relativity of simultaneity.

Nevertheless, in the Einstein interpretation, there is no way to distinguish any of the measured lengths as somehow "more real" than the others. If we look at the rotating ring, it has become shorter, so at least one of its pieces l_n (see Fig. 9.3) should have become shorter than before, actually all of them for reasons of symmetry. But if we look at the l_n from the corresponding S'_n, each l_n has its resting length l'_n, all the other l_n have shortened, the l_n opposite the resting l'_n the most of all. Although the ring has shortened, none of its parts has shortened or each has shortened more than necessary, depending on the reference system. The calculation of the total length, however, always gives the correct, experimentally verifiable result.

(Not all representatives of EI will agree that the rotating ring shortens, in which case there is an experimental way to distinguish between EI and LI, see ch. 24.1).

This discussion is intended to show: One part of the always real lengths speaks for real quantities to be attributed to the bodies, another part for apparent, pure measured quantities. The Einstein interpretation has no concept to distinguish between them, all measurement results are real in the same way. One must ask: Does the Einstein interpretation describe reality as it is, or does it inadmissibly declare as equal what is merely measured as equal, does a body have length in itself or is it real only relative to an inertial system?

One can object to these questions: For the Einstein interpretation, there is a four-dimensional space-time continuum, there is not space per se and time per se. Therefore, there is no length per se for bodies; what there is for all bodies is their four-dimensional distance. Δs (s. ch. 11.3):

$$\Delta s = \sqrt{\Delta x^2 + \Delta y^2 + \Delta z^2 - c^2 \Delta t^2} \tag{9.1}$$

Thus, the following applies to a moving length that rests in the system S':

$$\begin{aligned} \Delta s &= \sqrt{\Delta x'^2 + \Delta y'^2 + \Delta z'^2} \\ &= \sqrt{\Delta x^2 + \Delta y^2 + \Delta z^2 - c^2 \Delta t^2} \end{aligned} \tag{9.2}$$

where Δx, Δy, Δz are the $\Delta x'$, Δy', Δz ' corresponding lengths in the system S and Δt is the time difference indicated by the clocks in S when measured simultaneously in S' and this Δs has the same value for all inertial systems.

But this is nothing characteristic for lengths of a body, because any distance from two points has the same value when viewed from any inertial system, provided one measures the distances at the times corresponding to the Lorentz transformation. Invariant are also not the results of the length measurements, which are carried out simultaneously within the different systems, but the results of the measurements simultaneously in perhaps one system, but in the others then at the correspondingly transformed times, i.e. no longer simultaneously. For example, a length Δl at rest in S has simultaneously measured in S', in S' the distance

$$\Delta l' = \Delta l \sqrt{1 - v^2 / c^2}$$

While

$$\Delta l' = \frac{\Delta l}{\sqrt{1 - v^2 / c^2}}$$

if I measure the start and end points in S' in such a way that there is simultaneity in S. In other words, the invariance of the four-dimensional space-time distance also does not allow a definition of a spatial, possibly four-dimensionally measured length that is independent of the inertial system; one remains dependent on the simultaneity of the respectively chosen inertial system.

Therefore, it is consistent to say: For the Einstein interpretation of relativity theory, there is no length of a body per se. All measured different lengths are real in the same way.

9.2. Length measurements in the LI

An excellent inertial frame is available for the Lorentz interpretation. All measurement results are predicted in the same way as for the Einstein interpretation, but with the difference that relative to the excellent inertial frame the measurement results represent real physical quantities, relative to other inertial frames the measurement results contain synchronisation components, i.e. they are (partly) apparent. Due to the principle of relativity it is not possible to decide by measurements which inertial system is the excellent one, but it is "didactically" permissible in a discussion to regard any particularly suitable one as the excellent inertial system and to introduce real effects "in thought".

If a body or a scale is at rest in the excellent system, its length measured there is equal to the length at rest and this is its length. If a body moves with the speed v relative to the excellent system, it is Lorentz-contracted, its real length is now shorter by the factor $1/\gamma$. In any inertial system, on the basis of the relativity principle the same numerical value is measured for the scale at rest as in the excellent system and the same value is measured for the scale moving with v, shortened by the factor $1/\gamma$. In these cases, the measured value contains an apparent component that follows from the arbitrary synchronisation rule of the clocks. The measured value no longer reflects the real (= actual) length of the scale.

Since - also due to the principle of relativity - it is not possible to determine the distinguished system through measurements, one does not know the real actual length of a rod. All we know is that it exists, and if v is known relative to the excellent system, it is calculated from the measured rest length divided by the Lorentz factor.

The peculiarities of rotation can also be interpreted casually. In the case of a circular motion of a ring relative to the distinguished system, all ring elements l_n are Lorentz-contracted. Relative to an arbitrary inertial system the measured values are the same, the real portion contained in them is different and not known. If one considers a ring element l_n of circumference by itself, it may be just the one that is at rest relative to the excellent system, the others have a correspondingly higher velocity relative to the excellent system, or the center is at rest and none of the pieces l_n - but in any case the mean of the velocities is greater than that of the center and overall there is always a smaller circumference for the rotating than for the non-rotating ring. The rotating ring has and maintains a contracted length for any length of time. The amount measured relative to an S is exactly the amount the ring would have in real terms if S were the excellent inertial system.

Just as experience shows, the rotating ring gives a real length, its parts l_n have a constantly changing apparent proportion. A single part l_n considered by itself can have any contracted length, it can be interpreted as apparently shrinking relative to a system of rest, but not all parts l_n together - all parts together have something real, like the rest length itself.

The interpretation of measured lengths as real and apparent quantities is feasible in the Lorentz interpretation in accordance with experience. The four-dimensional space-time continuum in the Lorentz interpretation means the totality of measured lengths and times, it is not reality itself. Real space and real time, since the excellent inertial system cannot be determined by measurements, cannot be quantitatively determined, but the shape of the laws of nature can be recognised. Because of the principle of relativity, they do not differ from the formulae that only describe the results of measurements within any inertial system.

9.3. Linearly accelerated bodies

For further clarification of the question whether length contractions cause real changes in length or whether they are only measurable effects that have their cause in the relative motion of inertial systems, some thought experiments are to be carried out.

It is to be shown first of all that stress forces can occur due to relativistic length contraction, DEWAN et al. [69], [106], likewise, J. S. BELL [18], who vividly describes the great problems theoretical physicists can have with this.

Let us consider a rocket R_1, which is constantly accelerated in an inertial system S with the amount b, see Fig. 9.2. Then the following applies

(9.10) $$x_{R1} = 1/2\, bt^2$$

The length l of the rocket undergoes a Lorentz contraction, for which the following applies:

(9.11) $$l = l'/\gamma$$

The rest length is l', since we want to denote the momentary inertial system at rest relative to R_1 by S'. The length of the rocket, viewed from system S, appears shorter and shorter as time increases, while the rest length l' remains constant in S'.

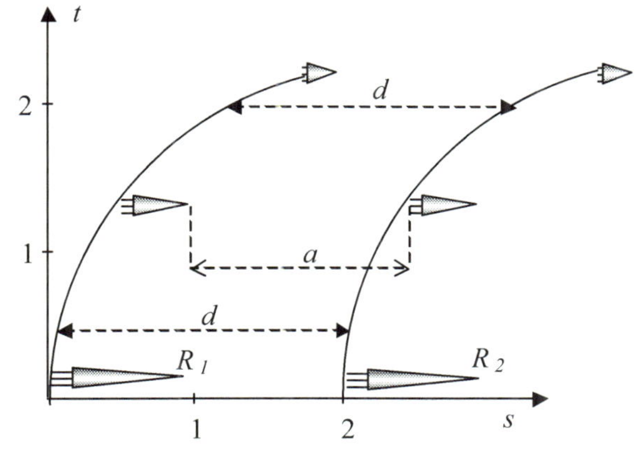

Fig. 9.2 Different Lorentz contraction of lengths and distances.

Two rockets R_1 and R_2 are accelerated at a constant rate of b in S. The distance d between the centers of gravity of the rockets remains unchanged, the distance a between the tip of R_1 and the end of R_2 increases as the rockets themselves contract.

The same applies to a rocket R_2, which takes off at a distance *d* from R_1 in *S:*

(9.12) $$x_{R2} = 1/2\, bt^2 + d$$

and of course, the length of this rocket has the same time dependence as that of the rocket R_1.

The instantaneous velocity is $v = bt$. The known formulas for free fall resp. constant acceleration were used [13].

In *S,* the difference between x_{R2} and x_{R1} is always *d,* i.e., the distance between the rockets R_1 and R_2 is always the same, as expected, since both rockets move and accelerate in the same way. (Distance here is always understood to be the distance between the centers of gravity of the rockets, not the distance between the tail and the tip of the rocket. Even more precisely: the distance between the engines). The lengths of the rockets become shorter as (9.11) and Fig. 9.2 show.

What does this process look like relative to *S'*? In S' the rockets are at rest at time *t* and the length of both rockets is equal to their resting length l'. But the distance d' between the rockets is no longer the original distance *d* but greater:

(9.14) $$d' = \gamma d$$

This is clear from earlier considerations: All distances in *S* appear contracted from S' and vice versa. Since the rockets are at rest in S' at time *t,* but have the fixed distance *d* as seen from *S,* this distance must be greater when measured in *S',* because it appears contracted in *S.* This is completely unproblematic in formulas:

$$x'_{R2} = \gamma(x_{R2} - vt)$$
$$x'_{R1} = \gamma(x_{R1} - vt)$$
$$x'_{R2} - x'_{R1} = \gamma(x_{R2} - x_{R1})$$

and it follows:

$$d' = \gamma d$$
$$d' > d$$

We now extend our thought experiment and connect the centers of gravity of the two rockets by an elastic ribbon and repeat the acceleration. Now there are two possibilities depending on the elasticity of the ribbon:

Relative to *S'*:

a.) The ribbon behaves like a rigid rod, it contracts and pulls the two rockets together to such an extent that the resting length of the ribbon remains the same, just as the resting length of the rockets also or

b.) the rockets keep their distance *d.* As soon as the acceleration of the rockets has ended, S' no longer changes its speed relative to *S,* but the ribbon is taut because in S' it now has the length γd. If the rockets are supported frictionless in *S',* they are slowly moved towards each other until the tension of the ribbon has disappeared. (The same formulae (9.11) and (9.14) differ, in (9.11) l' is constant, but in (9.14) it is *d.*) A similar consideration can be seen in Fig. 9.2, here the starting and end points of the rockets are connected by an elastic ribbon *a.* The tension is additionally caused by the Lorentz contraction of the rockets relative to *S.*

Relative to *S*:

a.) The ribbon contracts according (9.11) by a factor $1/\gamma$, the distance between the rockets becomes smaller.

b.) Similar to S', the ribbon shortens in S by a factor of $1/\gamma$ only after a certain period of time.

These considerations illustrate two important, physically distinct, relativistic effects.

1.) The tension forces in the ribbon arise because the ribbon is prevented from performing the Lorentz contraction. This happens relative to S.

2.) The tension forces in the ribbon arise because the ribbon is stretched. This happens relative to S'.

Both effects predict the same measurement results and fulfill the relativity principle. But who is right? The EI cannot answer this question. All inertial systems are equivalent. I can regard S or S' as correct at will.

For the LI is valid: Every inertial system can be the excellent inertial system. If it is S, then the Lorentz contraction is real and 1.) is valid, if it is S', then 2.) is valid, the ribbon is stretched. In this case the simultaneous start in S is in reality not simultaneous. 'Absolute simultaneity' is then the simultaneity in S', see fig.11.6.0.

Two quotes: "So, the contraction is only a consequence of the way of looking at things, not a change in a physical reality. So, it does not fall under the concepts of cause and effect." M. BORN [28]. Relative to S' BORN is right, relative to S not, because in S the Lorentz contraction is connected with forces.

The assumption of an excellent inertial system is "unbearable for the theorist" - EINSTEIN [90], ch. 5. But here it is really helpful and explains a contradictory fact.

9.4. Circularly accelerated bodies

In a similar form as in the previous chapter, some thought experiments on bodies moving in a circle are to be carried out in order to gain illustrative material for the question of the extent to which relativistic length contractions are real [187].

To avoid the more difficult discussion of the Ehrenfest paradox, s. ch. 7.5, two borderline cases will be examined, s. figs. 9.3 and 9.4:

a) Rotation of a circular ring

b) Rotation of circle segments connected to the pivot point via radial elements (spokes).

For the circular ring, the circumference shortens due to length contraction. Accordingly, the circle segments shorten, while the spokes only change their thickness but not their length.

The Lorentz contractions that occur during rotation are counteracted by the centrifugal force of the hollow cylinder. The following applies to them:

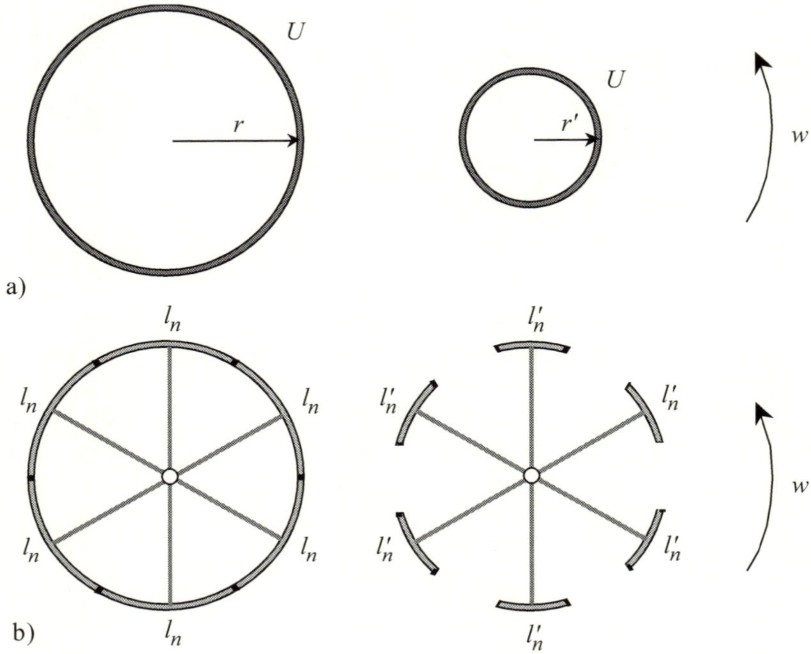

Fig. 9.3 Different Lorentz contraction of a closed and segmented circular ring

a) A thin rigid circular ring, set in rotation, shortens by the factor γ.
b) A circular ring divided into segments (wheel with spokes) set in rotation does not change radius and circumference, but gaps appear between the segments. (Rotation speed $w = 2\pi r f$, f= revolutions per second, (γ= 2), $w = v = 0.866c$.

Fig. 9.4 Rolling wheel

If the segmented ring rolls as a wheel in the negative x-direction, the segments are contracted differently and the spaces between the segments are enlarged differently. (γ= 2, $w = v = 0.866c$)

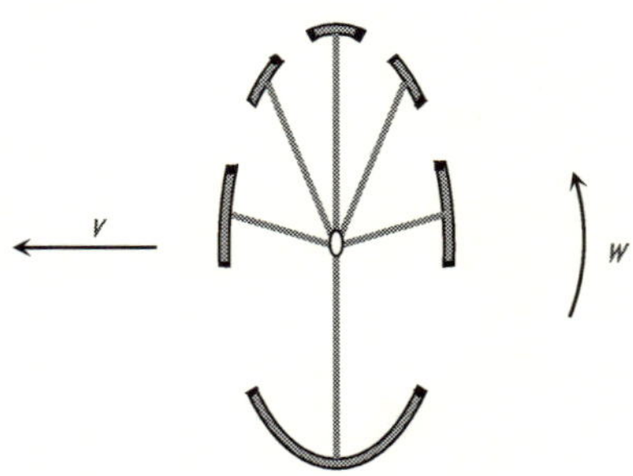

$$Z = mw^2 / r$$

(9.20)
$$= \frac{m(0)}{\sqrt{1-w^2/c^2}} \frac{w^2}{r}$$

$m(0)$ = rest mass, m = dynamic mass of the hollow cylinder of radius r moving at velocity w.

For a constant w, possibly close to c, the term

$$\frac{m(0)}{\sqrt{1-w^2/c^2}}$$

is also constant. For the limit value r towards infinity, the centrifugal force Z is zero, i.e. in practice, the centrifugal force can be kept negligibly small for each finitely large section by suitably choosing a sufficiently large radius and that's why its influence can be disregarded in the following considerations. (Also, the forces can be compensated by charging the sections and performing the rotation in a weak magnetic field).

We further assume that the height of the hollow cylinder is small in its dimensions and thus approximately a one-dimensional circle is formed. Then the rotating circular ring whose center rests in the inertial system S, has the circumference

(9.21)
$$U' = 2\pi r' = 2\pi r(1-w^2/c^2)^{1/2}$$

Both circumference and radius are shortened by the same factor.

To explain this, think of the circle at rest divided into n equal parts of length l_n , Fig. 9.3 a. Each piece l_n has velocity w at the end of acceleration and therefore a different length l'_n as viewed from S. How large does l'_n become? The velocity of l_n is tangential to the circumference of the circle, perpendicular to the respective radius. The direction of the velocity changes continuously. At any given time t, each l_n has velocity w in a particular direction. At any given time t, however, l_n is also at rest in an inertial system S'_n, which moves with velocity w in the same direction as l_n.

This applies to all n sections l_n, all S'_n have the same velocity w relative to S, their directions are different. Since l_n is at rest in S'_n at time t, it has a length shortened by $(1-w^2/c^2)^{1/2}$ as seen from S, like any other rod l'_n at rest in S'_n.

$$l'_n = l_n\sqrt{1-w^2/c^2}$$

Use has been made of the fact that l_n is the rest length of l_n in S at the beginning of the acceleration and in S'_n at time t.

Psychologically, one could object:

l_n does not move in a straight line and uniformly at time t, l_n is accelerated. That's right, l_n has a certain location at time t, a certain velocity of amount w and a certain acceleration of amount w^2/r . But for the Lorentz contraction, only w matters, that is unanimous opinion and follows from the Lorentz transformations. Quite analogously, clocks moving on circular paths suffer time dilation

$$t'_n = t_n/\gamma$$

(s. ch. 3.2), the acceleration is also irrelevant for them, which has additionally been proven experimentally [14]. For circular accelerations, the time dilation is the same as for a rectilinear motion of the same speed.

Ideal scales and ideal clocks are assumed, i.e. those that are not damaged by acceleration. In thought experiments, but usually also in practice, this is fulfilled. Moreover, as has been shown, one can keep the accelerations arbitrarily small by choosing r very large.

Something different is the question of what the conditions look like in a reference frame at rest on the circular ring. In such an accelerated reference frame, the geometry of space becomes non-Euclidean. This need not be known here and will be dealt with later (ch.10) for other reasons.

Now we carry out another, similar thought experiment. As shown in Fig. 9.3 b.), a circular ring divided into n parts is to be accelerated, each l_n is connected to the pivot point via a spoke of length r. Each l_n has a rocket that is ignited at the same time as the others and ejected at a later, common time.

At time t, the situation is as in Fig. 9.3 b.). The l_n have the velocity w and the following applies

$$l'_n = l_n/\gamma$$

because each l_n is at rest in a certain inertial system S'_n at time t. This S'_n moves with the same speed in the same direction as l_n at time t. The lengths of the spokes do not change because they are moved perpendicular to their length. The Lorentz contraction only shortens lengths in the direction of motion, not those perpendicular to it. In the formulas (4.1) of the Lorentz transformation, $z' = z$ and $y' = y$. The fact that the spokes are accelerated and subject to centrifugal forces does not matter, as shown in the thought experiment before, because the forces can be compensated.

As the l_n shorten, gaps grow between them. The size of a gap $l_ü$ is the difference between l_n and l'_n:

$$\begin{aligned} l_ü &= l_n - l'_n \\ &= l_n\left(1-\sqrt{1-w^2/c^2}\right) \end{aligned} \tag{9.26}$$

because the circumference of the circular disc, as it is determined by the spokes of radius r, remains

$$\begin{aligned} U &= 2\pi r \\ &= n l_n \end{aligned} \tag{9.27}$$

because r does not change or should not change.

In these considerations, the circular arcs l_n were treated as straight, which is all the more justified the larger n is chosen.

We see: Viewed relative to S, bodies accelerated on circular paths shorten by the Lorentz factor $(1 - w^2/c^2)^{1/2}$, but the distances between them increase. On the other hand, if we consider bodies at rest in an inertial system and their distances from inertial systems moving relative to them, both decrease by the Lorentz factor in the same way if the relative speed increases. The situation becomes even more complicated for a rolling wheel, Fig. 9.4 and rotating discs, Fig. 10.6. The EI may be asked, how does the four-dimensional space-time continuum recognise the difference between physical rods and distances between them? Between elastically connected and solid bodies? In any case, as expressed by KRAUS [187] there are two types of Lorentz contraction, that of rigid scales (bodies) and that of distances between them - in accordance with the considerations of ch. 9.3, but there is no explanation from the EI for the difference. This is discussed in more detail in ch. 10.5 – 10.7 (Ehrenfest paradox, Sagnac effect) and ch. 11, 24.1-3..

Fig. 9.5 Three absolute effects

LI: All three absolute effects can be explained, since an excellent reference frame can be assumed.
EI Each section of the moving circular ring is untruncated relative to its co-moving inertial frame and thus - contrary to observation - all are untruncated. The EI lacks a selection criterion that invalidates this paradoxical argument. This is also shown in the following chapters, e. g. the Sagnac effect.

As a similar problem, consider the absolute effects of Fig. 9.5. In order to disprove the Newtonian proof of an absolute space, EINSTEIN included Mach's principle (see chapter 11.7) into his considerations. This would have to be extended in order to be able to explain the additional, relativistically caused, absolute effects, Fig. 9.5 b.) and c.) - despite many efforts, however, not even the actual Mach's principle can be verified, not even with general relativity, see the author's contribution in [328]. So, if we assume that Mach's principle does not allow for an explanation, then EI has a problem, but not LI, because the following applies to LI: The faster a length element moves, the shorter it is. Regardless of whether the resting circular ring and its center are at rest in an absolute sense - which is not recognisable because of the principle of relativity - the rotating circular ring moves faster overall and is therefore shorter. For the EI, the absolute effects of Fig. 9.5 are correctly calculated for a fixed, given inertial system, but relative to the co-moving inertial systems in Fig. 9.5 b.) and c.) this is not true. In this case, the rotating ring is - contrary to observation -

untruncated. Analogously for the rotating clock. The obvious reason: Here the EI lacks a selection criterion that invalidates this paradoxical argument.

10. Paradoxes

10.1. Introduction

Paradoxes are apparently contradictory assertions. In contrast to genuine internal contradictions (antinomies) a paradox also called a paradoxy or paradoxon, can be explained logically without contradiction.

The paradoxes of special relativity owe their origin primarily to the following arguments: If two lengths move relative to each other, each of the other appears to be shorter; if two clocks move relative to each other, the other shows a shorter time, but at most one of the two lengths can actually be shorter than the other, at most one of the two clocks can actually run slower.

The question is not whether the normal concepts of length and time fit the principle of relativity. but whether results predicted by relativity theory contradict themselves. However, only measurement results are predicted. Therefore, one must reformulate the above arguments and ask: Are measured lengths and measured times for the same scales and the same clocks for one and the same inertial system predicted differently by the theory? Allowed would be relative to different inertial systems.

The most important approach to solving the paradoxes consists in a correct application of the Lorentz transformations. In particular, this takes into account the relativity of simultaneity. In the case of the twin paradox, also accelerated motion is to be investigated.

The paradoxes of length contraction for rectilinear movements, such as the garage problem, and for circular movements, that of EHRENFEST, as well as the famous twin paradox for time dilation are discussed. The latter in detail on the basis of the special and general theory of relativity, because in the technical literature there are various solutions that contradict each other. Precisely here lies a possibility to distinguish between the relativistic and Lorentz interpretations, as the author believes. The reader should be able to decide for himself on the basis of an investigation discussed from all sides, but without prolixity. The paradoxes concerning the 'Thomas precession' - they occur in a sequence of Lorentz transformations in different spatial directions (not only in the direction of the x, x' -axis) - are examined by the author in [328].

Anyone who has understood the following considerations will recognise the solution idea that is applied again and again: Make a real experiment out of the paradox and calculate it relative to different inertial systems. If these predictions are free of contradictions, the paradox is considered solved. What still remains of the paradox is considered only "philosophical", since a contradiction with physical experience, i.e. with the experiment, is excluded. Negatively formulated, this means that one avoids the philosophical-scientific-theoretical question of what, for example, length of moving body actually is. This reproach does not apply to LI because of chapter 9.

The following contradictions remain in the author's opinion:
Bell's spaceship paradox: contradiction in the cause of Lorentz contraction, Sagnac effect: speed of light is not equal to c everywhere, twin paradox: antinomian predictions about the age of twins, Ehrenfest paradox: Lorentz contraction is something real but there are inertial frames in which length *dilation* is supposed to occur.

First and foremost, the following is about correctly applying the Lorentz transformations.

10.2. Garage problem

A garage has the resting length l_G of 2 m, a car resting length l'_A of 8 m. The garage is at rest in *S*, the car in *S'*, *S* and *S'* move relative to each other with the speed *v*. *v* is 290 000 km/s, and thus $\gamma = 4$.

The paradox is: viewed from *S*, the car has the length

$$l_A = l'_A/\gamma$$
$$l_A = 2\ m$$

Because of $l_G = l_A$ the car just fits into the garage. Viewed from *S'*, the garage has the length

$$l'_G = l_G/\gamma = 0.5 \text{ m}$$

It's $l'_G < l'_A$, the car doesn't fit in the garage.

The solution is: Assume that the garage has two doors T_1 and T_2, T_2 is open, T_1 is closed, so that the car can drive into the garage unhindered, but will bump into door T_1. Just before the top of the car touches gate T_1, it is opened. This would be the case in *S* at the time t_1. Shortly after the rear of the car is in the garage, gate T_2 is closed, which is the case in *S* at time t_2 .

In *S* then applies $$t_1 = t_2$$
because the whole car just fits into the garage, door T_1 is opened and door T_2 is closed at the same time.

It is valid in *S'* because of the Lorentz transformations:

(10.1)
$$t'_1 = \gamma(t_1 - vx_1/c^2)$$
$$t'_2 = \gamma(t_2 - vx_2/c^2)$$

(10.2)
$$t'_1 - t'_2 = \gamma\left(t_1 - t_2 - \frac{v(x_1 - x_2)}{c^2}\right)$$
$$= -\gamma v l_G/c^2$$

Here x_1 and x_2 are the *x*-coordinates of the gates T_1 and T_2, t'_1 and t'_2 the t_1 and t_2 in *S'* corresponding times, see Fig. 10.1.

As you can see

(10.3) $$t'_1 < t'_2$$

t'_1 is earlier than t'_2.

The following happens when viewed from S': The car drives into the garage and approaches T_1, T_1 is opened, the car drives on, the rear passes T_2, T_2 is closed, the top of the car has already left the garage at this point. What happens simultaneously in S happens one after the other in S' (relativity of simultaneity). and in both inertial systems the process is described correctly, from S the car just fits into the garage, from S' it does not.

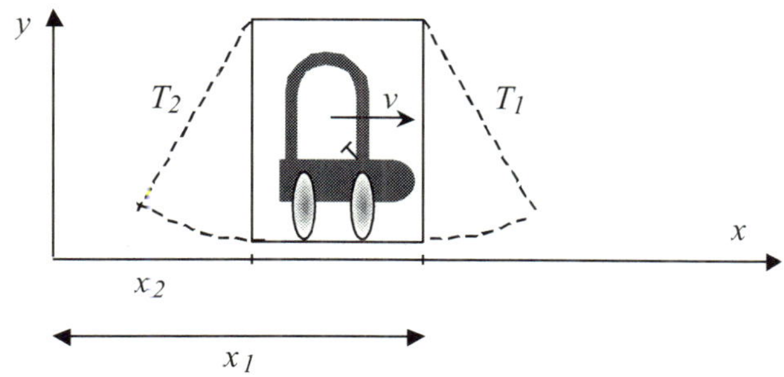

Fig. 10.1 The garage-car paradox

A car approaches with v, it experiences a Lorentz contraction and therefore fits into a garage that is too short; but for the car driver, the garage moves, it experiences a Lorentz contraction, the garage becomes even shorter and is therefore too small for the car in any case. T_1, T_2 Garage doors, x_1, x_2 start and end of the garage.

10.3. Wood's paradox

The Wood's Paradox [408] arises from the famous experiment by A. H. FIZEAU around 1850 to determine the speed of light. The principle is sketched in Fig. 10.2, as far as it is relevant here. Two gear wheels mounted on an axle at a distance L rotate with the angular velocity w_α. A beam of light falls from the left onto gear wheel Z_1, it is chopped into flashes of light by the teeth. A flash of light travels the distance L in the time $\Delta t = L/c$ and hits exactly a gap of gear wheel Z_2 - how does an observer in S', which moves relative to S with v, see the situation? He says that L is contracted by the factor $1/\gamma$, the speed of light remains constant at c. The flash of light is therefore faster at gear Z_2 and does not yet find a gap there.

Since a rotating gear wheel represents a clock (you only have to count the gaps that pass by and then know what time it is), the gears rotate slower by a factor of $1/\gamma$ when viewed from S'. This increases the effect even more.

To solve the paradox, it is shown: When a flash of light arrives at the gear wheel Z_2, it rotated by the angle α

(10.4)
$$\alpha = w_\alpha t$$
$$= w_\alpha L/c$$

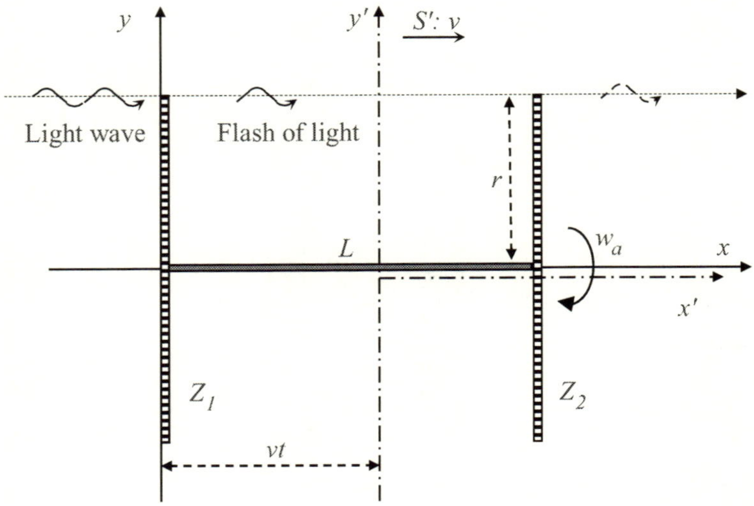

Fig. 10.2 Wood's paradox

Gear wheel Z_1 and Z_2 rotate around a common axis at a distance L. A flash of light hits a gap in Z_2. Viewed from S', L is shortened, the running time of the light flash is shorter, and Z_2 also rotates more slowly: the light flash does not hit the gap.

regardless of whether it is calculated relative to S or S'.

$\Delta t, \Delta t'$ are the transit times of the light flash in S and S' respectively.

Question 1: When does the flash of light arrive at Z_2?

In S: Start $\quad x_1 = 0, \quad y_1 = r, \quad z_1 = 0, \quad t_1 = 0$

Destination $\quad x_2 = L, \quad y_2 = r \quad z_2 = 0, \quad t_2 = L/c$

With the help of the Lorentz transformation applies in S':

Start $\quad x'_1 = 0, \quad y'_1 = r, \quad z'_1 = 0, \quad t'_1 = 0$

(10.5) Target
$$x'_2 = \gamma(L - vL/c) = \gamma L(1 - v/c)$$
$$y'_2 = r$$
$$z'_2 = 0$$

(10.6)
$$t'_2 = \gamma(L/c - vL/c^2) = \gamma L(1 - v/c)/c$$

In S', too, the light has propagated with c, as the quotient of (10.5) and (10.6) shows.

Question 2: How long has Z_2 been rotating in S' when the flash of light arrives at Z_2?

Not only for the duration $t'_2 - t'_1$, because in S' gear Z_2 at the start has the time t'_{Z2}

(10.7)
$$t_{Z2}' = \gamma(0 - vL/c^2)$$
$$= -\gamma vL/c^2$$

This applies because of the relativity of simultaneity. (In S $t_1 = 0$ also applies at the position of Z_2)

In S', Z_2 rotates for the duration of $\Delta t'_{Z2}$

(10.8)
$$\Delta t'_{Z2} = t'_2 - t'_{Z2}$$
$$= \gamma L/c \qquad \text{(note sign)}$$

In addition, the gear wheel Z_2 turns slower by a factor of $1/\gamma$:

$$w'_\alpha = w_\alpha/\gamma$$

(If you want to apply the Lorentz transformations here too, which would be very correct, think of a clock at Z_2, it goes as fast as Z_2 turns. The clock at Z_2 in turn goes as fast as an identical clock at Z_1. The calculation for this clock: see ch. 4.4)

Overall, the angle of rotation is α

(10.9)
$$\alpha = w'_\alpha \Delta t'_{Z2}$$
$$= w_\alpha L/c$$

as in S.

When the flash of light arrives at the gear wheel Z_2, it has rotated through the same angle relative to S and S'.

10.4. Lid paradox

In an inertial system S, a lid approaches an inspection shaft at rest in S from the left. As the lid moves, it is shortened and falls into the inspection shaft if both fit at rest. An observer resting on the lid sees the inspection chamber approaching. For him, the inspection chamber is shortened and the lid is therefore too big to fall through. [118]

To solve the paradox, specify the experimental procedure:

We assume, see Fig. 10.3, that the shaft approaches in S with the speed w perpendicular to the x-axis the lid approaching with v from the left. At the time t_0 they meet. The lid fits exactly and slides into the shaft.

Viewed from S', the lid rests, the shaft opening approaches and is shortened. The lid is thus larger than the shaft opening. The exact calculation shows, as can also be seen in Fig. 10.4, that the shaft opening approaches the lid at an angle, the right lid edge and the right shaft opening meet earlier than the left lid edge and the left shaft opening. (Relativity of simultaneity, in S both events are simultaneous.) The lid is caught from right to left, although it is larger than the shaft opening.

Now we have to calculate. First in S. At the time t_0 the left and right shaft edges meet with the left and right lid edges at the position x_l and x_r

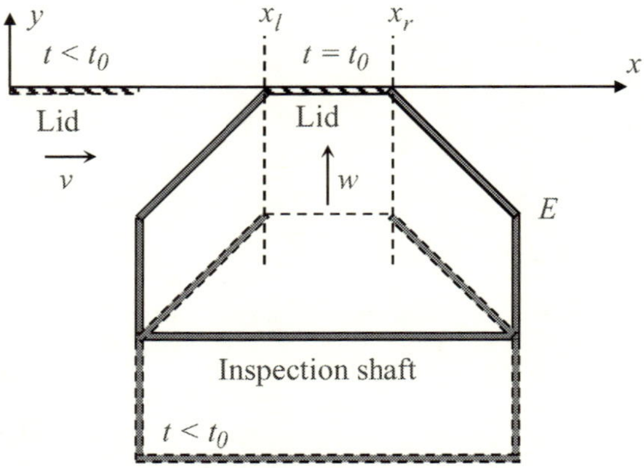

Fig. 10.3 Lid paradox

Lid and inspection shaft approach each other. At the time t_0 the lid sits on the shaft. For an observer resting on the lid, the shaft approaches, the shaft opening is shortened and the lid does not fit. *($\gamma = 2$, $w = v = 0.866\ c$)*

- Fig. 10.3 - and slide past each other. How this looks in *S'* is shown by the multiple application of the Lorentz transformations. (An elegant but intellectually more complex solution can be found in [59]).

t_0, x_r corresponds in *S'*:

(10.10) $$x'_r = \gamma(x_r - vt_0)$$

(10.11) $$t'_r = \gamma(t_0 - vx_r/c^2)$$

t_0, x_l corresponds in *S'*:

(10.12) $$x'_l = \gamma(x_l - vt_0)$$

(10.13) $$t'_l = \gamma(t_0 - vx_l/c^2)$$

t_l' is unequal t_r' and shows the relativity of simultaneity.

Considered in *S'*, the left lid edge meets the left shaft edge at time t'_1 at location x'_l and the right lid edge meets the right shaft edge at time t'_r at location x'_r. t'_l is greater than t'_r i.e. the event t'_r occurs earlier, as Fig. 10.4 confirms. In *S*, both events occur at the same time t_0.

Where, at which position x'_L, is the left edge of the shaft in *S'* at the time t'_r? *We know: x_l is constant for all times in S and we therefore know from the four quantities*

(10.14) $$x'_L, t'_r, t_L, x_l$$

t'_r and x_l and can calculate the others - x'_L and t_L - from the Lorentz transformations. For t_L this is done as follows:

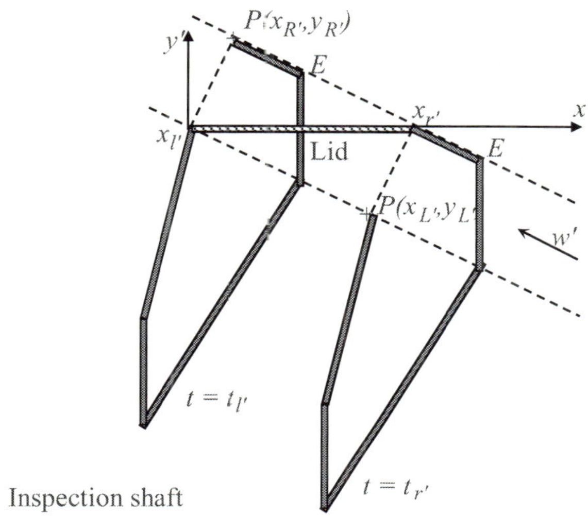

Fig. 10.4 Lid paradox

In S' the lid rests and is larger than the opening of the inspection shaft, the inspection shaft comes flying at an angle and catches the lid from right to left.
($\gamma = 2$, $w = v = 0.866$ c)

$$t'_r = \gamma(t_L - vx_l/c^2)$$

(to t_L dissolve)

$$t_L = t'_r/\gamma + vx_l/c^2$$

(for t'_r insert formula (10.11))

$$= \gamma(t_o - vx_r/c^2)/\gamma + vx_l/c^2$$

(γ shorten)

$$= t_o - v(x_r - x_l)/c^2$$

i.e. t_L is earlier than t_0.

The fourth quantity x'_L of (10.14) can be easily calculated, since t_L and x_L are now known:

(10.21)
$$\begin{aligned} x'_L &= \gamma(x_l - vt_L) \\ &= \gamma(x_l - v(t_0 - v(x_r - x_l)/c^2)) \\ &= \gamma(x_l - vt_0 + v^2(x_r - x_l)/c^2) \\ &= x'_l + \gamma v^2(x_r - x_l)/c^2 \end{aligned}$$

At what distance from the x-axis is the left edge of the shaft at the time t_L? It moves in S with the speed w and reaches the x-axis at the time t_0. It covers the distance until then:

$$y_L = w(t_L - t_0)$$

$$= w(t_0 - v(x_r - x_l)/c^2 - t_0)$$

(10.23) $$= -wv(x_r - x_l)/c^2$$

Furthermore: $$y_L = y'_L$$

Accordingly, we now ask: Where, at which point x'_R is the right-hand edge of the shaft in S' at the time t'_l? We know again: x_R is constant for all times in S and therefore we know from the four variables

(10.24) $$x'_R, t'_l, t_R, x_r$$

t'_l and x_r and can calculate the remaining quantities x'_R and t_R from the Lorentz transformations. For reasons of symmetry, however, in (10.21) and (10.23) we only interchange x'_l with x'_r, x_r with x_l and x_l with x_r.

(10.25) $$x'_R = x'_r - \gamma v^2(x_r - x_l)/c^2$$

(10.26) $$y'_R = +wv(x_r - x_l)/c^2$$

i.e. now the left lid edge and the left shaft edge meet after the right edges have passed each other and formula (10.25) and (10.26) describe the position of the right shaft edge in S' at this point t'_l. The larger lid in S' is captured by the smaller shaft from right to left, as graphically illustrated in Fig. 10.4.

(The more elegant solution [59] uses a formula for the slope of the straight line along which the shaft edges move:

(10.28) $$\frac{y'_L}{x'_L - x'_l} = \frac{-wv(x_r - x_l)/c^2}{\gamma v^2(x_r - x_l)/c^2} = \frac{-w}{\gamma v}$$

and agrees with our account except for the naming).

With the parameters chosen in Fig. 10.3 and 10.4, the lid collides with the shaft at the corner E. Because of the relativity of simultaneity, this happens in S when the lid has completely slid into the shaft, in S' earlier.

The idea we applied to solve the lid paradox is: The solution immediately obvious in S due to the sensibly chosen geometry in Fig. 10.3 is converted to S' with the Lorentz transformations and the result is drawn. It must be taken into account that the simultaneous event "lid fits" in S means two different events "left sides touch" and "right sides touch" in S'.

10.5. Ehrenfest paradox

The Ehrenfest paradox deals with rotating circular discs, special cases had already occupied us in ch. 9.4 and similarly ch. 9.3 (Bell's spaceship paradox). A knowledge of these special cases facilitates the understanding of the more complex Ehrenfest paradox. In Fig. 10.5, the circumference of a rotating circular disc should undergo a Lorentz contraction, the radius should not, because it is perpendicular to the direction of motion, but both together do not fit the circular formula.

(10.30) $$U = 2\pi R$$

EHRENFEST [88] describes it like this:

"A body behaves in a relatively rigid manner, i.e. it deforms continuously during any movement in such a way that each of its infinitesimal (i.e. arbitrarily small) elements exhibits at each moment for an observer at rest just that Lorentz contraction (relative to the state of rest) which corresponds to the instantaneous velocity of the center of the element."

"Some time ago, when I was trying to visualise the consequences of this approach, I came across corollaries that seem to show that the above approach leads to contradictions even for some very simple types of movement."

This includes "... the uniform rotation about a fixed axis. In fact: Let there be given a relatively rigid cylinder of radius R and height H. Let it be given a gradual rotational movement about its axis which finally remains constant. Let R' be the radius that it has for a stationary observer during this movement. Then R' would have to fulfil two contradictory requirements:

a) The periphery of the cylinder must show a contraction compared to the resting state:

(10.31) $$2\pi R' < 2\pi R$$

because each element of the periphery moves in its direction with the instantaneous velocity $R'w_\alpha$ (w_α = angular velocity).

b) If one considers any element of a radius, its instantaneous velocity is normal to its extension; thus, the elements of a radius cannot show any contraction relative to the state of rest. It should be:

(10.32) $$R' = R \quad ..."$$

From (10.31) it follows naturally: $R' < R$

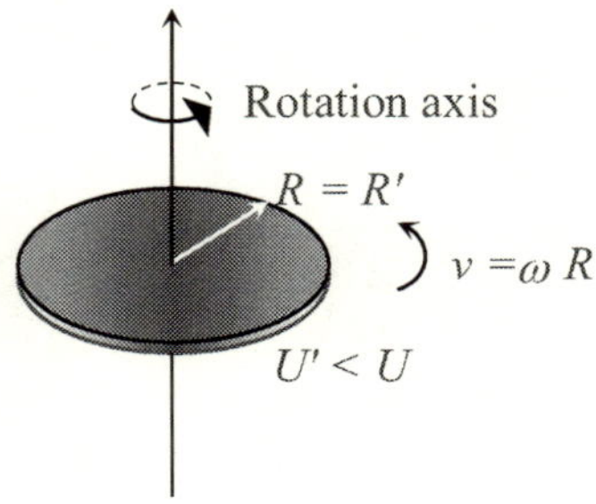

Fig. 10.5 Ehrenfest paradox

A disc with radius R and circumference $U = 2\pi R$ is set in rotation. Since the radius is always perpendicular to the direction of movement, its length does not change. The circumference is parallel to the direction of movement and undergoes length contraction, so that $U' < 2\pi R$ should apply.

EHRENFEST's paradox differs from the preceding paradoxes in that no different prediction is made here depending on the choice of reference system. In a moving inertial frame, the rotating disc no longer has a circular shape because a relative velocity is superimposed on the rotation. The radii are also contracted differently, but the circumference of the rotating disc is smaller than that of a non-rotating one, no matter which reference system is used and as long as centrifugal forces do not play a role. A corresponding calculation leads to the same result for any reference system, only the mathematical effort is different. We can dispense with this calculation, since the literature does not claim that the results are different for different inertial frames. A special case are the inertial systems that move with the rotating circumference: s. ch. 24.2. What is paradox now? With the considerations from chapter 9.4 only little. Only paradoxical in the sense of surprising, extraordinary is the thesis founded there that the Lorentz contraction can be something real and generate real forces. To solve the paradox, one must assume that with increasing rotation, the Lorentz contraction wants to shorten the circumference of the disc in real terms, and that this results in tension forces which try to compress the radii. This is then counteracted by the elastic forces in the disc since the radii only change their lengths under the influence of forces. Centrifugal forces act in the same direction, so that in real terms the circumference of the circular disc can rise. In any case, an equilibrium is reached in which the actual radius and the actual circumference fit together via the circular formula. (Thin discs could also bulge, as H. E. IVES [167] rightly assumes).

Neglecting centrifugal forces, the circumference U decreases because of Lorentz contraction, but to a lesser extent than predicted by special relativity because of the elastic counterforces of the circular disc. This is neither paradoxical nor antinomian. For the Lorentz interpretation, this can be seen first of all for the excellent inertial system, if the center of the rotating circular disc rests there. On the basis of the principle of relativity the same predictions apply to other inertial systems if the center of the rotating disc rests in them, otherwise one would have an experiment such as the one sought in chapter 7. The same is to be expected from the explicit calculation, after all, the Lorentz transformations are valid for all inertial systems as far as measurable effects are described. As a variant, construct a ring-shaped disc from small elements while it is already rotating, see Fig. 10.6. If we disregard the centrifugal forces, the rotation is stress-free [150]. Stress forces only occur when the disc is braked, because then the Lorentz contractions of the ring elements decrease and the ring-shaped layers expand, the further out the stronger.

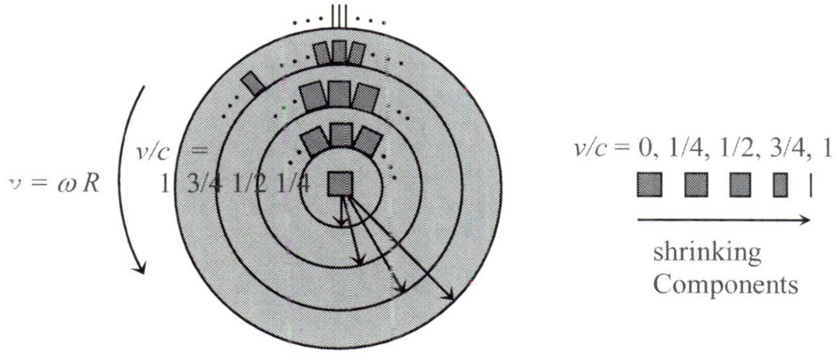

Fig. 10.6 Structure of a disc during rotation.

If you assemble a disc while it is rotating, relatively more components fit into the outer rings than into the inner rings, because the components contract with increasing speed [150]. Deceleration destroys the disc. (Here the Lorentz contractions are real, if an observer changes the reference frame, they are apparent).

For the Einstein interpretation, the argumentation is analogous, one only has to accept that Lorentz contractions lead to tension forces if any length elements are prevented from contracting, as is the case for discs set in rotation. There are real changes of (relative-) lengths during accelerations of bodies and apparent changes when changing reference systems - the Einstein interpretation only becomes contradictory if one disputes this difference.

One of the first solutions of the Ehrenfest paradox was made with the help of a Lorentz interpretation by H. A. Lorentz [202] 1921 and H. E. Ives [167] 1939, but several other physicists may have had the basic idea as well. Much later is a thorough, relativistic work by Ø. Grøn, 1979 [132] who correctly states: "For sufficiently high velocities, the material of the disk can break open radially due to Dewan-Beran stresses (see ch. 9.4)." W. Pauli [257] can also be quoted approvingly when he says:" ... Ehrenfest has shown that a rigid body cannot be set in rotation." That is to say: an ideally rigid body can suffer Lorentz contractions, but under the action of tension forces it does not change otherwise it would not be ideally rigid. It cannot therefore be made to rotate since it has to be cracked, stretched or compressed because the Lorentz contraction must take place. In the same way it could be said: "Ehrenfest has shown that a rotating rigid body cannot be slowed down." A thesis that Einstein also agreed with in letters [353].

A completely different question is how the circumference and the radius of a rotating circular disc in the co-rotating, i.e. accelerated, reference frame are related to each other. This is the subject of general relativity and C. W. BERENDA [19] has investigated it. This is dealt with in chapter 10.8 for other reasons and the consideration of C. MØLLER [235] is used. For the Ehrenfest paradox, R' and $U' = 2\pi R'$ - radius and circumference of a rotating circular disc - are measured quantities in the resting inertial frame S and the question was: If U' is measured shortened according to the special theory of relativity and then R' is also shorter, but then contrary to the relativistic prediction, or if R' is constant or even larger than R, then U' cannot have changed according to the prediction of the special theory of relativity. On the other hand, it can no longer be claimed that rotating discs and rotating inertial systems are equivalent. The rotation of ideally rigid disks cannot be changed at all and elastic disks change depending on the material. When changing between rotating reference systems, these differences do not exist.

It need not be a coincidence that the first solution of the Ehrenfest paradox was found with the help of a Lorentz interpretation, for it has didactic merits. The catastrophic variety of relativistic solution attempts also speaks in favour of such a thesis. T. E. PHIPPS [263] has summarised them in a table; in Fig. 10.7 they are listed, supplemented by our proposed solution. As can be seen, even the elementary questions - does the circumference and/or radius of the disk change relative to the inertial frame S'? - are altogether unanswered, there is no reference to the limiting cases of Fig. 9.3, for large radii straight lines and circles cannot be distinguished and the theoretical solutions would have to change into normal Lorentz contractions - all these deficiencies speak didactically for the Lorentz interpretation, if force effects caused by the Lorentz contraction have to be considered

Neither are there detailed, material-dependent calculations for rotating disks available yet, nor are there corresponding experiments. For the calculations, the relativistic version of Hoke's law, Ø. GRØN [133], together with the Lorentz transformations, should suffice, for the experimental verification an experimental setup as in [265], [201], [183], [383], [148], [353], [403], [385], [161], [306]. Given the abundance of proposed solutions, the author does not expect uniform agreement with his explanations. The mathematical and theoretical-physical treatments of the alternatives are impressive, each in itself justifies experimental verification, in view of the entire spectrum of solutions they should be carried out.

The Ehrenfest paradox suggests that Lorentz contractions can lead to real reductions in length. On the other hand, the rest lengths of scales remain unchanged and one could say that scales viewed from moving inertial systems only appear shorter, but are not. For the discussion of the question: What is real about Lorentz contractions? - it is of great importance for relativistic space-time theory - it is better, however, to consider the limiting case of a circular ring, as in Chapter 9.4. There, the Ehrenfest paradox has already been discussed from this point of view.

Attributes	I	II	III	IV	V	VI	VII
Freely moving scales at the edge of the disc experience a Lorentz contraction relative to S:	yes	yes	yes	yes	yes	no	yes
Disc circumference results in a Lorentz contraction relative to S:	no	yes	yes	yes	no?	no	yes
Disc radius contracts relative to S:	no	no	yes	yes	no	no	yes
Disc radius dilated relative to S:	no	no	no	no	yes	no	no
Tension forces along the circumference:	yes	no	no	no	yes?	no	yes
Radial tension forces:	no	no	yes	yes	yes	no	yes
Disc bulges:	no	no	no	yes	no	no	no
Gaps appear between freely movable scales at the edge of the disc:	yes	no	no	no	yes	no	yes

Fig. 10.7 Table on the Ehrenfest paradox

The Variety of „Solutions" to the Ehrenfest Paradox, taken from T. E. PHIPPS [263]. Solution IV (thin disc) and VII, as well as the limiting cases of Fig. 9.3 are represented here. There is no agreement on how the radius and circumference of a rotating disc change, so an experiment would be helpful.

The Ehrenfest paradox is generally passed over in relativistic textbooks, usually tacitly, often with the argument that circular motions are accelerated motions and therefore the subject of general relativity [all]. It should be noted in this regard:

a) U' and R' are measurable quantities for a rotating circular disc in the inertial system S in which the disc center rests. Even if U' and R' are supposed to be predictable only on the basis of general relativistic considerations, the argument could only be convincing if at least the result of these considerations (what value do U' and R' have?) is mentioned, which does not happen.

b) General relativity is responsible for accelerated reference systems, but not for accelerated movements relative to inertial systems. Thus FRENCH [115] states. : "Since Einstein based his general theory of relativity ... on the dynamical equality of an accelerated laboratory and a laboratory under the influence of a gravitational field, it is sometimes claimed or concluded that special relativity is not suitable for treating accelerated motion. However, this idea is wrong." Thus, general relativity is even competent for unaccelerated rectilinear motions - provided they are considered from accelerated frames of reference. [235]

c) On the one hand, it is correct to say that the Lorentz contraction is the consequence of a relativity of simultaneity and not the consequence of a force [59], but on the other hand this is not generally true, as the Ehrenfest paradox shows and the special cases in Chapter 9 have shown (see also the discussion with W. T. HERING, ch. 24.2).

10.6. Bell's spaceship paradox

s. ch. 9.3 and 24.3

10.7. Sagnac effect

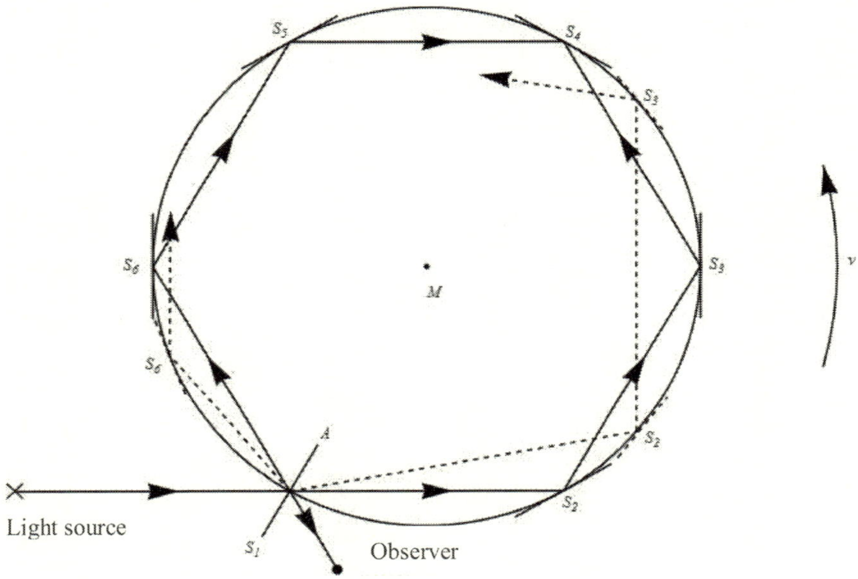

Fig. 10.7.1 Sagnac effect

Light from the source × is split into two coherent beams by the semi-transparent mirror S_1 at A and reflected at further mirrors S_2, S_3 ... S_6. At A, the beams meet again and interfere. (Ray path at rest:____, in motion: ......) Here $n = 6$, if you increase the number of mirrors, you approach a circular path. Then the mirrors S_i, $i = 1,2 \ldots n$, become a cylinder, the inside of which is mirrored, and the mathematical formulae become particularly simple. For finite values of n, on the other hand, the underlying inertial systems are particularly illustrative. Relative to the rotating disc, in contradiction to the experiment, a zero effect results for the EI, since light has the value c in all inertial systems in all directions, regardless of how the disc rotates. Relative to the laboratory system, both interpretations are consistent with observation.

In this experiment, a light source ×, the observer B, the mirrors S_2 ... S_6 and the semi-transparent mirror S_1 are located on a rotating disc, see Fig. 10.7.1. The light emitted by × is split by S_1 into two beams of opposite direction. After one round trip, both beams interfere at S_1, the interference is observable from B. For the basic understanding, a measuring arrangement with a finite number of mirrors as in Fig. 10.7.1 is preferable, for the quantitative calculation the limiting case of an infinite number of mirrors, because then the light in a mirrored cylinder moves circularly along the edge of the disk.

Relative to the laboratory system S, the time difference for the EI and LI is calculated according to the same formulae. For the LI, these formulae initially only apply to the excellent inertial system, but because of the principle of relativity, they describe what is measured relative to each inertial system. Light signals in the direction of rotation require back to S_1 the time t_+, in the opposite direction of rotation the time t_-, L is the circumference of the disc rotating with v measured in S, i.e. L is the Lorentz-contracted value. The following applies, see Fig. 10.7.2 + 10.7.3.

(10.35) $$ct_+ = L + \Delta L_+$$

$$vt_+ = \Delta L_+$$

(10.36) $$t_+ = \frac{L}{c-v}$$

Accordingly $$ct_- = L - \Delta L_-$$

$$vt_- = \Delta L_-$$

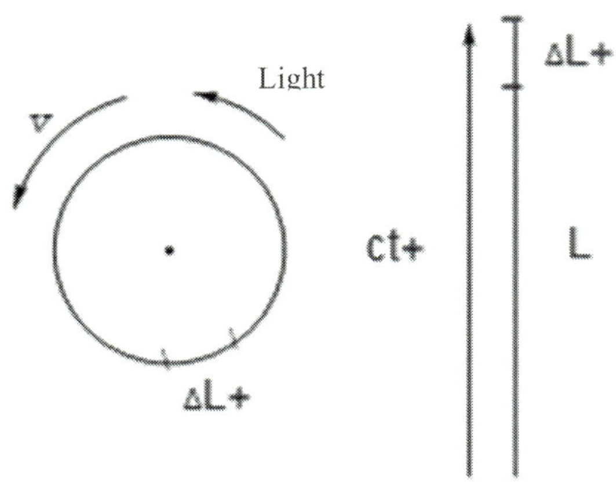

Fig. 10.7.2 The light path in the direction of movement and unrolled.

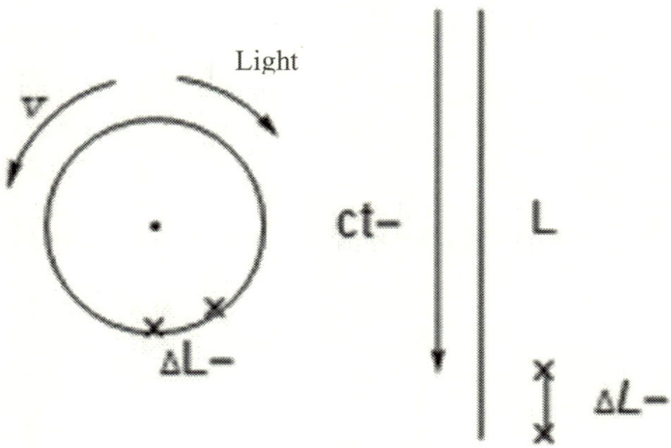

Fig. 10.7.3 The light path against the direction of movement and unrolled.

(10.37) $$t_- = \frac{L}{c+v}$$

(10.38) $$t_+ - t_- = \frac{L}{c-v} - \frac{L}{c+v} = \frac{2Lv}{c^2(1-\beta^2)}$$

Formula (10.38) is confirmed in many ways by experiment, SELLERI [331].

Relative to the coordinate system $S^{Scheibe} = S^{disc}$ which is connected to the disc, measurements are made with scales and clocks at rest there. The predictions for the measured interferences must not change. This is true for the LI, but not for the EI.

LI:

(10.39) $$t_+^{Scheibe} = t_+\sqrt{1-\beta^2}$$

(Clocks on the disc experience time dilation).

(10.40) $$t_-^{Scheibe} = t_-\sqrt{1-\beta^2}$$

(10.41) $$L^{Scheibe} = \frac{L}{\sqrt{1-\beta^2}}$$

(because of the Lorentz contraction)

What is the speed of light in the direction of rotation and in the opposite direction - relative to the edge of the disc, measured with clocks and scales at rest there?

(10.42) $$c_+^{Scheibe} = \frac{L^{Scheibe}}{t_+^{Scheibe}}$$

$$= \frac{L}{\sqrt{1-\beta^2}\, t_+ \sqrt{1-\beta^2}}$$ (because of (10.39) and (10.41)

(10.43) $$= \frac{c-v}{1-\beta^2} < c$$ (because of (10.36))

Accordingly

(10.44) $$c_-^{Scheibe} = \frac{c+v}{1-\beta^2} > c$$

(10.45) $$t_+^{Scheibe} - t_-^{Scheibe} = \frac{L^{Scheibe}}{c_+^{Scheibe}} - \frac{L^{Scheibe}}{c_-^{Scheibe}}$$

$$= \frac{2Lv}{c^2(1-\beta^2)}\left(1-\beta^2\right)^{1/2}$$

(10.46) $$= \left(t_+ - t_-\right)\left(1-\beta^2\right)^{1/2}$$ because of (10.38).

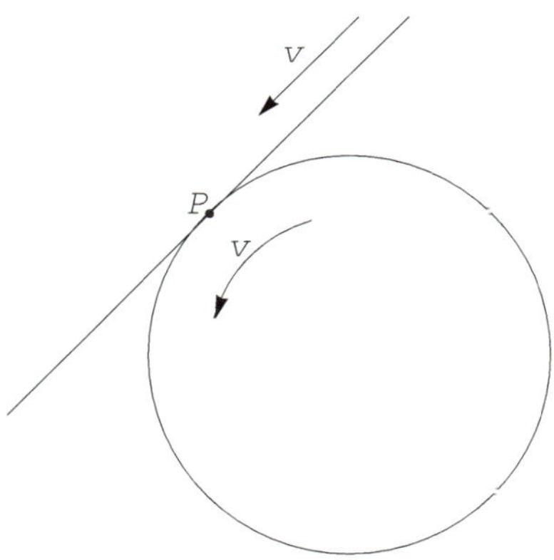

Fig. 10.7.4 Speed of light at point P

For LI, as it must be, except for the factor $1/\gamma$, the same result is obtained for S and $S^{Scheibe} = S^{disc}$. If S is not the excellent inertial system, the formulae describe what is measured because of the RP. For the EI, the formulae (10.43) and (10.44) are wrong. The speed of light has the value c for the EI in every point of the edge of the disk (and also in the rest of the disk), because:

Place the tangent line at any point *P* of the disc, see Fig. 10.7.4. A rod moving along the tangent line with the velocity *v* forms an inertial system in which *P* is at rest. In all inertial systems, light has the velocity c at all points and in all directions, thus also light in and against the direction of rotation. Since *P* is arbitrary, the following applies: Light signals along the circumference of the disc have the speed c everywhere, regardless of whether they are moving in the direction of rotation or in the opposite direction. Therefore, there are no differences in propagation time for the light waves in either direction and the Sagnac effect disappears - in contradiction to the experiment. In formulas, EI:

$$t_{+}^{Scheibe} - t_{-}^{Scheibe} = \frac{L^{Scheibe}}{c} - \frac{L^{Scheibe}}{c} = 0 \qquad (10.47)$$

This consideration does not apply to the LI, since the 'inertial system requirement', s. ch. 10.8.6, is violated, because a physical process is described relative to even an infinite number of inertial systems. In contrast, it is permissible to describe the Sagnac experiment relative to any inertial system, e. g. that for *P* of Fig. 10.7.4, without any contradiction to the experiment. There is no need to do the calculation because the RP holds. Is there an argument for the EI to prohibit the use of infinitely many inertial frames in this case? The author is not aware of any. On the contrary, to solve the Ehrenfest paradox, precisely these inertial systems are introduced by HERING, ch. 24.2.

A different kind of rescue attempt lies in the indication that rotating (i.e. accelerated) reference frames are present here, see CZERNIAWSKI, ch. 8.4 and [190, §89]. Very recent: PETKOV [261]. This leads to the correct formulae (10.44) but does not eliminate the paradox by formula (10.47). It is merely an additional approach that leads to a correct solution. To eliminate the paradox, the 'inertial system requirement' of the LI (or something equivalent) must be applied.

The importance of the Sagnac effect for the Lorentz interpretation (LI) has been thoroughly investigated by SELLERI [328], [331]. SELLERI uses the so-called inertial transformations in his calculations, which are particularly suitable for solving the relativistic paradoxes. Another convincing application example of the inertial transformations, which is not treated in this book, is aberration [331].

Exercise: Explain aberration relative to the inertial frame Earth.

Answer: It fails, see SELLERI [331]. The earth is, to a good approximation, an inertial system, continuously different in the course of a year. Since the telescope always rests in it, the prediction is: no aberration can be seen. But the inertial system requirement is not fulfilled. Relative to the sun, the earth, including the telescope, moves in changing directions, relative to the inertial system "sun" the aberration is calculable. Here, the 'inertial system requirement' is fulfilled.

10.8. Clock or twin paradox

10.8.1. Introduction

The clock paradox by LANGEVIN [191] has been dealt with again and again in countless technical articles [zwi] up to the present. For example, [219] lists over 200 citations on the subject, and this is by no means complete. There is a psychological problem in this; remember, these are experts who in their daily work have to deal with far more complicated subjects than the application of the Lorentz transformations to a straight-line outward and return journey of a space traveller from different inertial frames. There must be a clear answer after a few pages of calculus: It can, or it cannot. The abundance of articles speaks for an unexplained remainder in this paradox, and this will be worked out in the following.

Psychologically, it can be understood why an inner contradiction, if it exists, has no consequences at all in practice: What happens in reality is correctly predicted by the theory of relativity - and without LI only by it. Namely: If a space traveller moves back and forth relative to S in time $2\Delta t$ to a place $x = L$ with the speed v , then the space traveller has aged by $2\Delta t/\gamma$ because of time dilation, i.e. less than the twin who stayed at home. There is no doubt about this theoretical prediction of the special theory of relativity and exactly this effect has been demonstrated by the experiments of HÄFELE-KEATING [143] and [2], see ch. 5.2. What is a flight around the Earth other than a space journey, apart from the difference that atomic clocks and flight personnel have remained younger by only 50 microseconds and not by 50 years?

Since doubts about this impressive prediction are eliminated, the theory of relativity cannot be wrong. It contains - as is to be discussed - a contradiction that must be eliminated. The contradiction means: a consistent application of the Lorentz transformations has the contradictory consequence that both twins are older at the same time and are also not older.

The aim of this chapter is first to analyse the twin paradox step by step on the basis of special relativity. There is a clear position up to which the paradox can be resolved classically.

One does not have to accept this position as a solution. Therefore, in the second step, the general theory of relativity is consulted (in a generally understandable form, without mathematical proofs, but with precise references to the literature). Alternatively to both, as ultima ratio, there is a comparison of the paradox with the situation in the Lorentz interpretation. The latter has the clear advantage of distinguishing one inertial system, and in case of doubt, the considerations apply relative to this system.

The author's aim is to present each of the three positions in an understandable and convincing way; the actual decision is up to the reader.

It should also be pointed out that the paradox of Thomas precession treated in [328] can be solved with the same principles as the twin paradox (inertial system requirement, exclusion of synchronisation effects).

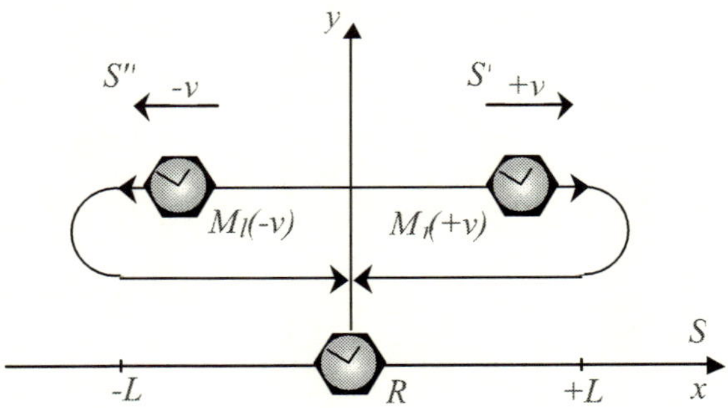

Fig. 10.8 Thought experiment on the clock paradox.

In S the clock R rests at $x = 0$, the clock $M_l(-v)$ moves to the left relative to S with $-v$ to $x = -L$ and from there back with $+v$ to $x = 0$. The clock $M_r(+v)$ moves to the right with $+v$ to $x = +L$ and from there back with $-v$ to $x = 0$. All clocks are at the time $t = 0$ at the position $x = 0$.

10.8.2. Preliminary remark on proper time

The basis of this discussion is the work of McCrea [225] with an addition by Ives [168], given the abundance of literature on this subject, earlier work with the same approach cannot be ruled out. In the following, it is always a question of comparing the proper times of clocks with the times predicted by any observers. Proper times are the times that a moving clock itself displays, i.e. measures.

First the designations: Clocks in the inertial frames S, S', S'' are examined, S' moves relative to S with v in the positive x-direction, S'' has the speed $-v$ and moves in the negative x-direction. Clocks at rest are designated R, $R(x)$, $R'(x')$, $R''(x'')$ are clocks at rest in S, S', S'' at the positions x, x', x'' . Moving clocks are designated with M.

The clock paradox assumes the following motion sequence, see Fig. 10.8: In S, the clock R rests at the position $x = 0$, relative to S, a clock M_l $(-v)$ moves to the left with the speed $-v$ to the position $x = -L$, turns around, and moves from $x = -L$ back to $x = 0$ with the speed $+v$. Correspondingly, a clock $M_r(+v)$ in S moves to the right with the speed $+v$ to the position $x = +L$, reverses, and reaches $x = 0$ again with the speed $-v$ (Instead of clocks, one can imagine twins or space travellers). The paradoxical states arise if one considers the movement times (travel times) of the clocks (twins) relative to each other: R sees the clocks $M_l(-v)$, $M_r(+v)$ constantly moving and conversely $M_l(-v)$, $M_r(+v)$ see clock R moving back and forth. Each predicts time dilation from the other, each clock expects the other clock to lag (stay young).

Acceleration objection:

The clocks $M_l(-v)$, $M_r(+v)$ are accelerated at the points $x = -L, x = +L$, the predictions of the accelerated clocks cannot be taken from the Lorentz transformations, general relativity is competent. This objection rightly exists if one wants to explicitly consider the acceleration phases at $x = -L, x = +L$. However, one can say: the acceleration phases are arbitrarily small relative to the total time of the journey and can be neglected. But this objection is not generally accepted and is therefore examined more closely here. The discussion of acceleration phases themselves is postponed to ch. 10.8.7, where general relativity has its say.

S	***S'***	***S''***
$x = 0$ $t = 0$	$x' = 0$ $t' = 0$	$x'' = 0$ $t'' = 0$
$x = -L$ $t = \frac{L}{v}$	$x' = -2\gamma L$ $t' = \frac{\gamma L}{v}\left(1 + v^2/c^2\right)$	$x'' = 0$ $t'' = \frac{\gamma L}{v}\left(1 - v^2/c^2\right)$
$x = +L$ $t = \frac{L}{v}$	$x' = 0$ $t' = \frac{\gamma L}{v}\left(1 - v^2/c^2\right)$	$x'' = +2\gamma L$ $t'' = \frac{\gamma L}{v}\left(1 + v^2/c^2\right)$
$x = 0$ $t = \frac{2L}{v}$	$x' = -2\gamma L$ $t' = \frac{2\gamma L}{v}$	$x'' = +2\gamma L$ $t'' = \frac{2\gamma L}{v}$
$x = -L$ $t = \frac{2L}{v}$	$x' = -3\gamma L$ $t' = \frac{\gamma L}{v}\left(2 + v^2/c^2\right)$	$x'' = +\gamma L$ $t'' = \frac{\gamma L}{v}\left(2 - v^2/c^2\right)$
$x = +L$ $t = \frac{2L}{v}$	$x' = -\gamma L$ $t' = \frac{\gamma L}{v}\left(2 - v^2/c^2\right)$	$x'' = +3\gamma L$ $t'' = \frac{\gamma L}{v}\left(2 + v^2/c^2\right)$

Fig. 10.9 Table: Coordinates of different positions in *S*, *S'*, *S''*.

Legend: If the clock in *S* at the location $x = 0$ shows the time $t = 2L/v$, it has the position $x' = -2\gamma L$ in *S'* and a clock resting at this location in *S'* shows the time $t' = 2\gamma L/v$

Elimination of the acceleration objection:

The movement of clock $M_l(-v)$ is resolved into the movement of two clocks R'_l and R''_l. The clock R'_l moves up to $x = -L$ like the clock $M_l(-v)$, i.e. it rests in S'. At the point $x = -L$ the clock R''_l comes to meet it, takes over the time from R'_l and flies on to $x = 0$. Correspondingly, the journey of clock $M_r(+v)$ is divided into two clocks $R''_r$and R'_rwhich meet at $x = +L$.

Now all acceleration phases are eliminated and the Lorentz transformations are applicable without restriction. Here i.e. LIEBSCHER [197] agrees. Next, the following comparisons of clock times can be made: The times indicated by R''_l, R'_r and R at the position $x = 0$ and the mutually predicted times. All times are calculated completely relative to S, S', S'', whereby for S' and S'', for reasons of symmetry, these calculations only mean swapping indices. *To avoid complicated notation, we will continue to speak of the clocks $M_l(-v)$ and $M_r(+v)$. These are now imaginary clocks which in reality consist of two real clocks meeting at $x = -L$, and $x = +L$ respectively.*

In the table Fig. 10.9, the essential positions in S have been converted into the corresponding coordinates of S' and S'' with the help of the Lorentz transformations. For example, the coordinate $x = 0$ at the time $t = 2L/v$ in S'' has the value $x'' = 2\gamma L$. One obtains these values by inserting the x, t values into the formulae (4.1) of the Lorentz transformations by elementary calculation.

Exercise: Check the table by your own calculation

10.8.3. Proper times relative to S, S', S''

We now calculate the proper times of the clocks R, $M_l(-v)$, $M_r(+v)$ relative to all three inertial frames S, S', S'' using the Lorentz transformations. The Lorentz transformations are valid relative to any inertial system. Regardless of which one is chosen, the times that can be read on each clock when it returns to $x = 0$ must be the same for each choice of inertial system. The proper time (or own time) of a clock is the time that a clock continuously displays. It is designated, as is usual in the literature, with τ so that we have the following quantities:

$\tau(R)$	the proper time of the clock R at the time $t = 2L/v$
$\tau(M_l(-v))$	the proper time of the clock $M_l(-v)$ at the time $t = 2L/v$
$\tau(M_r(+v))$	the proper time of the clock $M_r(+v)$ at the time $t = 2L/v$

relative to S.

$\tau'(R)$, $\tau'(M_l(-v))$, $\tau'(M_r(+v))$ are the corresponding proper times relative to S' and analogously for τ''. In other words, we calculate the proper times of R, $M_l(-v)$, $M_r(+v)$ relative to different inertial systems. The same value must always be predicted for each clock, otherwise there would be a contradiction with the experiment.

The inertial system S has a special position in this thought experiment, since all processes begin and end in S.

Let us first consider all processes relative to S.

(10.50) $$\tau(R) = \frac{2L}{v}$$

When R shows this time, all clocks are back at $x = 0$.

(10.51) $$\tau(M_l(-v)) = \frac{2L}{\gamma v}$$

(10.52) $$\tau(M_r(+v)) = \frac{2L}{\gamma v}$$

$M_l(-v)$ the time duration $2L/v$ moves at the speed of the amount v, it suffers a time dilation by the factor $1/\gamma$. The same applies to $M_r(+v)$, the sign of v does not matter, since v enters quadratically into γ.

Exactly the same predictions must result relative to S':

Relative to S', the clock R moves for the duration $t' = 0$ to $t' = 2\gamma L/v$ from $x' = 0$ to $x' = -2\gamma L$ (this corresponds to $x = 0, t = 2L/v$) at the speed $-v$. Clocks moving with v run slower by a factor of $1/\gamma$. Clock R shows the time as seen from S':

(10.53) $$\tau'(R) = \frac{2\gamma L}{v}\frac{1}{\gamma} = \frac{2L}{v}$$

The clock $M_l(-v)$ moves in S' with the speed

(10.54) $$w' = \frac{-2v}{1+v^2/c^2}$$

for the time $t_1' = 0$

(10.55) $$t_2' = \frac{\gamma L}{v}\left(1+v^2/c^2\right)$$

and then rests in S' until

(10.56) $$t_3' = \frac{2\gamma L}{v}$$

at the position $x' = -2\gamma L$

(Formula (10.54) corresponds to formula (4.13), relative to S' S moves with $-v$ and in S $M_l(-v)$ moves with $-v$.)

$M_l(-v)$ shows the time:

(10.57) $$\begin{aligned}\tau'(M_l(-v)) &= \frac{t_2'-t_1'}{\gamma(w')} + (t_3'-t_2') \\ &= \frac{\gamma L}{v}\left(1+v^2/c^2\right)\sqrt{1-w'^2/c^2} + \frac{2\gamma L}{v} - \frac{\gamma L}{v}\left(1+v^2/c^2\right)\end{aligned}$$

(in the first summand the time dilation for the velocity w', then formula (10.54) is taken into account)

(10.58)
$$\tau'(M_l(-v)) = \frac{2\gamma L}{v}\left(1 - v^2/c^2\right) = \frac{2L}{\gamma v}$$

The clock $M_r(+v)$ rests in S' for the period from $t'_1 = 0$ to

(10.59)
$$t'_2 = \frac{\gamma L}{v}\left(1 - v^2/c^2\right)$$

and has the speed

(10.60)
$$w' = \frac{-2v}{1 + v^2/c^2}$$

from

(10.61)
$$t'_2 = \frac{\gamma L}{v}\left(1 - v^2/c^2\right)$$

to

(10.62)
$$t'_3 = \frac{2\gamma L}{v}$$

$M_r(+v)$ shows the time:

(10.63)
$$\tau'(M_r(+v)) = \frac{\gamma L}{v}\left(1 - v^2/c^2\right) + \left[\frac{2\gamma L}{v} - \frac{\gamma L}{v}\left(1 - v^2/c^2\right)\right]\sqrt{1 - w'^2/c^2}$$

(in the second summand the time dilation for the velocity w', then formula (10.60) is taken into account)

$$\tau'(M_r(+v)) = \frac{2\gamma L\left(1 - \left(v^2/c^2\right)^2\right)}{v\left(1 + v^2/c^2\right)}$$

(10.64)
$$\tau'(M_r(+v)) = \frac{2L}{\gamma v}$$

Relative to S'' results analogously:

(10.65)
$$\tau''(R) = \frac{2L}{v}$$

(10.66)
$$\tau''(M_l(-v)) = \frac{2L}{\gamma v}$$

(10.67)
$$\tau''(M_r(+v)) = \frac{2L}{\gamma v}$$

Relative to S, S', S'', the same is predicted for each of the clocks

The author agrees with this solution by McCREA, but one can and must continue controversially: See the next but one chapter 10.8.5.

10.8.4. Proper times with change of S' and S''

There is another method to calculate the proper times of the clocks $M_l(-v), M_r(+v)$, which leads to correct results.

To do this, one calculates the proper times of the clocks when they return to $x = 0$ after the time $t = 2L/v$ as if an observer moving with them had read them continuously. Such an observer changes the inertial system when the movement is reversed:

Proper time of clock R:

$$\tau(R) = \frac{2L}{v} \tag{10.70}$$

Proper time from $M_l(-v)$:

It rests in S'' from $t''_1 = 0$ until it arrives in S at the point $x = -L$ at the time $t = L/v$. Then, according to the table 10.9 in S'', it has the time

$$t''_2 = \frac{\gamma L}{v}\left(1 - v^2/c^2\right) \tag{10.71}$$

It then rests in S' of

$$t'_1 = \frac{\gamma L}{v}\left(1 + v^2/c^2\right) \tag{10.72}$$

to

$$t'_2 = \frac{2\gamma L}{v} \tag{10.73}$$

because this corresponds to the values $x = -L, t = L/v$ and $x = 0, t = 2L/v$. The proper time of clock $M_l(-v)$ is the interval

$$\begin{aligned} \tau(M_l(-v)) &= \left(t''_2 - t''_1\right) + \left(t'_2 - t'_1\right) \\ &= \frac{\gamma L}{v}\left[\left(1 - v^2/c^2\right) + 2 - \left(1 + v^2/c^2\right)\right] \\ &= \frac{2\gamma L}{v}\left(1 - v^2/c^2\right) \\ &= \frac{2L}{\gamma v} \end{aligned} \tag{10.74}$$

Proper time of the clock $M_r(+v)$:

$M_r(+v)$ rests in S' for the same length of time as $M_l(-v)$ in S'' and in S'' for the same length of time as $M_l(-v)$ in S', so it has the same proper time

(10.75) $$\tau(M_r(+v)) = \frac{2L}{\gamma v}$$

(10.74) and (10.75) agree with (10.64). Both predictions are correct. The special feature of this method is the following: For the calculation of the proper times the inertial frame is changed, in the first method, ch. 10.8.3, it remains the same. But in S' and S'' only the information is used that the clock displays, which is directly next to the clock $M_l(-v)$ and $M_r(+v)$ respectively. In other words, it doesn't matter whether the clocks in S' and S'' are correctly synchronised, it is enough that they have no technical defects, because you only read the time running on one clock.

According to this method, the proper time of a clock can also be determined experimentally. Move the clock to $x = L$ and back to $x = 0$ and obtain the value of (10.64). As a check, see how long the clock stays in S' and S''. The values (10.71) and (10.72) - (10.73) should be readable. This method corresponds very well to the situation of a space traveller who changes from the inertial system S' to the inertial system S'' during his reversal and always reads off the time of a clock carried along. This procedure also provides no contradiction.

10.8.5. Contradictory conclusions in the clock paradox

We now consider the following behaviour, which leads to a contradictory statement. LIEBSCHER [197], who had still agreed to the elimination of the acceleration objection and also argues very thoroughly in other respects, does not comment on this. In S' the clock $M_r(+v)$ sees the clock $M_l(-v)$ for the duration $t' = 0$ to

(10.80) $$t' = \frac{\gamma L}{v}\left(1 - v^2/c^2\right)$$

moving with the speed

(10.81) $$w' = \frac{-2v}{1 + v^2/c^2}$$

Then the clock $M_r(+v)$ changes to S'' and the clock $M_l(-v)$ changes to S'. In S'', $M_r(+v)$ now sees the clock $M_l(-v)$ for the duration of

(10.82) $$t'' = \frac{\gamma L}{v}\left(1 + v^2/c^2\right)$$

to

(10.83) $$t'' = \frac{2\gamma L}{v}$$

moving with the speed

(10.84) $$w'' = \frac{+2v}{1 + v^2/c^2}$$

Throughout its proper time from $2L/(\gamma v)$, it sees $M_l(-v)$ moving with the amount $w = -w' = w''$. Clock $M_r(+v)$ thus predicts a time dilation for $M_l(-v)$ with the value:

(10.85) $$\frac{2L}{\gamma v}\sqrt{1-w^2/c^2} = \frac{2L(1-v^2/c^2)^{3/2}}{v(1+\frac{v^2}{c^2})}$$

$M_r(+v)$ says from $M_l(-v)$, it runs lower by a factor of $\sqrt{1-w^2/c^2}$.

As the very analogous calculation shows, conversely $M_l(-v)$ says from $M_r(+v)$, it goes slower by the same factor $\sqrt{1-w^2/c^2}$.

One can continue this game, after all, both $M_l(-v)$ and $M_r(+v)$ see the clock R constantly moving and calculate a time dilation for R. $M_l(-v)$ sees R moving away with v and returning again. This again applies to the duration of the proper time of $M_l(-v)$. The proper time is multiplied by the factor $1/\gamma$ for the time dilation and the adversarial time interval of $M_l(-v)$ for R *is* predicted:

(10.86) $$\frac{2L}{v}(1-v^2/c^2)$$

$M_r(+v)$ makes the same prediction, because its proper time is the same as that of $M_l(-v)$ and also $M_r(+v)$ sees R always moving with v.

The - incorrect - objection for these examples is usually that both processes are asymmetrical - incorrect, because with respect to $M_l(-v)$, $M_r(+v)$, formula (10.85), the process is symmetrical.

Numerical example for the correct proper times of the clocks R, $M_l(-v)$, $M_r(+v)$, formulas (10.50) - (10.75), and the contradictory ones, formula (10.85) and (10.86):

Let it be $v = 0.866\ c = 2.6\ 10^5$ km/s

$L = 0.866$ light years $= 2.6\ 10^5$ km

Then, formulae (10.50) to (10.75):

$$\tau(R) = \tau'(R) = \tau''(R) = 2L/v = 2 \text{ years}$$

$$\gamma = 2$$

$$\tau(M_r(+v)) = \tau'(M_r(+v)) = \tau''(M_r(+v))$$
$$= \tau(M_l(-v)) = \tau'(M_l(-v)) = \tau''(M_l(-v))$$
$$= 2L/(\gamma v) = 1 \text{ Jahr}$$

The clock $M_l(-v)$ is in S'', formula (10 71): 0.5 years and in S', formula (10.73) - (10.72): 0.5 years. It sees $M_r(+v)$ moving with $w'' = 0.9897$ c, formula (10.84), and predicts the wrong time of 0.1429 years for $M_r(+v)$, formula (10.85), and the wrong time of 0.5 years for R, formula (10.86). (In (10.85) one can explicitly insert $w''$and gets the same numerical value).

How can the contradiction be resolved? The following objections are ruled out:

a) The clock $M_r(+v)$ is said to be constantly at rest, but when it changes from S' to S'' it is accelerated. The clock $M_r(+v)$ is therefore not in an inertial frame, but in an accelerated frame of reference, therefore the general theory of relativity is responsible. This is of course correct, the clock $M_r(+v)$ considered as a real clock rests in an accelerated reference frame, it is wrong to apply unchecked the formulas for time dilation derived from the Lorentz transformations in such a reference frame. See ch. 10.8.7 for the application of general relativity. However, it does not help because of b).

b) How must one argue if the clock $M_r(+v)$ is again realised by two clocks R'_r and R''_r where R''_r takes over the time of R'_r when they meet? General relativity is then not competent. It also does not help that with (10.85) a prediction is made of which one knows that it is wrong, that one knows the correct prediction and that this is even confirmed experimentally. It only follows that the antinomy has no practical significance, since one obtains the correct value in other, even simpler ways and one is also aware of this but this is not an argument to solve the antinomy.

c) Could the antinomian formulae (10.85) have their cause in a non-observance of the relativity of simultaneity? $M_l(-v)$ and $M_r(+v)$ change their inertial frame S' and S'' simultaneously relative to S, but not relative to S' or S''. If $M_l(-v)$ changes at the point $x = -L$, $M_l(-v)$ says, $M_r(+v)$ has not yet changed. As soon as $M_l(-v)$ has changed and is in S', $M_l(-v)$ says, $M_r(+v)$ has long since changed before me. This is exactly what the above derivation expresses quantitatively and the calculation inevitably results in formula (10.85).

Serious objections against the derivation of (10.85) are not in sight, one must regard it as correct; a wrong result arises when a physical process - the movement of $M_r(+v)$ - is not described from a freely chosen but then unchanged inertial system, but in alternation to two or more inertial systems. The principle of relativity and thus the Lorentz transformations apply to any inertial system, but an *arbitrary change is not permitted.* This restriction is called the **"inertial system requirement"**. In Chapter 10.8.6, we will discuss whether it can be derived from the EI or the LI.

References to the antinomian formulae (10.85) and (10.86) as well as attempts to solve them in the manner indicated here are not known to the author from the literature Agreeing with the result, in the well-known textbook by G. FALK, W. RUPPEL [109] the kinematics of the twin paradox is described as contradictory. The acceleration objection is convincingly rejected by LIEBSCHER [197, p. 98], for example.

Let us summarise: *Clock movements, including accelerated ones, are described without contradiction relative to any inertial frame. Each inertial system has equal rights. The description becomes antinomic if it is done from changing inertial frames.*

10.8.6. The inertial system requirement and the first solution of the clock paradox

The discussion of the clock paradox on the basis of the special theory of relativity led to the so-called inertial system requirement. Every physical process, especially for accelerated moving clocks, must not be described relative to successively changing inertial systems. Otherwise, antinomian predictions arise for the clocks $M_l(-v)$ and $M_r(+v)$ in formula (10.85). Which inertial system one chooses is of course not subject to any restrictions. Formally, it is correct to say: changing reference systems do not represent an inertial system as a whole and the Lorentz transformations only apply to inertial systems.

But on the basis of Einstein's interpretation, it is not reasonable not to describe the times of clocks in successively changing inertial frames. For the clock $M_l(-v)$ there is real, directly measurable, only the time relative to the instantaneous inertial frame, in which the clock rests. The predictions relative to this inertial system for other, moving clocks are predictions for equally real time intervals. If one imagines the clocks $M_l(-v)$, $M_r(+v)$ as space travellers, each of the space travellers sees and experiences himself as the one who ages faster than the relative moving one.

In the Lorentz interpretation, there is an excellent inertial frame; synchronising clocks with light signals is a practical convention that only fits reality for the excellent inertial frame, because there the speed of light in all directions is in reality equal to c, in the other inertial frames it is only made so.

The antinomian formulas - as they become explainable with a synchronization effect - can therefore also be seen as a thought experiment to distinguish between the relativistic and Lorentz interpretations. It is not a normal experimental proposal to measure something. It is the compelling prediction of a measurement result that cannot be: Compelling for the Einstein, not for the Lorentz interpretation of relativity.

The synchronisation effect (as the cause of the antinomian formulas) is best seen like this: Consider for

$$x = +L$$

$$t = L/v$$

i.e. at a certain place in space, clocks in *S*, *S'*, and *S''*. They show

the time in *S*: $$t = L/v$$

and in *S'* the time: $$t' = \frac{\gamma L}{v}\left(1 - v^2 / c^2\right)$$

and in *S''* the time: $$t'' = \frac{\gamma L}{v}\left(1 + v^2 / c^2\right)$$

With the infinitely fast change from *S'* to *S''*, the clock must change its display from t' to t'' in an infinitely short time. *This time change cannot be real time, it is a synchronisation effect* that can be excluded with the inertial system requirement. However, synchronisation effects do not exist for an interpretation in which the speed of light is actually c everywhere and is not only made c or measured that way, i. e. not for the Einstein interpretation.

For the Lorentz interpretation, the clock paradox can be solved directly: Only the excellent inertial system is decisive, only the movement relative to it causes a real time dilation. The movements relative to the other inertial systems have a special position because of the Einstein clock synchronisation, which is only expedient (see Chapters 5 and 6, in particular formula (5.5)). If one does not take this special position into account and only looks at the same formulae of the Lorentz transformation, the contradictory relationship (10.85) results. The clock paradox arises because two (or more) inertial systems, equivalent in all points, are assumed. In the Lorentz interpretation, this is inadmissible and this explains the inertial system requirement and the distinction between real time and synchronisation effects.

If one remains with the description of an experiment in a single inertial frame, synchronisation effects are certainly excluded. On the other hand, there are many processes that can be described arbitrarily in parts one after the other in different inertial frames. This can also be formulated quantitatively: If we regard S as an excellent inertial system, then the change of $M_l(-v)$ from S'' to S' at the point $x = -L$ means a synchronisation effect ΔS from

$$\Delta S = t'' - t' \tag{10.88}$$
$$= \frac{2\gamma L}{v}\left(v^2/c^2\right)$$

Consider, the change from S' to S'' is supposed to happen infinitely fast for $M_l(-v)$, yet $M_l(-v)$ gains time ΔS. No time has passed, but $M_l(-v)$ has a gain of ΔS. In the opposite direction it becomes negative. From this it can be deduced: One and the same physical process may be observed in parts one after the other relative to different inertial systems, provided that the synchronisation effects may be neglected. This should be fulfilled for many experiments, even at high speeds. For $M_l(-v)$ and thus for the twin or clock paradox as a whole, the condition is not fulfilled. *The clock paradox owes its existence to the inadmissibly neglected synchronisation effects.*

Numerical example:

A space traveller moves with $v = 0.99c$ to $x = -L = 25$ light years and back. Then

$$\Delta t = \frac{2L}{v} = \frac{2\times 25\times 3\times 10^6 km/s\times 1\, year}{0.99\times 3\times 10^6 km/s}$$
$$= 50.5\, years$$
$$\Delta t' = \frac{2L}{\gamma v} = 7.12 \text{ years}$$
$$\Delta S = \frac{2\gamma L}{v}(v^2/c^2) = 350.9\, years$$

The gain in time of ΔS = 350.9 years is not real, it is a synchronisation effect. It is not negligible; Δt, $\Delta t'$ apply, the space traveller has aged 7.12 years, at home 50.5 years have passed.

Synchronisation effects do not play a role in the calculation of proper time in Chap. 10.8.4. There it was already pointed out that it is unimportant whether the clocks are synchronised, but in the prediction for other proper times, e. g. of clock R, they do play a role, as is easily understood, because the proper times of other clocks are calculated using synchronized clocks. The displayed times in S' and S'', formula (10.88), are taken as real times. Synchronization effects are not distinguished from real elapsed times.

In any case, we can state: *If one observes the inertial system requirement, there is no clock paradox, it does not exist even if synchronisation effects of clocks are negligibly small.*

The somewhat different justification of the inertial system requirement, recognised later by the author [vor14]: 'All inertial systems are equal' means: 'One may regard each inertial system as the distinguished inertial system'. *However, it is a logical contradiction to consider two different inertial systems moving relative to each other during the same physical process as the distinguished inertial system.* This is discussed largely independently of this chapter below.

10.8.7. Second solution of the clock paradox

The clock paradox does not occur if one observes the inertial system requirement (IS requirement), but what can it be derived with? Why is it more than an ad hoc principle? The IS requirement was: Every physical process must not be described relative to successively changing inertial systems. Otherwise antinomic predictions arise, e.g. for the clocks $M_l(-v)$ and $M_r(+v)$ in formula (10.85).

The EI can say concerning the IS requirement, changing inertial systems are altogether no longer an inertial system (IS). And the formulas of the SRT are valid only for IS. This is not convincing. In the twin paradox (TP) two inertial systems with a clock resting in it and a clock moving in it are considered one after the other and then the proper times of both clocks are calculated. These are elementary calculations, which can be doubted badly. There is no argument for the EI, why the IS requirement must be considered.

The LI gives an answer.

First of all, the derivation of the antinomian formula (10.85) violates the IS requirement. Formula (10.85) is the sum of two predictions. The first is made for the clock $M_l(-v)$ relative to S'' and the second relative to S'. In these two time intervals, $M_l(-v)$ sees the clock $M_r(+v)$ moving with velocity w. The proper time predicted for $M_r(+v)$, formula (10.85), is antinomial to that for $M_r(+v)$, formula (10..64), and violates the IS requirement because two IS (S' and S'') are involved.

For the LI, there is an excellent IS (distinguished IS) that proves helpful in justifying the IS requirement. In order to do so, one has to visualize some properties of a distinguished IS:

- The values measured relative to the distinguished IS are the absolute values of objects, also called true or real values. (e.g. absolute length of a rod, in EI there is no such length).
- Because of the relativity principle (RP) all IS are equivalent, any IS can be taken as the distinguished IS without changing the formulas and measurement results. The RP only makes statements about measurement results, not about the absolute value of a measurand. Example: The measured length of

a rod has different values in different inertial systems. Its absolute length is the length measured in the excellent inertial system.

- If you know the distinguished IS, you get absolute values also by a LT of the values of any other IS into the distinguished IS.
- But it is also true: two different IS cannot both be the distinguished IS, because they move relative to each other and only one can be (absolutely) at rest. For the twin paradox, this means the prediction relative to S' and the prediction relative to S'' cannot both be absolute measurements at the same time. (e.g. concerning the length of a rod, at most one of the length measurements gives the absolute length).
- From measured values relative to different IS the real quantities can be determined by Lorentz transforming them into the distinguished IS.
- If one Lorentz-transforms for the antinomic formula (10.85) the measured value relative to S'' into the measured value relative to S', one gets the correct prediction of the proper time of the clock $M_l(-v)$ formula (10.58) and clock $M_r(+v)$, formula (10.64).
- For the LI there is the world of measured values and the reality. Both are different because scales change when they are moved. For the EI, the set of measurements is reality. The LI can designate an IS, the EI cannot.
- All measurement values relative to S (or S' or S'') are absolute measurement values if S (or S' or S'') is the distinguished IS. Therefore, it is convenient if all measured values are converted into one and the same IS. All these measured values are absolute measured values, if it is the distinguished IS.

From this, the IS requirement can be derived in two steps

1) The IS requirement is proved for the special case of an excellent IS. (Two IS cannot be the excellent IS at the same time).

2) Then the IS requirement must be observed also for any IS, because one may consider any IS as the excellent IS because of the RP. Measurement results are predicted in all IS with the same formulas.

This also solves the twin paradox. It arises because of the non-observance of the IS requirement, which in turn arises from the existence of an excellent IS.

Example:

A rod at rest in S has the length $L_{rod;}$ the length $L'_{rod;}$ measured relative to S' is smaller by the factor $1/\gamma$. Both measurements taken as absolute measurements at the same time contradict each other. A rod cannot have the length $L_{rod;}$ and $1/\gamma\ L_{rod;}$ at the same time. The measurement result becomes contradiction-free, if the measurement result relative to S' is Lorentz-transformed into that relative to S (or vice versa).

Because of the great importance of the twin paradox independently of the previous considerations two essential arguments.

1.) Let us assume that S' is the absolutely resting IS: Then every measurement relative to S' is a statement about something real. But the measurement relative to S'' is not a statement about something real. This difference is not considered in the twin paradox. (The measurements relative to S'' become statements about something real, if they are Lorentz-transformed to S').

2.) All IS are equivalent, each can be considered the excellent IS without changing the readings, but not two at the same time.

Summary: The twin paradox is solved, if one can take as a basis an absolutely resting IS as in the case of the LI. In this situation one can derive the IS requirement. For the EI the IS requirement is an addition assumption.

10.8.8. The general relativistic solution

By considerations based on general relativity, the antinomian equations (10.85) in ch. 10.8.5 can be solved for EI and LI. Seen from S as well as from the clocks $M_l = M_l(-v)$ or M_r, the moving clocks run slower than the clock R at rest.

In the following we take as a basis the considerations of C. MØLLER [235]. A similar, well comprehensible discussion is contained in the textbook by G. FALK, W. RUPPEL [109]. In any case, they have two strengths:

a) The problem definition corresponds to ours of ch. 10.8.2 as in Fig. 10.8. This is not self-evident [313.1].

b) The mathematical formalism is glorious, it performs what is desired without contradiction.

Not convincing is:

a) The acceleration fields introduced by C. Møller are supposed to act like gravitational fields, but they lead to negative proper times, i.e. in certain acceleration situations to clocks going backwards.

b) The actual twin paradox is independent of an acceleration phase, see ch. 10.8.2 Therefore, the twin paradox cannot be solved in a general relativistic way. At best, general relativity shows another way to predict the correct result, the counterarguments of ch. 10.8.2 remain.

Let's look at the details first.

We will try to illustrate C. MØLLER'S ideas in such a way that you can make your own decision about their applicability. Knowledge of general relativity is not assumed, but this also means that we do not derive any of C. MØLLER's formulae. The general relativistic derivation can be dispensed with since it does not pose a problem in this context. The problem lies in the question of whether the correct application of the general relativistic derivation, which is formula (10.92), can be regarded as a solution to the clock paradox. This is the case if one considers negative proper times to be meaningless.

In Fig. 10.10 the thought experiment of Fig. 10.8 is specified for the case of a real accelerated clock M_r. In the inertial frame $S(x, y, z, t)$ the clock R is at rest during the entire journey of M_r. Along L_a to point A M_r is accelerated, along L_b to point B it has the constant velocity v, in B the deceleration along L_c to point C begins. Here M_r briefly has zero speed in S. Further acceleration brings M_r back to B, from there it returns to A with constant speed $-v$, there M_r is decelerated again to clock R and the round trip can begin again.

All accelerations are hyperbolic of the amount g. An acceleration is called hyperbolic if it is constant in the moving inertial system of M_r but then changes constantly relative to S. Such an acceleration is perceived as constant by M_r and means a rocket of constant thrust. This is mathematically convenient. A constant acceleration relative to S, on the other hand, means a constantly increasing thrust force in the respective inertial system of M_r and for M_r itself.

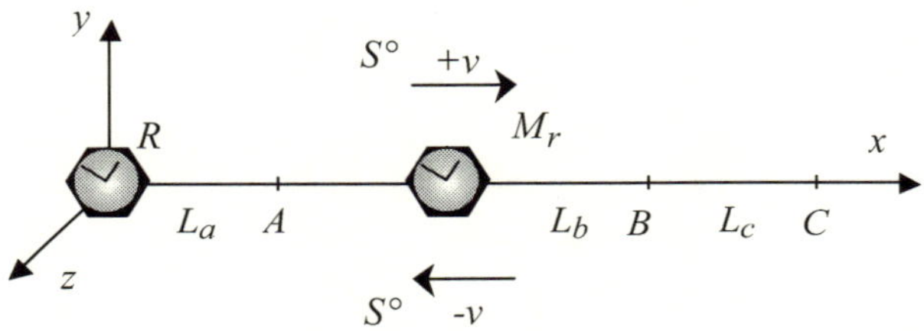

Fig. 10.10 The clock paradox taking into account the acceleration phases.
The clock $M_r = M_r(+v)$ is accelerated to $+v$ up to A, between A and B the speed of the clock remains constant, from B to C the clock M_r is decelerated, from C to B it is accelerated to $-v$ and from A to the origin of the coordinate system it is decelerated again until it comes to rest at clock R.

We introduce some designations:

In the inertial system S:

Δt_a = acceleration time from M_r along L_a to $+v$
Δt_b = movement time of M_r along L_b with the speed $+v$
Δt_c = deceleration time from M_r along L_c to zero speed
$\tau(\mathrm{R})$ = proper time from the start of M_r to the return of M_r
$\tau(\mathrm{M_r})$ = proper time of M_r from the start at R until return to R

$\tau_a(\mathrm{M_r})$ = Increase in proper time from M_r along L_a
$\tau_b(\mathrm{M_r})$ = increase in the proper time of M_r along L_b
$\tau_c(\mathrm{M_r})$ = increase in the proper time of M_r along L_c - the corresponding increases in the proper time of R are the intervals $\Delta t_a, \Delta t_b, \Delta t_c$

In the accelerated system $S° = S°(x°, y°, z°, t°)$:

In the accelerated reference system $S°$, M_r is constantly at rest at $x° = 0$, the clock R moves away from M_r in the negative $x°$ direction and then returns. There are only the two, here equivalent reference systems S and $S°$, therein lies the elegance of the general relativistic solution.

The quantities considered in $S°$ are marked with the ° symbol.

The simple relationships apply:

$$\tau(\mathrm{R}) = 2(\Delta t_a + \Delta t_b + \Delta t_c)$$
$$\tau(M_r) = 2(\tau_a(M_r) + \tau_b(M_r) + \tau_c(M_r))$$

because the proper times along the subsections L_a, L_b , L_c have the same values for the outward and return movement of M_r.

In S, because of time dilation:

(10.90) $$\tau_b(\mathrm{M_r}) = \Delta t_b\sqrt{1 - v^2/c^2}$$

If we assume that the acceleration force of M_r becomes infinitely large, but M_r should only ever be accelerated to v, the acceleration time Δt_a decreases more and more and finally reaches zero. Also, the proper time $\tau_a(\mathrm{M_r})$ becomes zero, because the proper time of a moving clock is always smaller than that of a stationary clock. (Accelerations have no influence on the proper time, only the constantly changing speed during acceleration. The factor $1/\gamma$ becomes increasingly smaller and the increase of proper time correspondingly. Overall, this is always less than the proper time of the clock at rest. Of course, this can also be seen from the formulae themselves, see MØLLER [235] page 294).

The same applies to proper times during movement along L_c .

For this limiting case, g becomes infinite:

(10.91) $$\tau(\mathrm{M_r}) = \tau(R)\sqrt{1 - v^2/c^2}$$

This is formula (10.52).

Now we apply the same considerations to the accelerated system $S°$ and show why the antinomian formula (10.86) does not result.

For movements as in Fig. 10.11, if they are hyperbolic, the following results from the general theory of relativity according to the considerations of *C.* MØLLER [235] for the proper time of $\tau°$ of "freely falling particles":

(10.92)
$$\tau° = \left(\frac{c}{g} + \frac{\overset{\circ}{x}_0}{c}\right)\tanh(g(t° - \overset{\circ}{t}_0)/c)$$
$$= \left(\frac{c}{g} + \frac{\overset{\circ}{x}_0}{c}\right)\frac{v}{c}$$

Thereby is

$t°$	the coordinate time, i.e. the time coordinate in S°
$t°_0$	the initial value of $t°$
$x°_0$	the initial value of $x°$

g the acceleration of M_r in the respective system of rest
x° the spatial coordinate in S°, measured in the x-direction
v is the final velocity reached during hyperbolic acceleration.

x° is not identical with the coordinate of an inertial system because the length contraction changes continuously due to acceleration. Only in the special case that M_r is at rest in S does apply to the clock R:

$$x^\circ = -L_a - L_b - L_c$$

This special case in particular is very important, because M_r is at rest in S when M_r reverses movement in C. Then the clock R is in S° at the position $x^\circ = -L$.

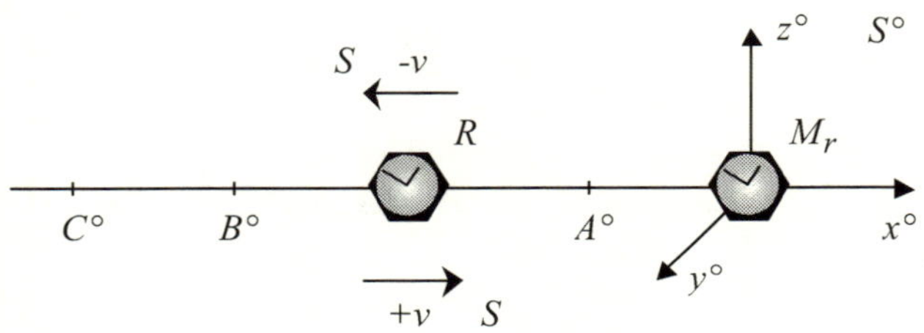

Fig. 10.11 The clock paradox observed from an accelerated clock.
In the system S° M_r is at rest, R moves away from M_r and back again with acceleration, between A° and B° R has a constant speed.

The formula (10.92) only applies to "freely falling particles", the proper time for other cases is not needed. For these general relativistic considerations, a free-falling particle is the clock R in S, because it is not accelerated relative to S, i.e. it is in an inertial frame. With formula (10.92), an accelerated moving space traveller can predict the clock times in an inertial frame (the clock R) relative to the coordinates of the accelerated reference frame in which he himself is at rest.

From the general formula (10.92) one obtains by substituting:

$$\mathring{\tau}_a(R) = \frac{c}{g}\tanh\left(\frac{g\tau_a(M_r)}{c}\right)$$

$$= \frac{v}{g}$$

(10.93) $$\mathring{\tau}_b(R) = \tau_b(M_r)\sqrt{1 - v^2/c^2}$$

because v is constant and time dilation applies. Formula (10.90) inserted:

(10.94) $$\mathring{\tau}_b(R) = \Delta t_b(1 - v^2/c^2)$$

Application of (10.92) for the phase L_c :

$$\overset{\circ}{\tau}_c(R) = \left(\frac{c}{g} + \frac{L}{c}\right)\tanh(\frac{g\tau_a(M_r)}{c})$$
$$= \left(\frac{c}{g} + \frac{L}{c}\right)\frac{v}{c}$$

For the limiting case g infinite, $\Delta \overset{\circ}{t}_a$ and Δt_c become zero, but

$$\overset{\circ}{\tau}_c(R) = \frac{Lv}{c^2}$$

remains unequal to zero. For Δt_b this results because of $L = L_b$:

(10.95) $$\Delta t_b = L/v$$

Thus one has

$$\overset{\circ}{\tau}(R) = 2\left(\overset{\circ}{\tau}_a(R) + \overset{\circ}{\tau}_b(R) + \overset{\circ}{\tau}_c(R)\right)$$
$$= 2\left[\frac{L}{v}\left(1 - v^2/c^2\right) + \frac{Lv}{c^2}\right]$$
$$= 2\Delta t_b$$
$$= \tau(R)$$

Formula (10.94) is the antinomic prediction (10.86), it is transformed into the correct solution (10.91) in the general relativistic approach by the proportion (10.95) and the clock paradox from the point of view of the acceleratedly moving space traveller is solved.

We now show a situation with negative proper times which detracts from the beauty of this solution.

The clock M_r may, when R is between points A and B at point D at $x = -L/2$, interrupt the uniform movement and decelerate to $u = 0$, and then immediately accelerate to v again. That is, the clock M_r initially behaves as if it wants to turn back, but then continues the movement. Only at point B does it turn back. What can be expected?

a) The proper time of M_r does not change for the limiting case of an infinitely large deceleration -g and infinitely large acceleration g, completely analogous to the previous considerations.

b) The proper time of R must not change either, because in this limiting case nothing has happened, the movement of M_r appears as if it has remained unchanged. This is also shown by the calculation:

In the deceleration phase, the proper time of R:

$$\overset{\circ}{\tau}_v(R) = -\frac{L}{2c}\tanh(-g(\overset{\circ}{t} - \overset{\circ}{t}_0)/c$$
$$= -\frac{L}{2c}\left(-\frac{v}{c}\right)$$
$$= +\frac{Lv}{2c^2} > 0$$

In the acceleration phase the same calculation applies with positive g:

$$\overset{\circ}{\tau}_b(R) = -\frac{Lv}{2c^2} < 0$$

The sum of the proper times is zero, but one after the other R runs forward a bit, then backwards, because R cannot "know" whether M_r will move on or turn around or do something completely different.

In various textbooks [313.1] the phase *A-B* is omitted, perhaps for this reason, but the clock paradox is then deprived of the actual phase, a constant motion. For the special case this creates, the objection of negative proper times is no longer directly applicable.

If we repeat these considerations for a finite acceleration g, the mathematical relationships become more complicated, but qualitatively nothing changes.

Let us summarise: The general relativistic solution is mathematically brilliant and free of contradictions, but it requires negative proper times. Negative proper times cannot be real. Moreover, the objections of chapter 10.8.2 remain. Negative proper times are once again the subject of the following chapter.

10.8.9. The twin paradox (ZP) in WIKIPEDIA

The article about the ZP in WIKIPEDIA [397.1] was included by WIKIPEDIA in the list of articles worth reading. It is very clearly formulated and suggestive; despite everything, the author cannot agree with the proposed solution. There are two reasons for this:

a) In WIKIPEDIA it is claimed that during the reversal phase the twins age so differently that the ZP can thus be solved. However, ch. 10.8.2, Elimination of the acceleration objection, has shown that the ZP must be solved independently of acceleration effects.

b) Furthermore, the effects of reverse acceleration are not correctly taken into account, as can easily be seen together with the following preliminary remark.

Consider in Fig. 10.8 an observer B_L at rest at location $+L$ relative to R. B_L observes the change of the space traveler from S' to S''. What time passes for B_L? The proper time τ of the space traveller changes during the reversal from τ_1 to τ_2, the proper time of the observer from t_1 to t_2 It holds:

(10.100) $$\int_{\tau_1}^{\tau_2} d\tau = \int_{t_1}^{t_2} \sqrt{1-\frac{v^2}{c^2}}\,dt$$

$$\tau_2 - \tau_1 = \int_{t_1}^{t_2} \sqrt{1-\frac{v^2}{c^2}}\,dt$$

(10.101) $$> \sqrt{1-\frac{v_{\max}^2}{c^2}} \int_{t_1}^{t_2} dt$$

$$> \sqrt{1-\frac{v_{\max}^2}{c^2}}\,(t_2 - t_1)$$

In the numerical example from WIKIPEDIA, $v_{\max} = 0.6c$. This, inserted into (10.101), yields $t_2 - t_1 < 1.25\ (\tau_2 - \tau_1)$.

In words: For the reversal phase, the change in the proper time of B_L is less than 1.25 times the change in the proper time of the space traveller. The important consequence: Since B_L is at rest in the same inertial frame as the twin on Earth, for the twin on Earth its proper time also changes by a maximum of 1.25 $(\tau_2 - \tau_1)$ during the reversal phase. If $\tau_2 - \tau_1 <<$ total journey, as assumed in WIKIPEDIA, this also applies to the proper time of the twin on Earth. Without influencing the overall result, both twins (almost) do not age during the reversal phase.

The authors of the WIKIPEDIA article ignore this simple fact, albeit very well in their argumentation and with a lot of detailed knowledge. They claim: During the reversal phase the twin on Earth ages significantly more than the space traveller and thus also significantly more than the observer B_L. Using the WIKIPEDIA NUMERICAL EXAMPLE:

travel speed: 0.6c, distance: 3 Lj (i.e. $L = 3$ Lj in Fig. 10.8), outward and return travel time: $t = s/v = 5$ years, together 10 years. $\sqrt{(1 - v^2/c^2)} = 0.8$, the space traveller ages by 5 x 0.8 = 4 years each way, 8 years in total. Since time passes more slowly for the space traveller on Earth, the twin on Earth has aged by 4 x 0.8 = 3.2 years, the one travelling there and back by 6.4 years. Up to this point, all the calculations in WIKIPEDIA are correct. But there is a contradiction: on the one hand, the twin on Earth has aged 10 years, on the other hand, only 6.4 years.

Now follows the incomprehensible assertion: "During the reversal phase, an additional 3.6 years pass on Earth [according to the space traveller]." No matter what arguments are brought forward for this in the article (read them up), the following objections of the author refute them:

Because of (10.101), both twins age by the same amount during the reversal phase - except for a factor of 1.25 - and not one almost not at all and the other by 3.6 years.

If the proper times actually change differently, S would no longer be an inertial system after the change, because the clocks would be adjusted differently. It gets even more absurd, because if the proper time of a clock in S changes, then so does the position of the clock relative to S' - it can be calculated with the help of the Lorentz transformations. Moreover, in order for the time and position to match, the Earth's twin must move faster than the speed of light relative to S'. The consequence: The change of anyone anywhere into another inertial system does not change any clock times anywhere. In WIKIPEDIA, synchronisation effects are wrongly interpreted as real.

The space traveller turns back (twin on earth becomes 3.4 years older), the space traveller decides to go further into space after all (twin on earth becomes 3.4 years younger again) and during the later, renewed turn back, the twin on earth ages again by 3.4 years. If the twin has died by 3.4 years during the first ageing, it comes back to life afterwards and dies again later. In a discussion with one of the authors [397], this is exactly what was said to be true.

The WIKIPEDIA ARTICLE is - at least at present - one-sided. As long as it is not expanded to include the arguments for the formula (10.101), a normal reader cannot form his own opinion, he is indoctrinated.

The ZP exists independently of what happens during acceleration when turning around, see ch. 10.8.2. The WIKIPEDIA article does not address this.

The WIKIPEDIA article refers under web links to further literature with proposed solutions of the ZP that differ from the main article. Obviously, even in the WIKIPEDIA environment there is no agreement on how the ZP should be solved - an important fact that should be pointed out.

In the meantime (2021), the WIKIPEDIA article changed and is no longer recognisable. The ZP is not resolved.

11. Philosophical contributions to Einstein (EI) and Lorentz interpretation (LI)

„Dreifach ist des Raumes Maß
Rastlos fort ohn Unterlass
strebt die Länge, fort ins Weite
Endlos gießet sich die Breite
Grundlos senkt die Tiefe sich"
Friedrich v. Schiller

"*Threefold* is the measure of space:
Restless away without break
the *length* extends, away into the infinite
Endlessly flows the *width*
Bottomless sinks the *depth*"

11.1. The terms one-dimensional and multidimensional

Dimension or extension refers to the property of real and imagined space to have length, width and height. Our aim is to extend these ideas as clearly as possible to four-dimensional and later curved spaces, but not to give a mathematically exact (see e.g. [293], [352]) or philosophically general definition (see e.g. [173]) of the concept of dimension, because our real problem lies less in the insight that space and time unified must be four-dimensional than in understanding the arguments why a four-dimensional space-time structure should fit reality better than the classical idea of space and time. In this sense, the terms one-dimensional and multidimensional are not defined in detail in philosophical-scientific terms, but only illustrated in order to synchronise the use of language.

Starting from normal, three-dimensional space, one first generalises the term space to (point) manifolds, which are then called space again, and says: Space can be everything that can be illustrated by points. Straight lines and surfaces are one-dimensional and two-dimensional spaces and not only part of normal space, because they represent (point) manifolds in themselves. Time, visualised as time points, forms a one-dimensional (point) manifold, i.e. a space. Space and time together form a - as we will see - four-dimensional point manifold, again a space, which is called a four-dimensional space-time-continuum. Since numbers can be represented as points, one also speaks of the spaces formed by the numbers; the same applies to vectors. It is not decisive whether objects are points, but whether they can be visualised by points or thought of as points.

Spaces have different dimensions, see Fig. 11.1. The number of dimensions can be described quite well as the number of coordinates with which points in a space are fixed. Normal space has three dimensions because a point in space is fixed by its *x,y,z* coordinates. As soon as one is able to define coordinate systems in a space, one makes statements about its dimension and the physically important spaces have coordinate systems, e.g. in the form of the inertial systems. They and the other coordinate systems discussed here must have the following two properties:

a) Coordinates are assigned to each point which unambiguously determine the position of the point in space.

b) The number of these coordinates is minimal (and therefore indicates the dimension of the space).

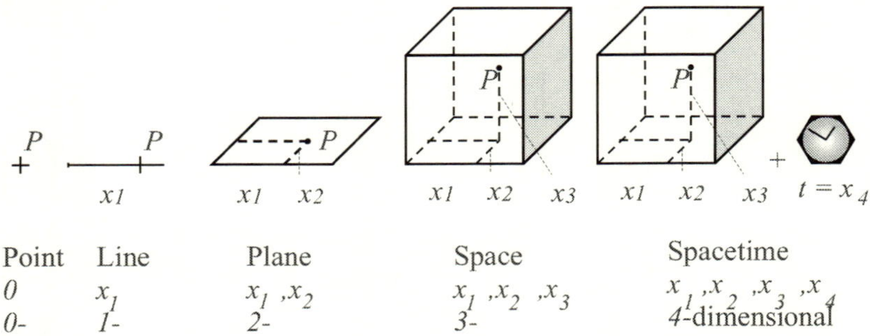

Fig. 11.1 Dimension concept

From a mathematical point of view, space-time points are located in a four-dimensional space, regardless of what the real space looks like.

Condition b) is intended to prevent, for example, assigning the coordinates $(x,y,x+y)$ to a point in the plane and then concluding that the plane has three dimensions. Another, but here meaningless, requirement to avoid falsification of the dimension is to require that points that are close to each other also have coordinates that are close to each other.

As a mathematician, one sees more precisely what was said: The dimension of an affine space is the number of its linearly independent vectors [293].

Another illustrative method for the dimension of a space is described by POINCARÉ [173]:

"The most important of all the theorems on the analysis of position is the assertion that space has three dimensions.... What do we mean by the theorem of the three-dimensionality of space? If we want to divide space into parts, we need cuts, which we call surfaces; to separate surfaces, we need cuts, which we call lines; to separate lines, we need cuts, which we call points. We cannot go further with division because a point that is not a continuum can no longer be divided. The lines that can be divided by cuts that are no longer continua are continua of one dimension. Surfaces that can be divided by one-dimensional continua are themselves two-dimensional continua. Finally, space that can be separated by two-dimensional continua is itself a continuum of three dimensions."

... and coordinates are respectively the distances of the points (or their projections) from these three sections. A similar construction occurs in ch. 13 for three-dimensional curved spaces.

Physical (= real) space therefore has three dimensions, because through three coordinates, the x,y,z-coordinates, every place (or location) in physical space is exactly defined, whereby a place in physical space is something where material particles (bodies) can be located. In special cases (Berlin has a latitude of 53° and a longitude of 13°) it may be fewer coordinates, in which case the missing one is implicitly given (... on the surface of the earth). Time in itself is a one-dimensional space, because any given point in time is described precisely by a numerical indication, the time - whereby the date is added to the time if the period of time extends beyond one day. Here, too, we have a coordinate system, namely a definition of the beginning of the time calculation.

What is four-dimensional?

A manifold (space in the more general sense) is four-dimensional if each of its objects (points in the more general sense) is determined by four independent quantities (coordinates in the more general sense).

Examples:
The location of a particle in space and its time at which it is at a certain location is a four-dimensional manifold.

The location of the particle and its rest mass are also a four-dimensional manifold although location and mass are not comparable quantities in themselves. This also applies to location and time, at first glance they are not comparable with each other either.

Of course, one can take this circumstance as an opportunity to reject the term four-dimensional manifold for place and mass or place and time, because the two clearly do not belong together. But that does not matter, the concept of space and dimension is used in this most general form - the same applies to the concept of point.

To further illustrate dimensionality, form the manifold of all particles, determined by their location, the times at which they are there and their rest masses. This is a five-dimensional space. One could object that a particle can only have a certain mass at a location and not any number of masses, as should be the case if the mass at rest is to represent a dimension. In order to comply with this, one can imagine all possible masses at a location or one can only roughly specify the location (to an accuracy of one kilometer) and there is place for many particles of different rest masses. Two particles with the same location and time coordinates but different masses represent different points in this five-dimensional space.

Such examples are intended to show that describing something as four-dimensional is by no means necessarily profound. All it means is to call objects points and to describe them by four independent properties that can be expressed in numbers.

A few more mathematical examples (Fig. 11.1): The totality of the quantities denoted by (x_1) are the points on a straight line. They form a one-dimensional space, the totality of the points (x_1, x_2) are the points of the plane, they form a two-dimensional space, accordingly the totality of the (x_1, x_2, x_3) form a three-dimensional space. To do this, it is sufficient to define a coordinate system on a straight line, in the plane or in space as desired and to regard the x-values as the coordinates of the points. And what does (x_1, x_2, x_3, x_4) mean? They form the totality of the four-dimensional points and thus a four-dimensional space. For each arbitrarily chosen value for x_4, e.g. $x_4 = 4.5$, the remaining x-values can be chosen as for a three-dimensional space. One can therefore also say that the four-dimensional space consisting of the points (x_1, x_2, x_3, x_4) contains an infinite number of spaces (x_1, x_2, x_3) that differ in the x_4 -coordinate, just as there are an infinite number of planes (x_1, x_2) in the (x_1, x_2, x_3)-space that differ in the x_3 -coordinates. Thus the space (x_1, x_2, x_3, $x_4 = 0$) contains all points of the (x_1, x_2, x_3) to which the coordinate $x_4 = 0$ has also been added. This space decomposes the four-dimensional space (x_1, x_2, x_3, x_4) into the two parts (x_1, x_2, x_3, $x_4 < 0$) and (x_1, x_2, x_3, $x_4 >= 0$). A space that can be separated by a three-dimensional continuum is itself four-dimensional - in continuation of POINCARÉ's argument above.

As much freedom as one has to construct multidimensional manifolds, more meaningful is their decomposition into submanifolds of smaller dimension. Thus the rest masses are a (one-dimensional) space by themselves and the spatial space is a three-dimensional one, while time is again a one-dimensional space. When is this possible in a meaningful way?

Typically, in physical (=real) space, every x,y,z coordinate in a rotated coordinate system can be described by x',y',z' coordinates. If one rotates around the z-axis, $z' = z$ and x, y lie in the same plane as x', y'. In Fig. 16.6 or with a collection of formulas [35] one sees

$$x' = x\cos(\alpha) + y\sin(\alpha)$$
$$y' = -x\cos(\alpha) + y\sin(\alpha) \tag{11.1}$$

There is no binding rule that determines what the x-, y- or z-axis should be in space. All directions in space are of equal value. This makes it sensible to speak of real space as a whole and to grant it the property of having three dimensions. A decomposition of real space into x-space, y-space and z-space cannot be justified.

Let's move on to the four-dimensional space-time manifold. The classical decomposition is: The space-time manifold decomposes into three-dimensional space and one-dimensional time. Places are determined with scales, points in time with clocks. According to the classical view, it is not possible to influence the time coordinates by rotations and displacements in space and vice versa. The four-dimensional manifold space-time (spacetime) is therefore more appropriately described in the classical conception as a three-plus-one-dimensional manifold.

Quite correspondingly, it is appropriate to call the four-dimensional manifold space-rest mass as a three-plus-one-dimensional manifold, since changes in the x-, y- or z-coordinate have no influence on the rest mass of the particle.

The property of being three-plus-one-dimensional is also expressed in the transformation formulae. In classical physics, the Galilean transformations, formula (4.6), apply. They apply to bodies (particles) and to the transition between inertial systems. They only indirectly describe space and time. The decisive statement for space and time lies in the relationship $t = t'$. No matter where one is in space, time t exists independently of it. This physically justifies speaking of space and time independent of it.

The situation is different for the Lorentz transformations. All coordinates x,y,z,t depend on each other, t no longer has any independence. (The formula (4.1) is specially chosen, it only shows a dependence of x and t). If one interprets this situation carefully, it means that the totality of the measured x-, y-, z- and t-values form a four-dimensional space, because coordinates are first of all only measured values. If the measured values are to be regarded as lengths and times on the basis of further considerations, they also form a four-dimensional space, the four-dimensional space-time continuum.

11.2. A first philosophical argument on the space-time continuum

This and the following chapters will attempt to use philosophical considerations to distinguish between a four-dimensional space-time continuum and a three-plus-one-dimensional space-time.

Here, an argument presented by A. MÜLLER [241] will be analysed. Among the numerous philosophical objections to a four-dimensional space-time continuum, this one is the most plausible among many doubtful ones. Whether it is compelling depends on the philosophical standpoint. A pragmatic physicist, for whom only experimental findings count, will certainly not be convinced. He will even find it presumptuous to want to criticise the fundamentals of an experimentally outstanding theory without even proposing an experiment for it. But this discussion is also significant for a pragmatist; it shows him where the often intense rejection of the space-time concepts of relativity theory by humanities scholars has its root, namely in the disregard for simple philosophical insights.

Characterised by a keyword, Müller's argument is called:
Psychic processes are time-like, but not space-like. They "are not located in a cube of $1.5 cm^3$", BAVINK [15]. But if there is time independent of space, time is something independent for itself and so is space and there is not only the unity of space and time.

The reflection of A. MÜLLER is phenomenological and comparatively simple. First of all, a few facts: Soul and consciousness are not measurable objects of physics, although both exist. The surgeon Virchow's saying is well known. "I have dissected many corpses, but I have never found a soul." Mental processes, although they exist, do not exist as measurable objects in physical space.

Similarly, a mirror image exists independently of feelings but is virtual, i.e. it does not lie in physical (= real) space like, for example, the photography produced by a camera. Of the mirror image, there are only some wave fields in space and they are not where one sees the mirror image. Also virtual is the magnification with a magnifying glass. Nevertheless, one can indicate distances between individual parts in mirror images. It is enough to look at the image of a scale.

However, all these processes and states occur in a real time, it can be measured with physical clocks. The time at which one sees an object and at the same time its reflection is one and the same, the time in consciousness is not outside or before or later than the real time that can be measured with clocks.

Feelings also run in time. Is the time in which they are experienced different from the time in which physical processes take place, e.g. the strike of lightning in a storm and the fear of it? Certainly not, it is not difficult to agree here. Furthermore, one should distinguish: The feeling of happiness is something different from the physical chemical process in the brain that underlies the feelings. Why certain physical-chemical processes lead to feelings of happiness, others to the sensation of "red light", is not clear. Why certain physical-chemical processes lead to feelings of happiness, others to the sensation of "red light", is incomprehensible, even if one assumes that such feelings and sensations necessarily arise as soon as certain physical-chemical processes take place. Or to put it another way, one will not be able to prove that the sensation "red light" cannot be the sensation "yellow light" when an electromagnetic wave of a certain wavelength hits the eye. The connection between colour and wavelength is only known through observation. It is not at all possible to deduce from a physical-chemical theory that there are sensations, because sensations are not objects in such theories.

These considerations, which are not physical but rather psychological in nature are intended to make clear: There are processes that really take place in time, just like physical processes, but which are not usual physical processes themselves. What does this mean for spatial relations? Do feelings and sensations have no particular place?

First of all: The person who has feelings is in a certain place in physical (= real) space. The physical-chemical processes that underlie feelings and sensations have a specific place, namely that of the atoms and molecules involved in them. The feeling of happiness, the sensation of "red light", they are not in a specific place in the physical sense. There are only fields, atoms, elementary particles etc. The feelings cannot be imagined sitting somewhere between the atoms - they have no place in the space of physics.

Now one can say that the feelings are in the body of the person, that is, somehow in space. However, this assignment is only made because one can localise the body of a person and one feels it that way. This becomes clearer when we look at perceptions. The red colour of a ripe fruit is assigned the same place as the fruit on the basis of the human imagination. In fact, there is nothing red there. Waves of different frequencies are reflected from the fruit, the colour red is only in my perceptual space. Sensation does not exist as a layer of colour in physical (= real) space. Insofar as sensations and feelings are spatially extended or can be localised in a space, this does not refer to physical (= real) space.

In this sense A. MÜLLER says [241]. : "There are processes which can well run in time but not in space, namely the psychic processes, ideas, feelings, impulses of will, fluctuations of attention, etc., have a temporal but not a spatial character. This guarantees an independence of space and time from each other. MINKOWSKI has completely overlooked this and that is why his other much-quoted statement is directly wrong: "Object of our perception are always only places and times connected. No one has noticed a place other than at a time, a time other than at a place".

Mental processes are something real and therefore not only sometime but also somewhere. However, they are real in a different way, their location cannot be determined by measurements and therefore Lorentz transformations cannot be applied to them. If these considerations are correct and if there are spatiotemporal realities of a different kind, Minkowski space cannot be space and time in the philosophical sense The EI is thus likely to face an insoluble difficulty; for the LI it is unproblematic, for it space and time are more than a four-dimensional spacetime for describing physical processes.

11.3. The four-dimensional Minkowski space of the EI

The explanations of the previous sections have created the prerequisites for discussing Minkowski's argumentation for a four-dimensional space-time continuum. This is based on the Lorentz transformations, formula (4.1). All properties of Minkowski space are derived from them. To see the differences, the Galileo transformations, formula (4.6), are also used.

In the Lorentz transformations, x, y and z denote the location coordinates and t the time coordinate. The totality of these values forms the four-dimensional space-time continuum (also called Minkowski space). No three-plus-one-dimensional space can be derived from the Lorentz transformations, because at the transition to an inertial system S', the new value x' as well as the new value t' depends on x as well as on t. Which time t' is measured in the system S' at the location x' depends on the location and time coordinates in S and not only on the time coordinate alone.

Numerical example: $v = 0.5\ c$. If the clock in S shows zero o'clock everywhere, it is also zero o'clock at the position $x' = 0$. A different time applies to the other positions:

(11.2) $$t' = -\gamma v x / c^2$$

Using the above figures, that is 1.9×10^{-9} seconds for each meter.

Space and time measurements are inextricably linked, one can say: Just as in three-dimensional space the spatial coordinates merge into one another when one changes the coordinate system, the same applies here including the time coordinates.

If one interprets the x,y,z and t coordinates relativistically as the real coordinates of the points in physical space, the consequence is inescapable: real space is a four-dimensional space-time continuum, indecomposable into a three-plus-one-dimensional manifold. This is MINKOWSKI's mental position, and will be discussed in more detail.

The following starting point remains for an alternative position: The *x,y,z* and *t* coordinates of the Lorentz transformation are first of all not coordinates of the points in space-time, they are only their *measured values* and are interpreted by MINKOWSKI in an obvious but inaccurately realistic way. This alternative thesis is pursued further in Chapter 11.4.

If you see time as what you yourself experience as time and as what a clock visibly displays, there is no universal time, because in every inertial system it is something different. Something faster when one is at rest, something slower when one sees oneself as moving. With this view of time, all that remains is the relativistic idea of looking at space and time as a whole.

MINKOWSKi reproduced his findings in 1908 in a lecture entitled "Space and Time" [229]. His statement is famous: "From now on, space and time should sink completely into shadows and only a kind of union of the two should retain their independence." A. SOMMERFELD [207] comments: "Objectively, of what Minkowski says here, even today, from the physical point of view, nothing can be taken back, ...; how one wants to position oneself epistemologically to MINKOWSKI'S conception of the space-time problem is another question, but, as it seems to me, a question that does not essentially touch the physical facts." MINKOWSKI continues: "The object of our perception are always connected only places and times. No one has noticed a place other than at a time, a time other than at a place. ... I want to call a point in space at a point in time, i.e. a value system *x*, *y*, *z*, *t* a world point. The multiplicity of all conceivable value systems *x*, *y*, *z*, *t* shall be called the world."

"The views of space and time ... have grown up on experimental physical ground". By this he means the Lorentz transformations, which he interprets relativistically: *x*, *t* and *x'*, *t'* are both real length and real time. A moving clock does not slow down and therefore shows one hour when in reality two hours have passed. The time at the place of the moving clock is as the moving clock shows it, time is slower there.

For the totality of these places and times, MINKOWSKI defines a mathematical model, a special four-dimensional space, the "world", which in itself reflects the properties of the Lorentz transformations. (Mathematicians call this a four-dimensional pseudo-Euclidean space of index 1, i.e. it is a space similar to Euclidean, the normal space, only deviating from it in one dimension - index 1 - namely time).

A space that is supposed to represent reality includes at least coordinate systems (to be able to locate its points) and a distance definition (to be able to express distances between points). The distance between two points is the simplest geometric property of figures, e.g. the length of a side in a triangle is the distance between its vertices, and therefore the distance definition is fundamental. It is also comprehensive, because any geometric figure can be seen as a set of points whose points keep certain distances from each other. Thus a circle or a sphere is the set of points that lie at one and the same distance *r* from a given point, the center.

In addition, a quotation from RASCHEWSKI [293] from which the meaning of the distance definition emerges: "The only question that one could still ask is whether ordinary space does not still have properties that are lacking in three-dimensional actually Euclidean space. But this cannot be because, roughly speaking, in the end all the properties of ordinary space are determined by the fact that the distance of points according to the formula

$$AB^2 = (x_1' - x_1)^2 + (x_2' - x_2)^2 + (x_3' - x_3)^2$$

is calculated, and because this formula coincides with the corresponding ... for the normal three-dimensional Euclidean space".

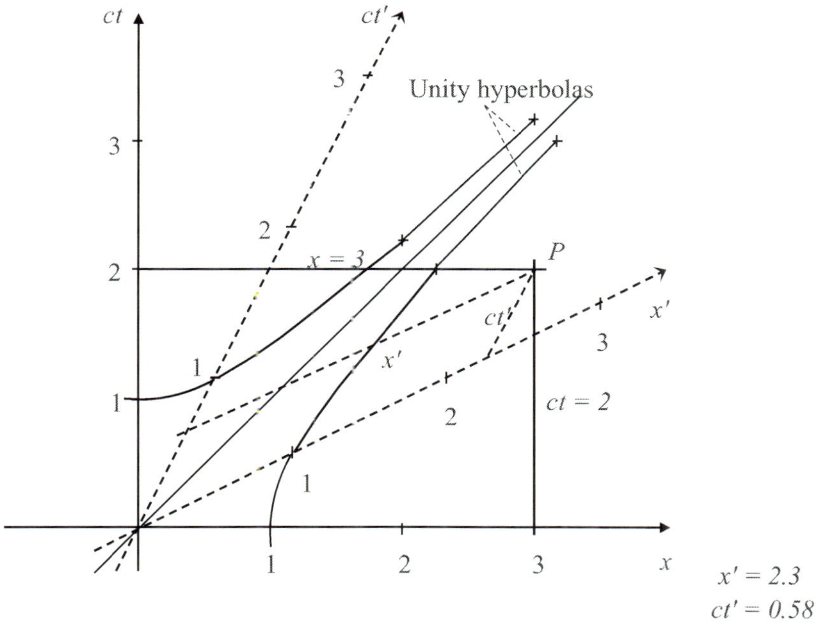

Fig. 11.2 Minkowski diagram for Lorentz transformations

The Lorentz transformations can be carried out geometrically. The *x*,*t* axes represent the inertial system *S*, the *x'*, *t'* axes the inertial system *S'*. The hyperbolas $x^2 -(ct)^2 = +1$, $x^2 -(ct)^2 = -1$ are the geometric locations of the unit points on the *x*, *x'* and *t*, *t'* axes. They define the length 1 on the axes. The slope of the *x'*-axis is *v/c* and for this example 0.5, the slope of the *ct'*-axis is the reciprocal of *v/c*. For the point *P* we get - and analogously for the general case - P(*x*=3, *ct*=2) = P(*x'*=2.3, *ct'*=0.58) in accordance with the calculation.

Coordinate systems in Minkowski space are defined by inertial systems, as in normal space. Inertial systems consist of spatial, mutually perpendicular *x*, *y*, *z* axes and clocks for measuring time. If times are represented by lengths, a coordinate system can be represented by four mutually perpendicular axes. The *x*,*y*,*z*-axes do not differ from normal space, the *t*-axis becomes another axis like those for *x*, *y*, or *z*. To better understand the difference between the spatial and temporal axes, it is sufficient to visualise the *x*- and *t*-axes, see Fig. 11.2. One does not see any difference to the classical representation of *x* and *t*. The difference arises when one wants to switch to another inertial system, i.e. change the coordinate system.

Before going into further detail, first the decisive argument for the existence of Minkowski space: its four-dimensional distance definition.

Distances are quantities that must be independent of a special coordinate system, they are invariants. In normal space

(11.3) $$\Delta t' = \Delta t$$

(11.4) $$\begin{aligned} \Delta l'^2 &= \Delta x'^2 + \Delta y'^2 + \Delta z'^2 \\ &= \Delta x^2 + \Delta y^2 + \Delta z^2 \\ &= \Delta l^2 \end{aligned}$$

Both because of the Galilean transformations. Thereby Δt, Δl, Δx ... are differences of times, lengths, coordinates as in ch.4.4 and 6.2. Thus is

$$\begin{aligned} \Delta x' &= x_2' - x_1' \\ \Delta x' &= (x_2 - vt) - (x_1 - vt) \\ &= x_2 - x_1 \\ &= \Delta x \end{aligned}$$

(11.4) is the PYTHAGOREAN THEOREM in space.

It is these mutually independent and invariant distance definitions for all inertial systems that justify the concepts of space and time in classical physics.

Because of the Lorentz transformations, the invariances (11.3) and (11.4) do not apply in Minkowski space, but instead

(11.6) $$\begin{aligned} \Delta s'^2 &= c^2 \Delta t'^2 - \left(\Delta x'^2 + \Delta y'^2 + \Delta z^2\right) \\ &= c^2 \Delta t^2 - \left(\Delta x^2 + \Delta y^2 + \Delta z^2\right) \\ &= \Delta s^2 \end{aligned}$$

Formula (11.6) is the four-dimensional distance definition of Minkowski space. It is independent of the respective inertial system and justifies the concept of a four-dimensional space-time continuum.

One proves formula (11.6) by substituting the Lorentz transformations by elementary calculation. For the *numerical example* of chapter 4.2 applies:

$$x'_1 = x_1 = t'_1 = t_1 = 0$$
$$x_2 = -300000\,\text{km}$$
$$t_2 = 0$$
$$x'_2 = -1.15 \times 300000\ \text{km}$$
$$t'_2 = +0.575 s$$

Thus
$$\Delta s^2 = -x_2^2 = -(300000\ \text{km})^2 \text{ (negative)}$$
$$\Delta s'^2 = c^2 t_2'^2 - x_2'^2$$
$$= (300000\ \text{km})^2 \left(0.575^2 - 1.15^2\right) = \Delta s^2$$

Without question, the distance formula (11.6) defines Minkowski space as a mathematical space. Whether it can also be the real space remains to be clarified. In the Minkowski diagram, see Fig. 11.2, the axes t' and t, x' and x have different positions in space. The transition from the inertial system S to the inertial system S' corresponds to a rotation of the x' and t' axes by the same angle in the opposite direction, as can be seen from the drawing. For any given point P, its x, t and x', t' coordinates can be read from the drawing. Therefore, Minkowski diagrams are first of all only a geometrical method of performing the Lorentz transformations. Quite correspondingly, there are diagrams to perform the Galilean transformations geometrically, see Fig. 11.3. Furthermore, both diagrams can be understood very simply as coordinate systems and their transformations within two-dimensional mathematical spaces. These mathematical spaces are referred to in LIEBSCHER [197] as Minkowski and Galileo geometry. It is therefore in any case justified to speak of Minkowski spaces in the mathematical sense.

However, the diagrams are not a direct representation of real, physical space. Here is a first consideration - more follow in chapters 11.5 and 11.6: *In the real world the x- and x'-axes coincide for the inertial systems S, S', they consist of the same real points, they glide past each other. In the diagrams, however, they are different.* If we compare this with the situation of the x, y-axes of a coordinate system of the normal plane, the difference is that here the coordinate axes are directly part of the real space. By extending the axes beyond the edge of the paper, all points of a plane of the real space are captured. In the Minkowski diagram, different straight lines correspond exactly to one real straight line, axes rotated by an angle correspond to the real movement of the inertial systems. The Minkowski diagrams are not a two-dimensional section through real space, they are only a geometric model of the Lorentz transformations.

The Minkowski diagrams cannot be used as direct proof of the existence of a four-dimensional space-time, since they do not represent a two-dimensional section of this four-dimensional space-time. However, they have an indirect significance for the proof of a four-dimensional spacetime. To this end, one can make the following argument in the sense of MINKOWSKI: As LIEBSCHER [197] shows, there is the normal, Euclidean geometry in the normal, Euclidean plane, furthermore the Galilean geometry, which results from the Galilean transformations - this geometry exists mathematically, but is not real (real are the Lorentz transformations) - and the Minkowski geometry, which results from the Lorentz transformations and has both mathematical validity and reality.

Examples:

1.) In Galilean geometry the length of a rod does not change when it is moved, in Minkowski geometry the length depends on the relative motion. In Euclidean geometry, there is neither time nor velocity, two lengths remain the same regardless of where they are in space and what time it is.

2) The length contraction is a physical law, but in Minkowski space it is a geometrical statement, which is illustrated in the Minkowski diagram. In Fig. 11.2 the x-, x'-coordinates of the point P are the different lengths of a stationary and a moving rod.

The Galilean transformations were considered as a proof of the reality of the normal Euclidean space and an independent time because their geometry fits, although the diagrams for Galilean transformations also do not represent a two-dimensional section through real space and independent time because the t and t' axes are different. The Lorentz transformations, on the other hand, are proof of the reality of Minkowski space if there is *no* absolute time. In this case, the only reality that remains is Minkowski space. Its reality guarantees the validity of the Lorentz transformations. Its existence is in accordance with the experiment, because the Lorentz transformations are experimentally confirmed and the existence of Minkowski space follows from the experiment, because only such a space contains the Lorentz transformations.

The shortening of moving bars, the deceleration of moving clocks, everything can be graphically represented in the Minkowski diagrams of Fig. 11.2, as LIEBSCHER [197]. explains in more detail. It is astonishing that such a mathematical space exists and the elegance with which much of the theory of relativity can be illustrated in it. If elegance is a proof of truth and if only mathematically identical formulae matter, then MINKOWSKI'S assertion is valid: The world is in reality four-dimensional, space and time for themselves are shadows.

From the mathematical properties of Minkowski space, if one puts it at the beginning, transformation formulae for the transition between coordinate systems follow. They agree with the Lorentz transformations, which are thus part of the Minkowski space. It is real because the Lorentz transformations apply. We do not live in three-dimensional Euclidean space - in it the PYTHAGOREAN THEOREM applies and the distance between two points is described by formula (11.4) - we live in a Minkowski space and the distance between two world points is described by formula (11.6).

These conclusions are compelling if an absolute time is regarded as experimentally excluded (and some other things, see ch.11.4 - 11.8.), because then the Minkowski space is the remaining reality. If there is an absolute time in the philosophical sense,

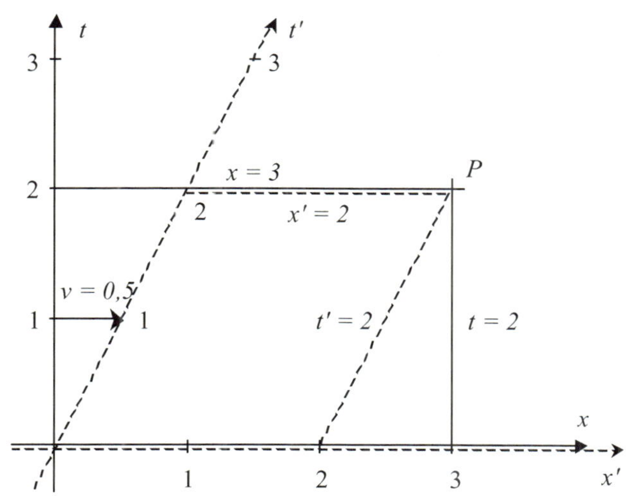

Fig. 11.3 Diagram for Galilean transformations

The Galilean transformations can also be carried out geometrically. *x*, *t* represent the inertial system *S*, *x'*, *t'* the inertial system *S'*. The gradient of the *t'*-axis, it is *l/v*, and its scaling results from the drawing. For the point *P* we get P(x=3, t=2) = P(x'=2, t'=2) and accordingly the construction proceeds in the general case. ($v = 0.5$, the units are arbitrary).

as the LI assumes, the Minkowski space can only be embedded mathematically-physically and in a three-dimensional physical (real) space together with an independent time, as was also true for the Galileo space. For the LI, therefore, the question of whether the Minkowski diagram and (the two-dimensional section through) the Minkowski space are the same is not decisive. With the considerations of RASCHEWSKI [293] (Euclidean and pseudo-Euclidean spaces have the same affine space) it is provable that the Minkowski diagrams establish a 4-dimensional spacetime (Minkowski space). Therefore, with some justification, textbooks do not distinguish between the two.

For the philosophical foundations of physics, the Minkowski space is of great importance. but how does the physicist himself see it? A. P. FRENCH [115] puts it this way: "Little is added to the fundamental statements of the space-time transformations (i.e. the Lorentz transformations) by such considerations..." On the other hand, for Einstein they formed the mathematical approach to the general theory of relativity and therefore have great significance in theoretical research.

11.4. Alternative interpretation of the Minkowski space of the LI

As compelling as the assumption of a four-dimensional space-time continuum as real space and real time appears in physics, because it explains the unconventional structure of the Lorentz transformations, there are objections on the part of philosophy. These objections state that the world cannot be imagined in the Minkowskian form and that in reality there can only be three-dimensional space and one-dimensional time. Some of these objections follow from the paradoxes, others e.g. from chapter 11.5 and from chapter 11.3.

If we leave open for the time being whether this is rightly asserted or merely proof of the inflexibility of some humanists in the face of compelling experiences in physics and mathematics, the question remains: What at least has been proven with MINKOWSKI'S considerations? Since all physical results underlying Minkowski space cannot be meaningfully doubted - the experiments decide - the existence of Minkowski space is proven at least in the following way:

The Lorentz transformations describe (only) the relationship between location and time measurements, not directly the relationship between space and time. I.e. the measured points in time are not the universal time, the location measurements are not the universal space, they only reflect this relationship. The Minkowski space is the four-dimensional space of the location and time *measurements.*

Of course, it is not enough to make such a claim, it becomes necessary to give the physical reasons why *measurements* of place and time should not be space and time themselves. In other words, a physical theory must be developed that shows how time and space can lose their claimed independence and decay or collapse into a four-dimensional manifold of space and time measurements. The alternative derivation of the Lorentz transformations, ch. 6, has shown that by assuming an excellent inertial system together with the principle of relativity, a three-plus-one-dimensional space in accordance with physical experience is conceivable. But this is not enough. It becomes necessary to say something about what an excellent system is supposed to mean physically and how it can be proven.

Philosophy owes such a physical theory, but it is equally clear that it is not the task of philosophy to do physics. It has the right to demand such a theory, but not the duty to develop it. In any case, it is correct to point out from the philosophical side that in the Lorentz transformations only physical measurement results are linked and to interpret them directly as space and time coordinates is hypothesis.

Anyone who, as a philosopher, on the one hand demands three-plus-one-dimensional space-time, but on the other hand does not dispute the experimental findings of special relativity and agrees to conceive of Minkowski space as the totality of space-time *measurements* in physics, cannot be refuted. For representatives of this philosophy, however, a physical theory is missing that explains why space and time appear in experiment as a four-dimensional space-time continuum and what an excellent inertial system should physically be. However, there are approaches to such a theory; they will be discussed in Chapter 11.7.

11.5. Inertial systems with the speed of light

Two mutually exclusive facts can be demonstrated by simple considerations:

a) In Minkowski space there are no inertial frames moving at the speed of light.

b) In real space, however, there are inertial systems moving at the speed of light.

From this, one must conclude that Minkowski space cannot be the space of reality. Minkowski space only includes the totality of the inertial systems described by the Lorentz transformations, but real space includes all existing inertial systems.

First of all, you can convince yourself that there are or can be particles in Minkowski space that move at the speed of light, e.g. photons (light particles), while other particles, e.g. electrons, can only come arbitrarily close to the speed of light. Particles moving with c fulfil the path-time relationship $x = ct$ in an inertial system S. They then have the coordinates in inertial systems S':

$$x' = \gamma x(1 - v/c)$$

(11.10) + (11.11)

$$= \frac{x\sqrt{1 - v/c}}{\sqrt{1 + v/c}}$$

$$t' = \frac{t\sqrt{1 - v/c}}{\sqrt{1 + v/c}}$$

as can be shown by inserting $x = c\, t$ into the Lorentz transformations. Its path in S' is then quite analogous

$$x' = c\, t'$$

as can be obtained by dividing the two equations (11.10) and (11.11) together with the relation $x = c\, t$

In Minkowski space, there is thus the obvious fact that one can describe the path of a particle moving with c relative to any inertial system S. From $x = c\, t$ follows $x' = c\, t'$.

Next, the question arises: Are there in reality inertial systems moving at the speed of light? The answer is that there are even an infinite number of them, as many as there are directions in space. Select any direction in space. Send flashes of light in this direction at certain time intervals. Then the photons thus created move in the same direction at certain spatial distances from each other. All these particles form an inertial system that moves at the speed of light. In it, there are temporal processes: the creation of light quanta as an effect is later in time than its cause, the excitation of atoms. Oncoming light quanta have a different speed than those at rest in this frame. Collisions with atoms are not experienced by all of them, and thus not all at the same time, etc. There is a rectilinearly, uniformly moving system, i.e. an inertial system.

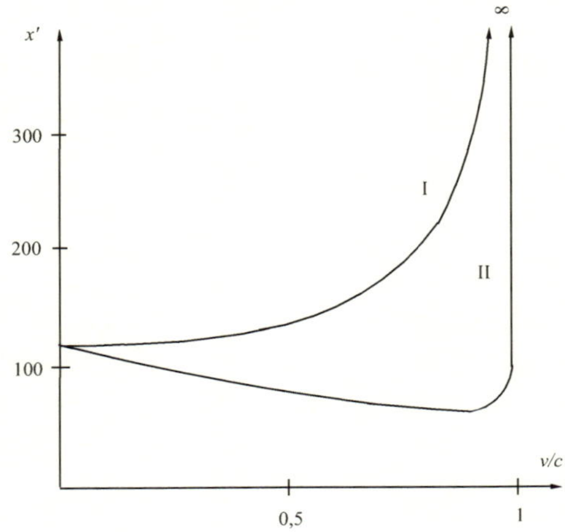

Fig. 11.4 Limit value $v \to c$

For $v = c$, the Lorentz transformation $x' = \gamma(x - v\,t)$ has a pole.

Examples: I: $x = 120$, $t = 0$

II: $x = 120$, $t = 100$

If there are inertial systems moving with *c* in Minkowski space there must be a Lorentz transformation that describes the transition into this system. Since in the Lorentz transformations the quantity *v* indicates the relative velocity of two inertial systems, put $v = c$ into the Lorentz transformations. However, the factor γ becomes infinitely large and x' and t' are no longer defined, i.e. this inertial system does not exist in Minkowski space.

In mathematics and physics, it often helps in such cases to examine the limiting value, in this case the limiting value of *v* against *c*. It turns out: It does not exist either, as sketched in Fig. 11.4. The calculation is not difficult either:

$$\begin{aligned}\lim_{v \to c} x' &= \lim_{v \to c} \gamma(x - vt) \\ &= (x - ct)\lim_{v \to c} \gamma \\ &= \text{undefined}\end{aligned}$$

because the limit value of γ is undefined as long as the special case $x = c\,t$ *does* not exist. For the special case $x = c\,\mathrm{t}$

applies (see 11.10): $x' = \gamma c\,t\,(1 - v/c)$

$$x' = ct\sqrt{1 - v/c}\,/\sqrt{1 + v/c}$$

and for $v = c$ results:

(11.20) $x' = 0$

and accordingly because of (11.11) for all t

(11.21) $t' = 0$

Formula (11.20) is a reasonable value for x', because $x = c\,t$ is the movement of the origin $x' = 0$ of an inertial system moving with *c*. The value $t' = 0$ for all t - formula (11.21) - is not plausible, the above considerations have shown that different points in time exist in an inertial system moving with c. $t' = 0$ for all t would mean time is standing still, which is in contradiction to reality. (There is no contradiction in the assumption that clocks moving with c stop showing the time.)

The previous considerations indicate: Even the consideration of the limiting case $v \rightarrow c$ does not deliver an inertial system, only the x-coordinate for points $x' = 0$ can be meaningfully transformed. On the other hand, it is plausibly explained that in reality there are not only particles moving at the speed of light c, but also such inertial systems. It is precisely these that classical relativity does not know, since they are not covered by the Lorentz transformations. *This means that the four-dimensional space-time continuum of special relativity does not fully describe reality.* What is described with the Einstein interpretation of special relativity is the totality of possible spatio-temporal measurements, not the space and time behind them, because, although inertial frames moving with c exist: Measurements cannot, of course, be made in them. Neither scales nor clocks can be accelerated to the speed of light because of $m(v) = \gamma\ m(0)$. It is not possible to stand next to a clock moving at the speed of light and immediately check what time it shows.

In this case, too, the considerations initially have no significance for experimental physics, because what is measurable is correctly described in the special theory of relativity, and what is measurable lies in Minkowski space.

11.6. Spatial objects and space, time sequences and time

11.6.1. Geometric objects in mathematical spaces, bodies in physical space

Everyone knows what a body is in the physical sense, namely what it is in everyday life - a piece of matter composed of atoms and molecules. A body can be moved around in space at will and still retain its shape, as long as it is rigid.

Just as a body is located in physical space, geometric objects are located in mathematical spaces - sphere, cylinder, cone, for example, in normal three-dimensional space. There are numerous geometric properties of objects in the different mathematical spaces - we will not discuss them here. We only ask a very simple question How are any geometric objects defined in any mathematical spaces? There are different answers to this, but they have one thing in common: a geometric object is or comprises a subset of this mathematical space. That is, one can say of any point in the space, this point belongs to the geometric object and that one does not.

Example: A sphere is the set of all points in three-dimensional space that have a distance from a point $M = M(x,y,z)$ that is less than or equal to the radius r. It is certain whether each point in the space lies inside or outside the sphere.

We can therefore say that geometric objects are defined subsets of a mathematical space. Of course, there are other properties as well. So we can ask what distinguishes spheres from each other and what they have in common, and possible answers would be, different sized radii on the one hand, but the property of being able to define all opposite pairs of points on the surface of the sphere as north and south poles on the other.

What are physical bodies in normal space? Without question, they correspond to geometric objects, i.e. they are subsets, defined sets of points in them. This does not say anything about how the point sets change when the body is moved.

Let us now consider the four-dimensional space-time continuum. Are bodies there also defined subsets? Let's go into an inertial system $S(x,y,z,t)$. A body is described by all the points it encompasses at any fixed point in time. In the case of a steel ball, these are all points that are up to 5 cm away from the center of the steel ball, for example. More precisely: points with such coordinates x,y,z at time t. Measurement inaccuracies should not be of interest. What does this body look like in S'? The answer is to apply the Lorentz transformations, the points remain the same, only the coordinates change.

In S' a new situation now arises: t' is no longer constant for a constant t (relativity of simultaneity, t' also depends on the spatial coordinate x. Our steel sphere only has the same points in the four-dimensional space-time continuum relative to S' if one measures in S' at different times. But what does it look like in reality? An observer in S' counts exactly the points to the steel sphere that it occupies for a given t'.

It follows from this: In classical relativity, bodies do not correspond to geometrical objects in the mathematical sense. As we see, bodies are only defined relative to inertial systems.

Let us look at a quantitative example:

A sphere is at rest in S, its center is the zero point. All points $P(x,y,z,t)$ belong to the sphere with $x^2 + y^2 + z^2 \leq r^2$.

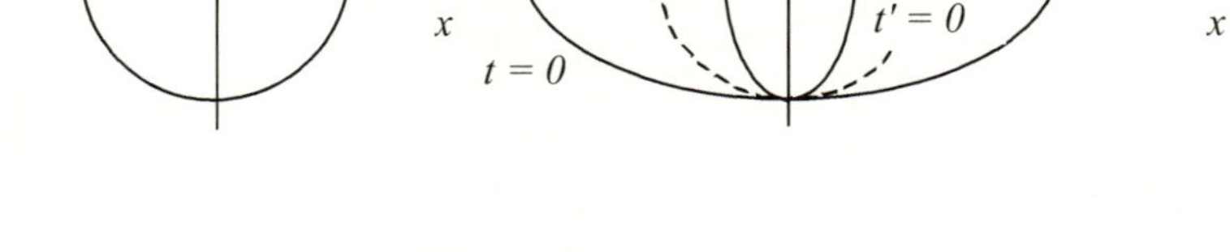

Fig. 11.5 A sphere in S appears stretched in S' at time $t = 0$, at time $t' = 0$ flattened. ($\gamma = 2$)

One expects: A defined set of points corresponds to a real body in real (three- or four-dimensional) space. In fact, the dimensions observable relative to inertial systems do not fulfil this condition.

Let us take the point P_1 with $x_1 = +r, y_1 = z_1 = 0$ and P_2 with $x_2 = -r, y_2 = z_2 = 0$. All points between P_1 and P_2 belong to the sphere, they form the diameter of the sphere on the x-axis. All points x' correspond to it in S' at time $t = 0$.

$$x' = \gamma x \text{ with } -r \leq x \leq +r$$

Here is $$|x'| > |x| \text{ because } \gamma > 1$$

($|x|$ is the amount of x)

For the same diameter, however, the following applies in S' at time $t' = 0$

$$x = \gamma x' \text{ with } -r \leq x \leq +r$$

$$|x'| < |x|$$

(This is the Lorentz contraction)

Fig. 11.5 illustrates the ratios for $v = 0.866\,c$, i.e. $\gamma = 2$. For $t = 0$ one has a stretched ellipsoid in S' for a sphere, for $t' = 0$ an oblate ellipsoid.

Of course, there is no contradictory measurement in either. The measurements are made at different times ($t = 0$ or $t' = 0$) and can have different results. $t = 0$ means in S' for t' an interval

(11.25) $$\frac{-vx'}{c^2} \leq t' \leq \frac{+vx'}{c^2}$$

Since the sphere is moving in S', this means that the same points on the sphere are observed over a certain period of time. The same applies for $t' = 0$ in S.

But this does not change the fact that the same body has different sets of points in S' for $t = 0$ or $t' = 0$ and correspondingly for S. The body is not a firmly delineated part of the four-dimensional space-time continuum. What means the whole sphere for S, means in S' the observation of the sphere's movement over a time interval.

It is not possible to define bodies of the theory of relativity as subsets of the four-dimensional space-time continuum. One can always name points that belong to it on the one hand and not on the other. *The term body is defined in classical relativity only relative to an inertial system.* Although it is something real, it only has the status of a measurable object, it is not a physical quantity uniquely defined for all inertial systems. In S', the sphere consists of different points than in S. This is not true in normal space.

"*Threefold* is the step of time
Hesitantly comes the *future*
Quick as an arrow the *now* has flown away
Eternally still stands the *past*"
Friedrich v. Schiller
„Dreifach ist der Schritt der Zeit
Zögernd kommt die Zukunft hergezogen
Pfeilschnell ist das Jetzt entflogen
Ewig still steht die Vergangenheit"
Friedrich v. SCHILLER

11.6.2. Relative and philosophical simultaneity

The Minkowski space includes present, future and past events, because its time parameter t may be zero, positive or negative.

Since our world consists solely of the totality of present events, certainly not of future ones, and since the past has at most contributed to the present, Minkowski space cannot be the space and time in which and in which we live. Only a part of Minkowski space could describe our world. Which part should that be if simultaneity is only defined relative to inertial systems?

Similar questions are: What is a future event in Minkowski space, what is a past event? Questions that can easily be answered only relative to an inertial frame. To clarify them, first some quantitative correlations:

At a great distance in space, a normally bright fixed star (the Sun) may explode as a supernova, visible from afar, and thereby transforms into a faint neutron star. In the inertial frame S

$$x_{Fix\ star} = -1 \text{ Million Light years}$$
$$x_{Earth} = 0$$

The supernova event is taking place in S now,

$$t_{Supernova} = 0$$

Let S', S'' be the inertial systems as in ch. 10.8 with $v = c/2$. Then applies $\gamma = 1.15$

$$x'_{Fix\ star} = -1.15 \text{ Million Light years}$$
$$t'_{Fix\ star} = 1.15\,(0-0.5c\, x_{Fix\ star}/c^2)$$
$$= +575\,000 \text{ Years}$$
$$x''_{Fix\ star} = -1.15 \text{ Million Light years}$$
$$t''_{Fix\ star} = 1.15\,(0+0.5c\, x_{Fix\ star}/c^2)$$
$$= -575\,000 \text{ Years}$$

This follows from the Lorentz transformations, formula (4.1). Three observers at rest in the origins of S, S', S'' fly past each other at the time $t = t' = t'' = 0$ and each says something different about the outburst of the supernova - which is not visible to them at this time:

For the S observer, the eruption is happening now, for the S'' observer it happened 575,000 years ago, and the S' observer predicts it will happen in the future. He still has 575000 years to prevent the eruption. Of course hopeless for normal space travellers, who cannot even cover 575000 light years in this time, since $v < c$ applies. In the Utopia world, however, as a navigator of a vehicle with faster-than-light speed, one is on the spot in time to avert the catastrophe. If clocks of S, S', S'' move past the explosion site during the explosion, the different clock times displayed by them are correctly predicted by all observers S, S', S'', and these are the clock times that an observer at the explosion site could read from the clocks. The event is the same in all details, including the coordinates, for all observers, only its simultaneity with other events is disputed.

Such future or past events are called by M. CAPEK [46]. in his Einstein interpretation of Minkowski space 'non-causal future' or 'non-causal past'. Because of the finite speed of light, the observer at the point $x = 0$ at time $t = 0$ in the above example is neither influenced by the supernova nor can he influence the supernova vice versa. If he sees the supernova in the telescope - then it can have a causal effect on him - or if he flies to it, it has long since passed.

Another type of future or past events are those that can be causally related to an observer. Causes (light waves, projectiles) can only exert effects on objects that are not too far away. If a light particle moves for the time duration t, it can only have an effect on objects at the distance x if $x \leq c\, t$. Such events form, in CAPEK's words, the "causal future" or the "causal past" and lie in the Minkowski diagram in the light cone, i.e. between the straight lines $x = -c\, t$ and $x = +c\, t$ in the direction of the ct axis of Fig. 11.2. The causal future of the observer at the origin of S has $t > 0$, the causal past $t < 0$. The point P in Fig. 11.2 lies outside the light cone, a light signal emanating from the origin never reaches this space-time point, since $x > c\, t$, only the later space-time point $P'(x = 3, ct = 3)$ is reached.

Causal past are facts like: "The books of the Greek philosopher Plato can be read by everyone", causal future: "a book completed in five years still no one" [46]. The causal future are points in Minkowski space, but they are not yet part of the universe. There is something absolute about the causal future (and correspondingly about the causal past) because: No other observer in the universe can experience it earlier than - in our example - the observer at the origin of S; cause and effect cannot reverse even for other observers; causal past is later past for other observers. Non-causal past and non-causal future, on the other hand, swaps against each other depending on the choice of reference frame, as the above example has shown.

After these preliminary considerations, at least two questions arise: Why does Minkowski space, in contrast to everyday experience, require two kinds of past and future? To which part of Minkowski space does reality correspond? The world to which neither the future nor the past belongs, since the latter represents nothing real (no longer or not yet)?

There are three possibilities for deleting from the Minkowski space future and past. To do this, choose an arbitrary time t, e.g. $t = 0$, which marks the present.

a) Assuming that an excellent inertial system exists, one may argue: All events that are simultaneous in the excellent inertial frame and for which $t = 0$, form our world; the events that are simultaneous to the other inertial frames are in reality not simultaneous because of the only purposeful synchronisation of the clocks, they belong to the past or future. In this case, the world is the three-dimensional normal (Euclidean) space, the different terms causal - non-causal past or future describe synchronisation effects.

b) There is no excellent inertial system.

b1) All inertial systems are equal, everything that is simultaneously at any inertial system at the time $t = 0$ is equally reality. The consequence: The entire Minkowski space is our world. Future, past (causal or non-causal) are always also present in some form and must therefore be added to reality.

b2) One removes all events that mean causal or non-causal future or past for a point at the time $t = 0$ - only this point remains - or one deletes only its causal past and future - then for distant points their present and future belong to the world.

The versions of b) are not immediately satisfying, but version a) is philosophically elegant. The same applies to the different concepts of past and future. The future and the past are experienced no differently at great distances than they are here, and for an excellent inertial system there is no need to discuss them.

The relativity of simultaneity expresses two possibilities:
a) There is no absolute simultaneity and therefore it is not measurable (Einstein interpretation).
b) There is simultaneity, but it is not measurable because on account of the principle of relativity there are no physical laws of that make it measurable (Lorentz interpretation).

Is it really impossible to say something about simultaneity even though it is not measurable? After all, there are physical quantities that are not measurable, but nevertheless have certain, quantitatively fixed values. One example of this is the one-way speed of light. As was shown in Chapter 5, its amount is not measurable; this is only possible for the two-way speed of light, but the light emitted by the sun approaches the earth in a precisely fixed form, the speed exists without being measurable [110]. In this sense, there is a scientific-theoretical argument for an inertial-system-independent simultaneity that is in some sense obvious, Fig. 11.6.0:

At any point in space - that is, for example, all points existing relative to any inertial system - I can imagine an observer and a clock at rest there that shows any time. Every observer has a clear idea of what the present is for him, because he experiences it. It is distinguishable at every moment from all other moments of time by the time indicated by the clock. A little later it is the time of a past event, also it is always certain what times future events will have. This definition of present is inertial system independent, for example, the observers of S, S', S'' have the same conception of present at the point $x = 0$ and therefore none says of the other that it is now already past or still future; their clock times indicated by their own clocks can, however, be different. It is important to keep the terms "really being", "present" clearly in mind. Why should "really be" only apply to a point in time, to "the present"? Hasn't Cologne Cathedral been "reality" and a symbol of eternity for several hundred years? How should the "reality" of Cologne Cathedral stand for a point in time, for the present? This is not a contradiction, the present is the Cologne Cathedral in the splendour of a celebration and always in an emotional variant. Eternity is the present in changing uniqueness, "really" always something different, always only a point in time.

On the other hand, the state of "being real" is completely independent of time or simultaneity. A bird grasps the "real" danger from the sneaking cat directly and not through the knowledge or feeling of a common time. The bird is "real", the sneaking cat is "real" and both are "real together" - there is no talk of simultaneity yet.

Unquestionably through one's own experience, a chain of time points belongs to every space point of any inertial system and always exactly one of these time points is present. This is recognisably and observably completely independent of all other space-time points of Minkowski space and expresses that "in thought" one can go anywhere and there must always be something there. The totality of these clearly defined present points forms our world, all other space-time points are the past or describe the future. Only these present points exist, they "are", and they are so "together". *All points that "exist together", the present points, are called "simultaneous".* In this way, simultaneity has been defined independently of physics, and it is up to physicists to measure it or not, but it is not okay to deny simultaneity on the grounds that it is not measurable. Since there are non-measurable physical quantities (one-way speed of light), this reference is not sufficient anyway.

It is not possible to explicitly specify the present points in Minkowski space; the relativity of simultaneity that prevails there can be understood as an expression of the fact that simultaneity cannot be measured by physical methods, i.e. to have the clocks of all observers display the same time for all present events, which are, after all, clearly fixed. In this sense, Minkowski space describes exactly what can be measured with the help of the natural laws of space and time and not space and time. Or formulated differently: In Minkowski space, the laws of nature express themselves directly. Thus, the principle of relativity through the existence of inertial systems that are equivalent in everything, and later the laws of general relativity as their limiting case. It is not without reason that EINSTEIN described MINKOWSKI'S findings as significant for him in the discovery of the general theory of relativity.

These considerations can be continued. Past events are also united or differentiated by whether they were once real together or not. For a deepening, compare with the work [46], a consistent Einstein interpretation of the Minkowski space, as well as with the references there to REICHENBACH, GRÜNBAUM and their critics, such as Hugo DINGLER. Corresponding more recent works can be found in the philosophical journals [84], [5]. Whether the concept of "being real together" can precede the concept of "being simultaneous", as is claimed here, is ultimately something one must decide for oneself.

As has been shown, not only the different time in inertial systems is a consequence of the relativity of simultaneity, but also the ambiguity about what reality, the world, is supposed to be as opposed to "no longer real" or "not yet real", although the state of "being real" is clearly established for every observer everywhere in the universe through their own experience. On the other hand, this ambiguity within Minkowski space reflects physical laws, especially the principle of relativity, they become geometrically descriptive through it.

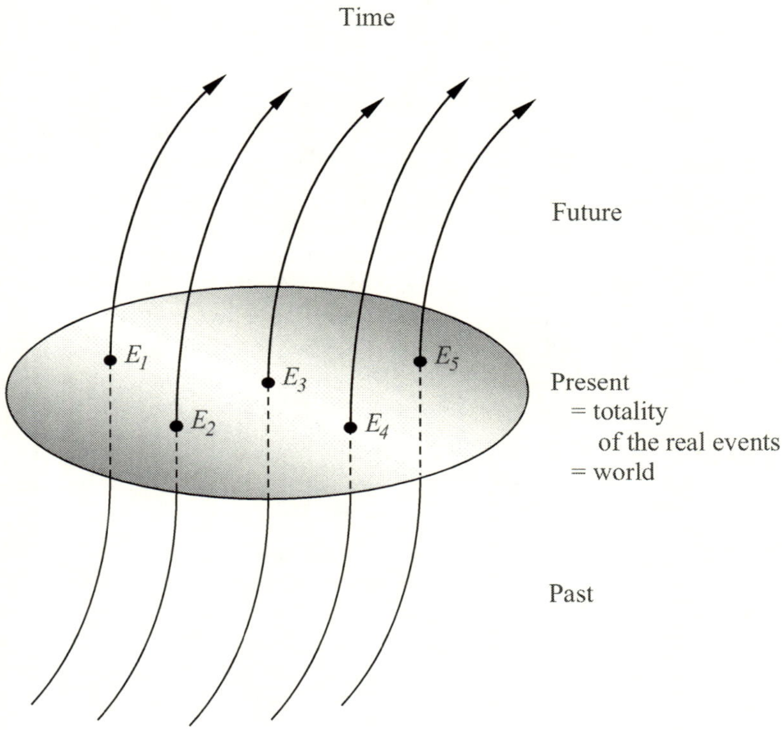

Fig. 11.6.0 Absolute and relative simultaneity

At every location x_i something is "real". If one uses the event E_i to designate everything that is real at the location x_i, then it follows: All events E_1, E_2, ... E_i, ... are "jointly real". For one "real" event at one location there is one "real" event E_i (and not several ones) at another location. "Being real together" can be defined as "being real at the same time" without using the concept of time in a circular way. It follows: Simultaneity exists and is absolute, but is not measurable (relativity of simultaneity).

'... especially since physicists have little understanding for logical-philosophical arguments."
EINSTEIN 1951

11.6.3. Minkowski space and real space

Analogous to the present argument of the previous chapter, there is an argument about spatial properties that the Minkowski space does not possess [vor26]. Consider a rod x at rest, passed by a rod x' moving with velocity v, see Figs. 11.6.1 and 11.6.2. An observer at the point P sees at any time, i.e. always, exactly one point X' opposite him on the rod x'. It follows in particular that if O' *is* opposite O - which is any point in time - then there is exactly one point P' opposite P, see Fig. 11.6.2.

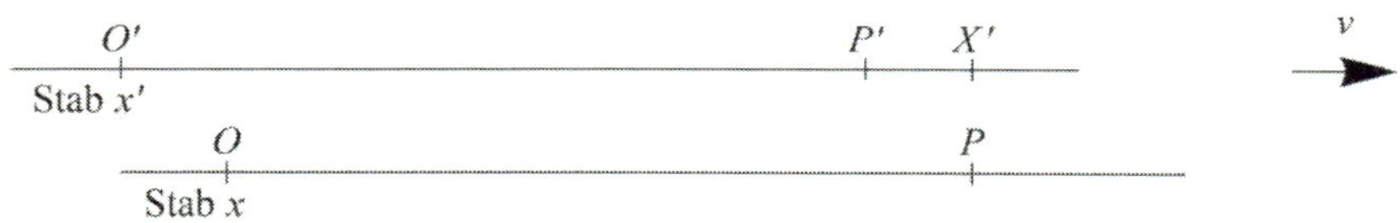

Fig. 11.6.1 Stab x' (rod x') moves with v past the Stab x (rod x). The observer in P on x always sees exactly one point X' opposite him on x'.

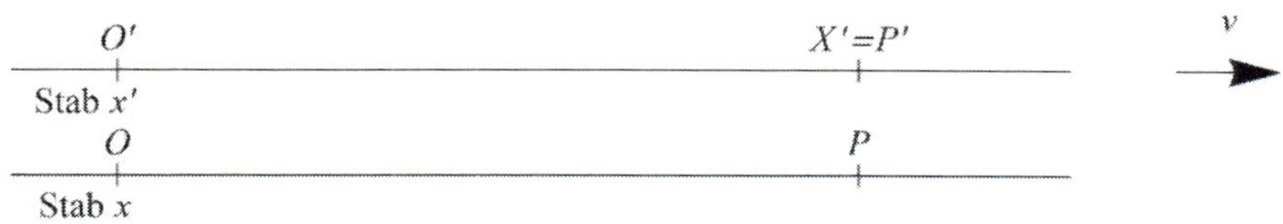

Fig. 11.6.2 So also when O' is opposite O. This point is called P'.

What coordinates does this point P' have? Neither EI nor LI know the answer, they only know on the basis of LT: If O and O' meet, then the following applies

For $t = 0$: $$x'_P(t=0) = \frac{x_P}{\sqrt{1-\frac{v^2}{c^2}}}$$

and for $t' = 0$: $$x'_P\left(t' = 0\right) = x_P\sqrt{1 - \frac{v^2}{c^2}}$$

Fig. 11.6.3 Minkowski space.

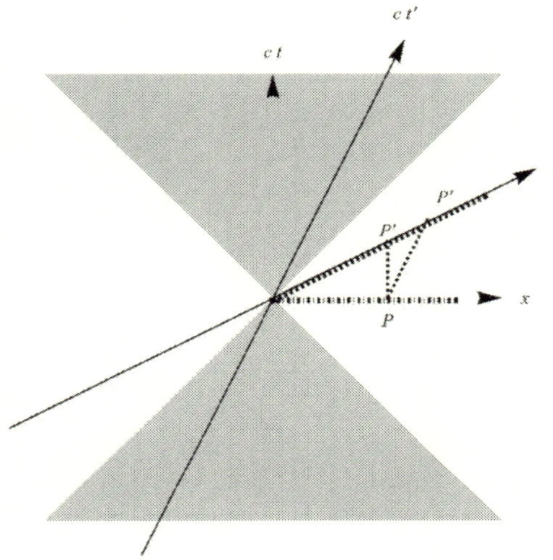

1.) Which point of x is opposite which point of x'? Except for the coordinate origin O, O' it cannot be seen. You have to construct it or calculate it with the help of the Lorentz transformations. The points $P'(t=0)$ and $P'(t'=0)$ are opposite the point P.

2) The rods x and x' have any lengths. Where do the rods lie in Minkowski space if the starting points O and O' are opposite each other? An observer at rest on x sees x' as the dashed line ($t = 0$) and an observer at rest on x' sees x as the dotted line ($t' = 0$). In contrast to real space, the rods have no defined position. In particular, what is opposite P is something different in every inertial frame.

But that is not the answer, at most one of the two points can be opposite x_P. What the LT calculates are the *measurement results* at the given times $t = 0$ or $t' = 0$, see Fig. 11.6.3. The LI would have the correct answer if the excellent inertial system were known, for the EI there is a contradiction because both predictions are equivalent and both must apply. Put a little differently: The Minkowski space does not know the term "opposite",(apart from the trivial case that rod x' is at rest relative to rod x), it only knows "being relatively simultaneous" (or "being relatively opposite", see Fig. 11.6.3) and the following question is not answered: What is opposite P if O' is opposite O? The eminent philosopher E. MAY [223] puts it like this: "There is no more than a three-dimensional manifold to which one could also assign the original meaning of "next to". The four- and multi-dimensional "spaces" are purely relation-theoretical schemes without any spatial meaning."

Exercise. One objection, representative of many others, is: According to special relativity, the rod x' is shortened relative to x. If you change the inertial frame and go over to x', the rod x is shortened. The statement "P is exactly opposite a point P' " is false. Refute this objection.

Note: In Minkowski space there are only measured lengths. To call all measured lengths real lengths is hypothesis.

In Fig. 11.6.2 P' is the point opposite P when O' is opposite O. It is to be shown that this indicates a reference frame which moves with $+w$ relative to rod x and with $-w'$ relative to rod x'.

Measured on x, OP has the length l_x, measured on x', $O'P'$ has the length l_x'. In the thought experiment of Fig. 11.6.2 OP and $O'P'$ are then of equal length, i.e.

(11.30) $$l_x\sqrt{1-\frac{w^2}{c^2}}=l'_x\sqrt{1-\frac{w'^2}{c^2}}$$

(11.31) $$v=\frac{w+w'}{1+\frac{ww'}{c^2}}$$ or

(11.32) $$w'=\frac{v-w}{1-\frac{vw}{c^2}}$$

Formula (11.31) results from (4.13) assuming: The distinguished inertial system moves relative to S with w, S' moves relative to the distinguished inertial system with w' and S' relative to S with v. Since l_x, l_x'. and v are known by measurement one has two equations with two unknowns w, w'.

With (11.32) one eliminates w' in (11.30) and obtains an equation for w which characterises the distinguished reference system. These formulae show mathematically perfectly, in Minkowski space an inertial system is distinguished by Fig. 11.6.2; LI and EI do not contradict each other, but the EI is extended by the LI, [vor26]. The LI distinguishes between real and measured length of a rod.

Or to put it another way: While for the EI the previously stationary and subsequently moving rod changes its length in real terms when the inertial system is changed, the LI says: The real length of the rod does not change, but the measurement is made with different (or changing) scales and clocks.

Since it was E. MAY who brought the rod argument to the author's attention, some quotations from MAY on this and the previous chapter: "The "relativisation" of time, partly glorified as a deed of genius, partly branded as a corrosive activity, is due to nothing other than the simple fact that the empirical determination of simultaneity is impossible. This has been emphasised with particular force by Th. HAERING... HAERING had also clearly recognised that the impossibility of an empirical determination of simultaneity applies independently of any possible outcome of the Michelson-Morley experiment, indeed retains its validity in general independently of any "experimental confirmation"."

"We have recognised time as a structure of order meaning "one after the other" and thus as one of the conditions of the possibility of all experience. The same applies to space, which must be described as a structure of order with the meaning "side by side".

The grasping of the beside happens quite immediately; it can be regarded with certainty as a primordial or ultimate meaning that admits of no further decomposition and is indefinable." [223]

Similar to this, with more concrete reference to the theory of relativity: The scientific-theoretical investigations of TETENS, [365].

"The king is dead, long live the king."
Francis. Ritual

11.7. Ether and quantum ether

"It followed from the MICHELSON-MORLEY EXPERIMENT and special relativity that the aether did not have the simple mechanical properties that had initially been assumed. An arguably somewhat hasty conclusion drawn by many was that the aether did not exist at all." HERRMANN, JOB [152]. EINSTEIN in his lecture "Ether and Relativity" [90]: "The nearest point of view ... seemed to be the following. The ether does not exist at all. ... However, a closer look shows that this denial of the ether is not necessarily required by the special principle of relativity." Correspondingly in [152]: "The existence of an ether was indeed quite seriously questioned for some decades, and some scientists would have liked to banish it from physics. In any case, soon after this partial banishment, it was officially reintroduced, but under a new name (one might almost think so that no one would notice the embarrassment)." "In the modern view, a field is an excited state of this new ether." [152] "The new ether bears the name vacuum.." "Those who have some sensitivity to language will take offence at the fact that something that exists has been given a name that expresses precisely a non-existence."

Nobel Prize winner LAUGHLIN (Theory of the fractional quantum Hall effect):

“The word ether has extremely negative connotations in theoretical physics because of its past association with opposition to relativity. This is unfortunate because, stripped of these connotations, it rather nicely captures the way most physicists actually think about the vacuum” .... “The modern concept of the vacuum of space, confirmed every day by experiment, is a relativistic ether. But we do not call it this because it is taboo.” This and further quotes see V. BALTZ [384.1].

Therefore, in the following we will speak of quantum ether or simply ether. It is the so-called 'non-empty vacuum' of quantum mechanics (QM) and quantum field theory (QF), to it apply the terms such as: Polarisation of the vacuum, zero-point energy of the vacuum

$$\sum_{w} \tfrac{1}{2}\hbar\omega_{w}$$

virtual particle generation, zero-point oscillation or vacuum fluctuation. Its enormous energy s. ch. 24.13. Electromagnetic waves have the uniform velocity c in all directions in the quantum ether.

Thus, the first demand of the LI is fulfilled, because of QM and QF there is a quantum ether. The second requirement is: motion through the quantum ether must cause the relativistic effects like Lorentz contraction, time dilation etc. In all completeness, this has not been proven (LORENTZ's contraction theory was a start), and remains an open problem, but fortunately, because of the principle of relativity, one does not need to know the mode of action of the quantum ether when applying LI, (so little that it sounds plausible to say it does not exist). The existence of a quantum ether additionally follows for the LI from the solutions of the paradoxes and the absolute effects of classical special relativity, see Fig. 9.5. The *rest of the chapter discusses this in more detail.*

An almost self-evident property is that the (quantum) ether admits Maxwell's equations. It then follows that for a preferred system, the speed of light is uniform in all directions and has the value c, moving lengths contract and moving clocks go slower - at least this is the content of one of the better-known ether theories, as proposed by PROKHOVNIK [287] and other researchers [new]. One of them, H. R. Brown [37], writes: "Rods and clocks in motion behave relativistically because the forces that are responsible for the cohesion of matter, forces which inevitably are described by quantum theory, satisfy equations that are Lorentz covariant."

The basis for these models, and also the partly quite different considerations [77], [258], [259], [111], [78], is NEWTON'S classical concept of space. This space corresponds to normal (Euclidean) space. Every place in space is defined by the three coordinates x, y, z, plus the time t, which is independent of them. Mentally, you can imagine the space built up of small cubes (points), then you immediately see that they cannot move as part of the space. Now there are two ideas. The obvious one, there are point-like masses moving through empty space, whereby it is unclear what absolute motion should be, and the other in the sense of QM and QF: In infinite space, quantum ether is everywhere and material particles are wave packets moving through the quantum ether. With this idea, absolute motion is everything moving in the quantum ether. The first idea fits better to EI - Euclidean space becomes Minkowski space, the second better to LI - with the quantum ether relativistic effects like Lorentz contraction could become physically explainable.

Absolute rest in empty space cannot be experimentally proven. Bodies that are "left at rest" and not subjected to any forces can be at rest or move in a straight line, uniformly (Newton's law of inertia). Due to the principle of relativity, there are no laws of nature that make absolute speed measurable.

Is the absolute Newtonian space more than a human idea to be able to better classify experiences? For NEWTON, there is no doubt about it, because apart from constant velocities, there are also accelerations that have an absolute character. An example is the famous bucket experiment. If a bucket filled with water rotates, the water surface is curved; if it does not rotate, it is smooth. In this way, a movement relative to absolute space is indicated, because if one imagines this experiment in empty space, there is only this explanation as the cause for the curved water surface: The bucket rotates relative to space, it must "exist" or better the 'non-empty' space, i.e. space together with quantum ether.

An alternative is offered by Mach's thesis: for it, the water surface would always be smooth in empty space because the fixed star sky is missing. Although Mach's thesis is about 100 years old, sounds very attractive to physicists and is therefore rarely absent from discussions, there is no halfway accepted physical theory that makes the thesis plausible, although the fixed star sky naturally exerts forces on rotating bodies [60]. EINSTEIN no longer believed in it: "...one should really no longer speak of Mach's principle at all" [139]. In any case, it is not sufficient in this form, it would have to be extended and explain the other absolute effects of Fig. 9.5 as well. The excellent inertial system, the system at rest relative to the fixed star sky? The LI could live with that.

The concept of ether is associated with the most diverse prejudices; by DUFFY [80], [81] they are analysed. Widespread and equally false is the claim that special relativity has abolished the ether (and not only tried to do so). The emotional appeal as headline of a journal article "Don't bring back the ether" [71] fits this. A too late plea, EINSTEIN could no longer read it and decided otherwise. See the quote above.

EINSTEIN comments on Newtonian space and the ether as follows [90]: "On the other hand, an important argument can be put forward in favour of the ether hypothesis. To deny the ether is ultimately to assume that empty space has no physical properties whatsoever. The fundamental facts of mechanics are not in harmony with this view. The mechanical behaviour of a system floating freely in empty space ... still depends on its state of rotation. ... In order to be able to regard the rotation of the system at least formally as something real, NEWTON objectifies space. ... NEWTON could just as well have called his absolute space "ether"."

Without doubt, physical theories [ne2] exist for which it is a gain to be allowed to presuppose an excellent inertial system. The author mentioned in [73] is none other than the famous physicist P. A. M. DIRAC. In his work on electromagnetic phenomena, he feels compelled to speak of an excellent inertial frame. This does not prove the validity of the Lorentz interpretation, but it once again becomes plausible that there are no difficulties in principle to physically interpret the concept of an excellent inertial system.

In such considerations, one has further theoretical leeway: Thus JAMMER [172] on the concept of force of special relativity: "... All these restrictions and objections (note: by special relativity) ... do not, however, affect the concept of force in any more profound way. One important point, however, should be noted: Due to the rejection of an absolute simultaneity of two events taking place far away from each other, one has to conclude in special relativity that remote effects are untenable as a legitimate physical concept."

Exactly this is no longer true with an excellent inertial system. The concept of remote action is legitimate again (but thus not yet existent), because for the LI there is philosophical simultaneity and not only relative simultaneity. (Long-distance effects, non-locality, infinitely fast interactions perhaps aids to physically interpret 10-dimensional spacetimes and rolled-up dimensions?)

Now to the question of why it is of fundamental importance to pursue these questions: In this book, LI and EI are two positions with equal rights. Nevertheless, no physicist can choose which of the two theories he wants to take, because there can only be either space and time independent of it or a four-dimensional space-time continuum (apart from completely different alternatives). If, however, the Lorentz interpretation applies, the difference in principle between non-relativistic and relativistic physics is abolished, because it is always permissible to neglect approximately the influences of certain physical quantities, but not fundamental facts about space and time. For example, it is correct to use the Galilean transformation for small velocities, but it is never correct to regard the (or a) four-dimensional space-time continuum as three-dimensional space and independent time in such an approximation. This is exactly what happens in the transition from relativistic to non-relativistic physics, but not in the transition from the Lorentz interpretation to non-relativistic physics, because space and time do not change in this process. The situation is similar when comparing special relativity with QM and QF. For LI, the Lorentz invariance of the laws of nature has physical reasons that lie in the properties of the fields and their particle-free ground state, the ether; for EI, the Lorentz invariance is an abstract consequence of a four-dimensional space-time. In a search for a unified world view of physics, this is a clear advantage for the Lorentz interpretation of relativity - not an argument for the correctness of the theory, but one to pursue it.

11.8. Physical quantities in the EI and LI

EI and LI differ in what they consider to be the actual physical quantities. In the EI, the actual quantities are the four-dimensional vectors and tensors, in the LI, what one would observe of them in the excellent inertial system. For example, for the EI, there is no electric and magnetic field strength as independent physical quantities, but only the twofold, obliquely symmetrical tensor of the electromagnetic field. The electric and magnetic field strength only exist as its components with different values in the different inertial systems. For the LI, the electric and magnetic field strength exist as actual physical quantities, their value is the value of these quantities in the distinguished inertial frame. For the LI, it is a (meta-)law of nature that the *measurement results* relative to arbitrary inertial frames are described by the twofold, skew-symmetrical tensor of the electromagnetic field. Quite analogously, for the EI there is the four-vector distance, for the LI 'length per se' as its *(x,y,z)* components in the distinguished inertial frame. Here the (meta-) law is that the *measurement results* relative to arbitrary inertial systems are described by a four-vector. *This will be discussed in more detail and with some repetitions in the rest of the chapter.*

The Lorentz transformations apply to both interpretations, clocks are synchronised with light signals in both interpretations, and thus the same amount of formulae and the same amount of experimental evidence apply in both interpretations. What is their difference? Their essential difference lies in what of the measured values and what of the formulae linking them are supposed to be statements about real physical quantities. Real means as much as being "in itself", existing independently of, not only relative to an inertial system.

Physical quantities are length, time, mass, temperature, electric and magnetic field strength, momentum, energy, etc. Physical quantities are measured on physical objects. Their measured values consist of the measured value and the unit of measurement. For example, the length of a rod is measured as 1.5 m, the time shown by a clock or the duration of a process as 10 s, the mass of a body as 20 g.

Length, in the Einstein interpretation, is a physical quantity that has meaning only relative to an inertial frame. The measured length of a scale has any value between nearly zero and the resting length depending on its velocity or the velocity of the inertial system in which it rests. Time, quite analogously, is a physical quantity relative to an inertial frame because clock times, as measurements of time, are only unambiguous relative to an inertial frame. Space is a physical quantity relative to an inertial system because the measured values of space, the distances, are only unique relative to an inertial system. The same can be said for electric field strength, magnetic field strength, etc.

What are these quantities in the Lorentz interpretation? For them, space, time, length, electric and magnetic field strength are physical quantities that are independent of inertial systems, but which can only be measured incompletely due to the principle of relativity. The principle of relativity states that all inertial systems are equivalent or – formulated completely equivalently - if there is an excellent inertial system it cannot be localised because of the principle of relativity. If there is an excellent inertial system, it is obvious to say that the distances, times, lengths, electric and magnetic field strengths measured relative to it are the real values of space, time, electric and magnetic field strength.

The situation is similar for the principle of the constancy of the speed of light. The real value of the speed of light is the value measured relative to the distinguished inertial system and it is the principle of relativity, which enforces the same measurement result relative to any inertial system with the difference that here light signals have a different real speed. (In the direction of movement of the inertial system, for example, $c - v$).

In the Einstein interpretation, all measurements are regarded as equivalent and as real is regarded what endures independently of all inertial systems. In the Lorentz interpretation, one has an excellent inertial system and the real measurements associated with it. Additionally, because of the principle of relativity, the other inertial systems that behave in the same way.

All further discussions have this one difference in mind: all inertial systems are in fact equal (Einstein interpretation) or all inertial systems appear to be equal but in fact are not, and the principle of relativity prevents this from being proven experimentally (Lorentz interpretation).

The fact that also for the Lorentz interpretation , $E^2 - H^2$, Δs^2 etc. is constant for all inertial systems expresses a physical law. Four vectors, tensors in general are not real physical quantities, as can best be seen for the four-vector *(x,y,z,ict)*. It is not the reality for LI, but *(x,y,z)* and *(t)*, the four-vector only describes the transformation properties of *measured values*.

The simplest example is the speed of light c. In the Einstein interpretation, it applies to all inertial frames and is a real physical quantity for them. In the Lorentz interpretation, c is only the measured value of the speed of light, not its real value. The measured value is the same in all inertial systems, the (one-way) speed of light itself *(c+v, c-v)* is different for different directions in an inertial system. Therefore scalars, four-vectors, tensors in the Lorentz interpretation is what can be measured from the real quantities in all inertial systems. It is the principle of relativity that makes all inertial systems to appear as the excellent inertial system, with the consequence that in every inertial system physical quantities cannot be measured as they are there, they are measured there as they are and appear there as if an excellent inertial system were present. One can express this as follows: If an inertial system were the excellent inertial system, then time would have the value of the time of a clock at rest in it, length would have the value of the length measurement, the electric field would have the measured field strength value in this system.

Within the Einstein interpretation, there are two types of measurements:

(a) Measured values of physical quantities relative to an inertial system, including length measurements of bodies or scales, time measurements (clock times) and, for example, measurements of electric and magnetic field strength.

b) Measured values which apply independently of inertial systems and which are calculated from the measured values in a). These include the space-time interval

(11.30) $$\Delta s^2 = (ct_1 - ct_2)_2 - (x_1 - x_2)^2 - (y_1 - y_2)^2 - (z_1 - z_2)^2$$

the quantity $E^2 - H^2$, the speed of light c, which is measured as constant c regardless of the inertial frame (and the other scalars and invariants of four-vectors, tensors of the nth degree)

The Einstein interpretation arrives at this double reality in the following way:

The principle of relativity is: All inertial systems are equal. It follows from this: Physical quantities relative to an inertial system are not quantities in themselves, not real physical quantities, since they have different values for each inertial system. The measurement of, for example, clock times with clocks at rest is not a measurement of time per se, since the measurement of such clock times is different in each inertial frame. The measurement of length is not a measurement of length per se, since length measurements turn out differently in each inertial frame and are all on an equal footing. Hence the thesis, space and time by themselves are not real physical quantities.

The thesis of the Lorentz interpretation is:

All inertial systems appear as equal, they cannot be classified as different from each other by measurements. It follows from this: There is no measurement procedure (for which, for example, MICHELSON-MORLEY's was planned) that determines a distinguished inertial system. It follows from this: With physical methods the distinguished inertial system is not recognisable. From this follows: Similarly as before, there are two types of real physical measurements:

a) measured values or measurements relative to any inertial system

b) Measured values or measurements relative to the excellent system, in these measurements the real physical quantities show up directly, i.e. as they are.

From a) follows: The qualitative properties of physical quantities can be seen in every inertial system, but it is not recognisable which of the real, measurable values they have in the concrete case.

Example: Length is a physical quantity because length can be shown relative to any inertial system, but the real length of a scale is not known because the real length of a rod must be measured relative to the excellent inertial system. Time is a physical quantity because clock times can be measured with clocks at rest relative to any inertial frame, but the real time interval is not known. Moving clocks, moving rods actually go slower or are shortened because it is measured that way relative to every inertial system, but their real values are not known because one does not know the excellent inertial system relative to which they would have to be determined.

Physical quantities relative to inertial systems are, apart from time and length, e.g. electric and magnetic field strength, velocity, momentum, energy. For the Einstein interpretation, they are physical quantities that only exist relative to inertial systems, so in a sense they are not physical quantities. For the Einstein interpretation, physical quantities are the four-vectors or tensors such as

Four-vector of length: $(c\Delta t, \Delta x, \Delta y, \Delta z)$

Four-vector of speed: $(u_t, u_x, u_y, u_z) = \left(\frac{dt}{d\tau}, \frac{dx}{d\tau}, \frac{dy}{d\tau}, \frac{dz}{d\tau}\right) = \gamma(v)(c, v_x, v_y, v_z)$

Four-vector of momentum:

$$\left(\frac{E}{c}, p_x, p_y, p_z\right) = m_0(u_t, u_x, u_y, u_z)$$

$$= m_0\left(\frac{dt}{d\tau}, \frac{dx}{d\tau}, \frac{dy}{d\tau}, \frac{dz}{d\tau}\right)$$

$$= m_0\gamma(v)(c, v_x, v_y, v_z)$$

or the twofold covariant tensor of the electromagnetic field.

In the Einstein interpretation, these quantities are the actual physical quantities, since they describe what is independent of the inertial systems. For the Lorentz interpretation, four-vectors, tensors etc. are the physical connections which ensure that the principle of relativity remains fulfilled for laws such as the conservation of momentum and energy or Maxwell's equations, and that the laws of nature retain the same form in all inertial systems.

The electric and magnetic field strength, the one-way speed of light in chapter 5, simultaneity etc. can be measured relative to any inertial system with the clock synchronisation there. The correct inertial system is among them, the problem is not the measurement but the evaluation of the measurement results. With this subtlety, the world remains in order even for the Lorentz interpretation: physical quantities are measurable physical quantities.

Pragmatically, it would be to say that Minkowski space is a mathematical-physical space that does not describe space and time or a space-time that might exist instead, Minkowski space describes only that of space and time that is expressed in the laws of nature and their physical measurements. Not everything of space and time has to be measurable, philosophical considerations can go beyond the measurable.

The explanations in this chapter can be made more precise and perhaps also shifted in emphasis and should only make the differences between the two interpretations in physical concept formation more understandable. From a physical point of view, the Einstein and Lorentz interpretations are equivalent in that they predict measurable relationships in the same way. A principle of relativity and a principle of the constancy of the speed of light apply to both interpretations, albeit modified. They differ in philosophy, in the theory of space and time that cannot be verified by measurements alone.

"The bad thing about searching for the truth is
that sometimes you find it."
R. DE GOURMONT, French. Writer

12. Comparison of the Einstein and Lorentz interpretations of special relativity theory

The LI is an extension of the EI with the same formulae and the same predictions for the measurement results (to be cautious: at least if the predictions of the EI are unambiguous). With restrictions, s. ch. 24.1, it is not possible to decide between the two experimentally, but only mentally. The question is not whether the EI predicts the right thing - that is the case - but whether it sometimes also predicts the wrong thing. Clearly experimentally verifiable differences exist in the transition to general relativity, s. ch. 21.5. The main differences within EI and LI of special relativity lie in the understanding of absolute effects related to accelerated motion, in the solvability of paradoxes and in the compatibility with simple philosophical insights into space and time. Specifically, EI has the following difficulties:

1. absolute effects: Newtonian bucket effect, clocks moving on circular paths (elementary particles), rotating circular ring, rotating circle segments, s. ch. 9 - all effects that are easy to understand if there is an excellent inertial system, still computable, but incomprehensible if there is not. No matter in which inertial system the rotating bucket, the circular accelerator, the rotating circular ring or the rotating circular segments are located, they show an absolute deviation from the non-rotating state.

Example: If a rotating ring is smaller than a non-rotating one in all inertial frames, then the Lorentz contraction must be something absolute *and not:* exist as an absolute effect on the one hand and not exist after a change of the observer's inertial frame on the other hand, see Fig. 9.5.

2. Antinomian conclusions from the paradoxes.

Bell's spaceship paradox - the ribbon breaks, because of Lorentz contraction (internal forces) or because the attachment points are accelerated differently (external forces).

Ehrenfest paradox - the disc tears because of the Lorentz contraction of the circumferential elements or because they are stretched.

Twin paradox - the space traveller remains young, but just as compellingly, the twin on earth remains so. Its solution needs an additional condition (IS requirement), which can be derived for the LI. For the EI it remains an additional requirement. See chapter 10.8.6 and 7.

Sagnac effect - Correct prediction relative to the laboratory system, a zero effect is predicted relative to the co-rotating reference frame. The one-way speed of light is not c everywhere.

3. Philosophical objections - Minkowski space does not know absolute, philosophical simultaneity, it does not know the concept of "oppositeness", see Fig. 11.6.1. The

Minkowski space does not know the terms "length in itself" but only "length relative to an inertial system". Minkowski space only knows what is measurable, but obviously there is more.

All are reasons for the author to choose the LI - knowing that the main thing, the formulas, are the same in both interpretations. Both interpretations remain equally valid. In the sense of **KIP S. THORNE** (Nobel Prize 2017) it is **"extremely useful to have both paradigms at one's fingertips."**

13. An exact definition of the Lorentz interpretation of special relativity

The following attempt at an exact definition of the LI of special relativity theory comprises three postulates. The first two define special relativity as far as the prediction of measurement results is concerned. Here the author follows BERGMANN-SCHÄFER [21], the variants of other textbooks apply analogously.

Postulate 1: Principle of Relativity (RP) of LI

(see chapter 6.6)

Postulate 2: Principle of the constancy of the speed of light of the LI.

(see chapter 6.6)

alternatively: The speed of light is independent of the speed of the source.

The alternative 2nd postulate is more physical and fits better with LI. The 1st postulate applies to measurement results, the 2nd postulate assumes Einsteinian clock synchronisation and also applies to measurement results. For both the EI and the LI, the relativistic equations correctly predict the measurement results. For the EI, the measurement results are also a statement about the real properties of physical objects, but this is not true for the LI, since the measuring instruments are also subject to relativistic changes and the measurement result reflects the overall effect of the changed measuring object and changed measuring instrument. To separate the two, a 3rd postulate is needed.

Postulate 3: There is an excellent inertial system.

All properties of physical objects and processes measured in the excellent inertial system are real properties of objects and processes. The deviation from the measurement results in other inertial systems shows the influence of the measuring instruments changed by movement. The introduction of an excellent inertial system could be avoided if it were definable within the EI what real and what apparent quantities are to be. This is not successful, as an example shows: For a moving rod there are different *measured* lengths, such as the intrinsic length and the lengths relative to S or S'. Here there is no difference between EI and LI. The differences lie in the question, what is the *real* length of a rod? The EI can offer:

(a) The real length of a rod is its intrinsic length, reality is then the four-vector belonging to a rod. (dx^0, dx^1, dx^2, dx^3) whose length $ds^2 = \eta_{\mu\nu} dx^\mu dx^\nu$ has the same value independent of the respective inertial system and is equal to its intrinsic length. This definition is too narrow, because the length measured relative to any inertial system is also something real, see EINSTEIN's thought experiment, chapter 9.1.

(b) All measured values are real. Then a rod has an infinite number of real lengths. This definition leads to contradictions in the solution of paradoxes and in simple philosophical considerations, see ch. 10 and 11.

(c) The EI even denies reality in the usual sense: "For us believing physicists ("gläubige Physiker ") the distinction between past, present [= reality] and future has only the significance of an illusion, albeit a stubborn one" EINSTEIN [121]. This contradicts the fundamental insight of DESCARTES: "cogito ergo sum", see ch. 3.1 and 11.6.2.

All statements of the EI about real quantities and reality are unsatisfactory, therefore it is obvious to extend the EI and to introduce the 3rd postulate.

Conclusions:

(a) The one-way speed of light has the value c in all directions in the excellent inertial frame of reference. (Postulate 2 and two-way speed of lights. ch. 7.3.)

(b) The real length of a rod is the length measured relative to the excellent inertial system

$$\Delta l = \Delta l_0 \sqrt{1-\beta^2}$$

Δl_0 = rest length, $\beta = v/c$, v = velocity of the rod.

(c) Electric and magnetic field strength are two independent physical quantities, the electromagnetic field tensor describes what can be measured in any inertial system. Thus, a charge at rest in an excellent inertial frame has only an electric field, while a charge in motion has an additional magnetic field, which, however, cannot be measured by the observer in motion, resting relative to the charge.

(d) Any inertial frame can be considered as the excellent inertial frame without being experimentally refutable.

(e) The inertial system requirement, ch. 10.8.6, applies.

(f) Minkowski space is a physical space and not, as for the EI, the (or a) four-dimensional spacetime in which we live.

(g) How the excellent inertial system is realised does not have to be known within the theory of relativity, assuming its existence is sufficient. One may suppose that it rests in the 'non-empty vacuum' (quantum ether) and that movement through the quantum ether causes absolute effects such as length contraction and time dilation, etc.

EI: "Why should I distinguish the system K, to which the systems K' are physically completely equivalent, in theory from the latter by the assumption that (there is an excellent inertial system)? Such an asymmetry of the theoretical edifice, to which no asymmetry of the system of experiences corresponds, is intolerable to the theorist. It seems to me that the physical equivalence of K and K' with the assumption that (there is an excellent inertial system), though not downright incorrect from the logical point of view, is unacceptable." EINSTEIN [90].

LI: Motion or rest in 'non-empty vacuum' (quantum ether) are, in contrast to 'motion or rest in empty space', meaningful physical states distinguishing K and K'. In any case, this cannot be ruled out. K and K' are not equivalent if it is true that moving meters change.

II. The Curved Space of General Relativity

II. The Curved Space of General Relativity

"Downhill the saints help you".
Theory of gravity

14. Overview

The aim of the following explanations is to define the concepts of curved, closed and expanding space of the general theory of relativity clearly and precisely in order to be able to understand the fundamental philosophical and physical consequences. The aim is not to guarantee formal mathematical or formal theoretical physical generality, but this can easily be made up for with the help of the specialist literature, [all]

Furthermore, it is not an introduction to the entire general theory of relativity, as e.g. in [95], [338], but a generally understandable presentation of the sub-areas of general relativity that have special significance for the concept of curved, closed and expanding space. For curved space, these are questions and results of the Schwarzschild metric, with the advantage that here the theoretical predictions are well confirmed experimentally. Closed and expanding space are illustrated by elementary theses on cosmology, which is naturally more speculative. Three-dimensional spherical space ($k = 1$) is closed, the properties of an expansion of space become particularly clear in the special-relativistic limiting case of hyperbolic space ($k = -1$)..

For all three selected areas, the Poincaré thesis is verified by way of example that both curved space and normal (Euclidean) space, together with additional mathematical-physical assumptions, are equally suitable for describing general relativistic phenomena. The same applies to closedness and expansion. Thus, the concept of the Einstein and Lorentz interpretation can be generalised to general relativity. The Einstein interpretation corresponds to the idea of a curved, closed and expanding space, the Lorentz interpretation corresponds to the idea of being able to physically explain the curvature of space through changes in scale in gravitational fields (analogous for the closedness). In the Lorentz interpretation, the expansion of space means nothing other than movement in space - here of galaxies after an explosion in compliance with Hubble's law.

Mathematically, both views are equally valid, but in the physical details they have different credibility.

.

15. The Significance of General Relativity for the Special Relativistic Space-Time Theses

As a result of the discussion of special relativity, two different space-time versions have emerged:
a) the four-dimensional space-time continuum, also known as Minkowski space.
b) the three-dimensional space and one-dimensional time with simultaneous validity of a principle of relativity.

The question arises whether general relativity is in harmony with both versions or whether it is suitable to bring about a decision between them. Without question, it is in harmony with the first version, it generalises the four-dimensional space-time continuum to the so-called normal hyperbolic pseudo-Riemannian space, i.e. to a curved space that can be described for not too large dimensions (locally) with a Minkowski space. But it also harmonises with the second version, because one can conceive of any physically curved space as a normal (Euclidean) space for which scales and clocks change in a location-dependent manner.

Fig. 16.1 illustrates the curved path of light rays near the sun. According to the general theory of relativity, space is curved near the sun, like flat paper in moist air. The curvature of space is something real, gravitational forces are not needed. The shortest connection between two points is therefore not a straight line as in normal space, but a curved line such as the great circle on the surface of a sphere. In curved space, light rays move along the shortest path between two points and one can define "straight" with this property in curved spaces, but since space itself is curved, "straight" has some other consequences than in normal space. This fits with thesis a).

Similarly, the general theory of relativity also allows the statement that space itself is not curved, but that the light rays are deflected by gravitational forces. There are then no ideal scales in the gravitational field, but all scales change the more the gravitational field becomes larger. However, these changes in scale, which lead to curved light paths, can be more easily dealt with mathematically if they are illustrated by curved spaces. The same applies to the clocks that move more slowly in gravitational fields, and in this way thesis b) remains justified.

Both interpretations are state of the art [338], [91], [393.1] and state that with the help of general relativity, for the first, no decision between the different interpretations of special relativity is possible. This would be the case if general relativity could compellingly prove the existence of four-dimensional, curved spaces, as is often wrongly believed. Therefore, it is expedient to generalise the concept of Einstein and Lorentz interpretation to general relativity, whereby the Lorentz interpretation explains all relativistic phenomena on the basis of a three-dimensional Euclidean space. The definitional details already emerge from an examination of the Schwarzschild metric. Moreover, it turns out that in application, some of the general relativistic formulae are physically interpreted differently by the two interpretations, so that several experimental possibilities emerge to distinguish between the two interpretations. In principle, however, each of the interpretations can adapt to the respective experimental result - albeit possibly with a loss of credibility.

16. Schwarzschild metric (SM)

16.1. Introduction

This chapter is central to the understanding of curved spacetime and explains the Schwarzschild metric as well as curved (Gaussian) coordinate systems associated with it.

The importance of the Schwarzschild metric lies in several points:

- it is experimentally well proven
- it describes curved space only for spherically symmetrical mass distributions, i.e. suns and planets, and thus makes statements about spatial areas in the order of magnitude of our solar system, but: Since the gravitational force increases towards infinity for small radii, and the magnitude of the gravitational force is proportional to the curvature of space, it can already be clearly seen here what is meant by the concept of curved space. [99]
- Just as clearly, the Schwarzschild radius also shows a possible limit of classical general relativity.

16.2. The experimental evidence for the validity of the Schwarzschild metric

In order to become familiar with the phenomena that have significance in general relativity, the experimental evidence for the validity of the Schwarzschild metric shall first be described. They relate to a curved spacetime and the first thing one would have to do is to introduce the coordinate system modified for this purpose. Thus, one finds oneself in a seemingly contradictory situation: On the one hand, one wants to represent observations of the Schwarzschild metric, and on the other hand, one already wants to use a non-classical coordinate system for this purpose. Strictly speaking, this is the situation of all physics: it needs a coordinate system to describe measurement results, but on the other hand it does not want to make any unproven assumptions about the representability of the observations.

These difficulties can be escaped for the time being, since the experimental predictions of the Schwarzschild metric are to be verified at great distances from the centers of gravity, and for great distances the coordinate system of the Schwarzschild metric does not differ from that of classical physics. This is not an unusual situation, for example, for many astronomical observations the fixed star sky is considered invariant, although all stars are moving relative to each other. In this sense FLIESSBACH [113] states.:"... therefore we can take the fixed star sky as an asymptotic frame of reference. In particular, ... calculated angular changes (for example for the deflection of light, perihelion rotation ...) are equal to observable angular changes with respect to the fixed star sky".

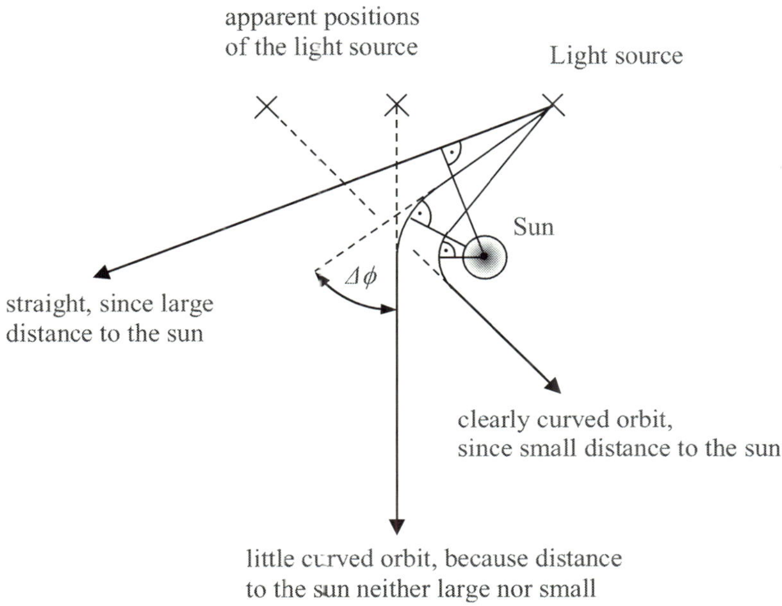

Fig. 16.1 Light moves crookedly near the sun

(In contrast to the drawing, angular changes of up to 2 arc minutes (2 × 1/60 Grad) are observed for the sun).

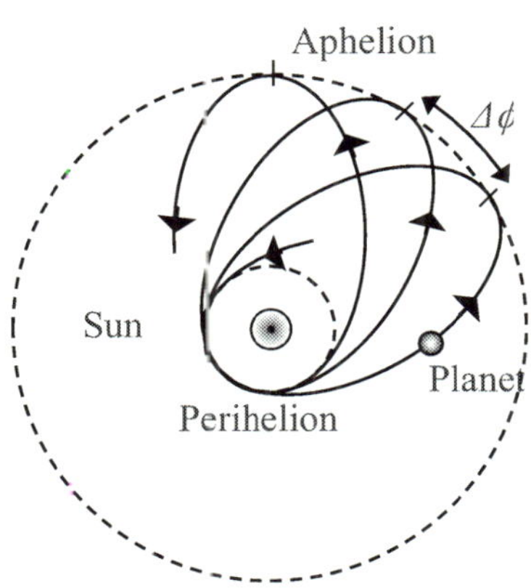

Fig. 16.2 Deviation of the planetary orbits from an ellipse.

The real planetary orbit appears as an ellipse revolving around the sun.

The calculation of the effects to be observed, on the other hand, must be carried out in a curved coordinate system.

The four best-known experimental confirmations of general relativity are, see Fig. 16.1-16.5:

a) Deflection of light rays in the gravitational field of the sun,

b) Deviation of the planetary orbits from an ellipse (perihelion rotation),

c) Delay of the radar echo from Venus when the Sun moves between the Earth and the planet,

d) Time delay of clocks in the Earth's gravitational field.

All four experiments are quantitative confirmations of the theoretical predictions for centrally symmetric gravitational fields as described by the Schwarzschild metric, but only in areas where the gravitational forces are weak.

The deflection of light rays in the gravitational field of the sun is also to be expected according to the Newtonian theory of gravity. Light consists of particles (photons), they have a certain energy and thus a certain mass because of the special theory of relativity and are therefore attracted to the sun. The gravitational acceleration g is independent of the size of the mass and is therefore the same for light particles as for planets. Near the sun, passing light rays are accelerated perpendicular to their path, i.e. they are deflected. If one observes the starry sky during a solar eclipse, one sees the stars near the sun in a different direction compared to the night sky, namely shifted outwards, see Fig. 16.1. However, the general theory of relativity predicts a greater deflection by a factor of 2 compared to the Newtonian theory, which the experiment confirms. (For the classical derivation, use the hyperbola equation, [41].)

Since not only the gravitational forces of the sun but also those of the other celestial bodies act on planets, a deviation from an elliptical orbit is again to be expected according to Newtonian theory. The result in both cases is an orbit that is no longer closed, as shown in Fig. 16.2, which appears as if the point of the planet's orbit farthest from the sun and the point closest to the sun (aphelion and perihelion) rotate around the sun. The perihelion rotation caused by the other planets - it is better observable than that of the aphelion - is clearly greater than the deviation of the predictions of Newtonian gravitation and general relativity for the case of undisturbed, elliptical orbits. The Newtonian calculations and their observation already took place before the development of the general theory of relativity and a generally known, inexplicable small deviation remained. For EINSTEIN it was, as he himself expressed it, one of his happiest experiences when he succeeded in predicting exactly the known deviation from his theory.

The numerical values for the various planetary orbits can be taken from [23].

With today's technical equipment, radar signals emitted by the Earth and reflected by the planet Venus can be received again and their transit times measured with atomic clocks, Fig. 16.3. If the signals are influenced because the Sun is approaching its orbit, their transit time increases measurably. The transit time has been observed daily for about a year. Fig. 16.4 shows the increase in transit time as the sun is approached, the maximum being reached when the radar signal grazes the surface of the sun. The path of the radar beam is slightly curved, as is generally the case for light in the experiment in Fig. 16.1; however, the reduction in the speed of light with increasing gravitational fields is measured here. The entire course of the deceleration curve shown in Fig. 16.4 - not only the maximum - agrees with the general theory of relativity.

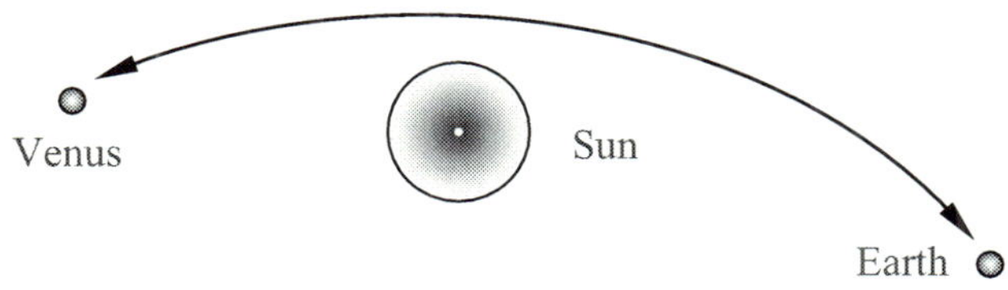

Fig. 16.3 Radar signals from Earth to Venus and back

Radar signals emitted by the Earth are reflected by the surface of Venus and received back on Earth. The transit time of the signals increases as the Sun approaches its orbit.

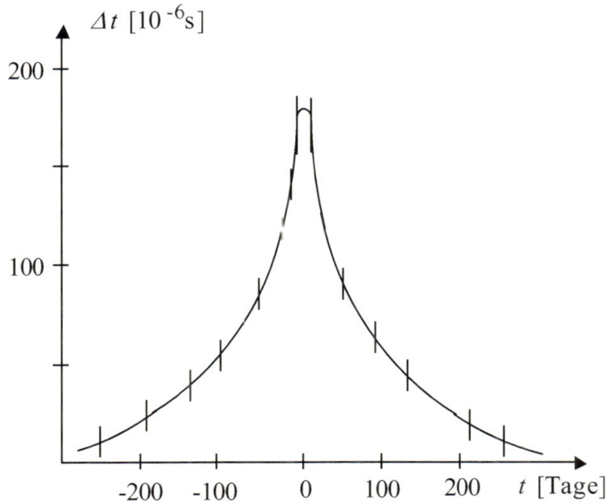

Fig. 16.4 The propagation delay Δt of the radar signals from Fig. 16.3

... as a function of time t has a clear maximum when the signals pass close to the sun.

The evidence for a slowing down of clocks in the Earth's gravitational field was provided by the experiment of HÄFELE and KEATING [143]. Clocks are transported to great heights and the time difference to the slower moving clocks on the Earth's surface is measured, see Fig. 16.5. With this experiment, the time dilation of the special theory of relativity was simultaneously proven. The measurement accuracy has been constantly improved and is far below one percent [2].

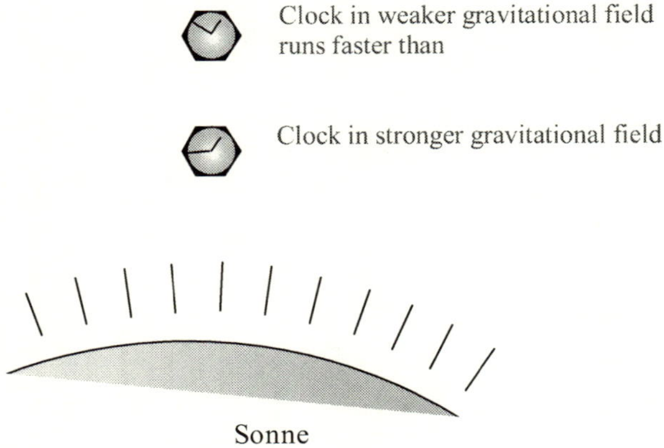

Fig. 16.5 Clocks go slower in gravitational fields

All the experiments described prove the validity of the Schwarzschild metric, as will be justified in the next chapters, and thus that of the entire general theory of relativity. In addition, the prediction of black holes [338], whose existence can be regarded as certain with constantly increasing probability (but they have different properties in both interpretations, see ch. 21), and of the expanding universe (ch. 22) are important qualitative confirmations of the theory.

16.3. Fundamentals of the Schwarzschild metric

The result of general relativity for centrally symmetrical fields, i.e. for spherical bodies, is expressed in the Schwarzschild metric. With it, times and length measurements are linked together, just as lengths for normal space are given by

(16.1) $$ds^2 = dx^2 + dy^2 + dz^2$$

and lengths and time intervals for the Minkowski space through

(16.2) $$ds^2 = dx^2 + dy^2 + dz^2 - c^2dt^2$$

Both relationships are also called metrics of these spaces. The term metric stands for measure, i.e. it specifies how the distance *ds* of two spatial or spatiotemporal points can be measured. (The distance *ds* initially refers only to points that are close to each other. However, this is not a restriction in principle, since any distance can be resolved into a sequence of points that are close to each other).

By introducing the Schwarzschild metric, one arrives at curved spaces without detours. To facilitate the understanding of the metric, it is analysed in several steps. First, parts of the metric are calculated in order to get to know its individual elements, then the metric is illustrated geometrically. Only then is its physical significance for curved space-time shown. (In the literature, e.g. [230], mathematical considerations about curved spaces are sometimes prefixed. Those who prefer this should first read the independently written chapter 18 on the mathematical concept of curved spaces).

Since the Schwarzschild metric is formulated in so-called spherical coordinates, first consider Fig. 16.6. There it is shown how the usual *x,y,z* coordinates of a point *P* can be replaced by the spatial distance *r* of the point *P* from the origin, as well as by the angles θ and φ. θ and φ indicate the direction in which the point *P* lies as seen from the origin. dr, $d\theta$, $d\varphi$ are the differences of the coordinates of two points lying close together, as is also true for dx, dy, dz. If P_1 $(r_1, \theta_1, \varphi_1)$, P_2 $(r_2, \theta_2, \varphi_2)$ are two such points, then, for example, $d\varphi = \varphi_1 - \varphi_2$.

With these coordinates, formula (16.1) is

(16.3) $$ds^2 = dr^2 + r^2(d\theta^2 + \sin^2\theta \, d\phi^2)$$

and formula (16.2)

(16.4) $$ds^2 = dr^2 + r^2(d\theta^2 + \sin^2\theta \, d\phi^2) - c^2dt^2$$

and the Schwarzschild metric has a very similar shape:

(16.5) $$ds^2 = \frac{1}{A}dr^2 + r^2(d\theta^2 + \sin^2\theta \, d\phi^2) - Ac^2dt^2$$

with $$A = 1 - \frac{2GM}{c^2 r}$$

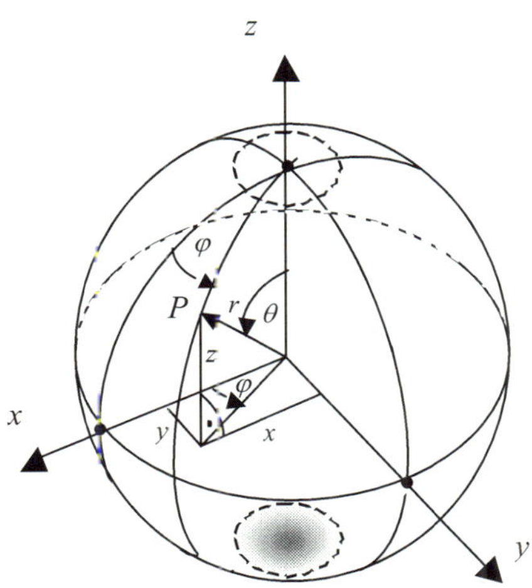

Fig. 16.6 Spherical coordinates in normal space.

The point *P is* fixed by its coordinates (x, y, z) or (r, θ, φ). Constants θ, φ correspond to latitude and longitude circles on the globe.

As you can see, the difference to the non-curved spaces lies here in the factor *A* before *dt* and its reciprocal before *dr* and also only this factor has something to do with the curvature.

It means:

r, θ, φ	spherical coordinates of a point *P*
$dr, d\theta, d\varphi$	the difference of these coordinates for two nearby points.
t	is the time, *dt* is the time difference between two closely succeeding events
G	is the gravitational constant
M	the mass of the spherical body (sun)
ds	the spatiotemporal distance between two near points

Setting *dt* equal to zero in formula (16.2) results in (16.1), the spatial part of (16.2). Correspondingly, *dt* equals zero in (16.6) leads to the curved space of the spacetime of (16.5):

(16.6)
$$ds^2 = \frac{1}{A}dr^2 + r^2(d\theta^2 + \sin^2\theta\, d\phi^2)$$

where *ds* is the spatial distance between two near points at a time *t*.

Numerical example:

The spatial distance of two points P_1, P_2 on the equator of the sun with P_1 ($\theta = 90°$, $\varphi_1 = 0°$, $r = R_{Sun}$) and P_2 ($\theta = 90°$, $\varphi_2 = 1°$, $r = R_{Sun}$) results in the following for normal space with formula (16.3)

$$ds^2 = r^2 d\phi^2 = (1.2 \times 10^7\ \mathrm{m})^2$$

since $dr = 0$, $d\theta = 0$, $d\varphi = 2\pi\, 1°/360° = 0.0175$ (radians) and $\sin(90°) = 1$ (constants: see table in the appendix.)

The same result follows for the Schwarzschild metric, since $dr = 0$ and both formulas do not differ further.

For a point P_3 at height $dr = 1$ m above P_1 (or P_2) $ds^2 = dr^2$ for the normal metric

and
$$ds^2 = \frac{1}{A}dr^2 = 1.000013\, dr^2$$

for the curved space of the Schwarzschild metric

In words: A vertically placed meter scale *dr* has shrunk by 6 10^{-6} m, root of *1/A*, at the surface of the sun compared to normal space.

Another property of the Schwarzschild metric arises from the examination of large radii *r*. In this case, both *A* and *1/A* become one because the magnitude *1/r* becomes zero and the Schwarzschild metric transitions to the metric (16.2) of the Minkowski space. This is a reasonable result, since for distant suns the gravitational forces disappear and special relativity applies. This is also illustrated by the geometrical considerations in the next chapter.

Finally, the case where in (16.3) the spatial distance is zero and dt is non-zero. This means dr, $d\theta$, $d\varphi$ are zero and the Schwarzschild metric holds

(16.10) $$ds^2 = -Ac^2dt^2$$

This situation describes a clock at rest in the gravitational field and is still needed in chapter 16.6.

Numerical example:

With the above values, $A = 0.999\,987$. Near the sun, a clock runs slower by the factor $\sqrt{0.999\,9871}$. If one second has passed on Earth, it is only 0.999 994 seconds near the sun.

16.4. Geometric model of the Schwarzschild metric

The Schwarzschild metric can be represented geometrically as a curved surface similar to a spherical surface. For this, however, one must limit oneself to two-dimensional sections through the three-dimensional curved space, since three-dimensional curved spaces could only be represented in a four-dimensional space.

To illustrate this idea, we first look for a geometric model for the metric (16.3) of normal space. The result is clear, two-dimensional cuts through three-dimensional (normal) space lead to the normal (Euclidean) plane. To obtain the same from the metric (16.3), set $\theta = 90°$ and consider only points with this θ, then $d\theta$ becomes zero and formula (16.3) reduces to

(16.18) $$ds^2 = dr^2 + r^2 d\phi^2$$

From Fig. 16.7 it can be seen that two points P_1 and P_2 each at a distance of r_1, r_2 from the origin in the direction φ_1 and φ_2 in the x,y-plane satisfy the relation (16.18). The plane itself is thus the geometric model for the metric (16.3), if one restricts oneself to the intersection $\theta = 90°$ and ds is the spatial distance between two near points, which can be verified in a drawing without mathematics by measuring.

Quite analogously, a model for the Schwarzschild metric can be found if one first restricts oneself to the intersection $\vartheta = 90°$. See Fig. 16.8. It is claimed that any ds measured in the curved surface, i.e. any distance between two near points, can also be calculated from the Schwarzschild metric obtained for $\theta = 90°$: [393.1]

(16.19) $$ds^2 = \frac{1}{A}dr^2 + r^2\,d\varphi^2$$

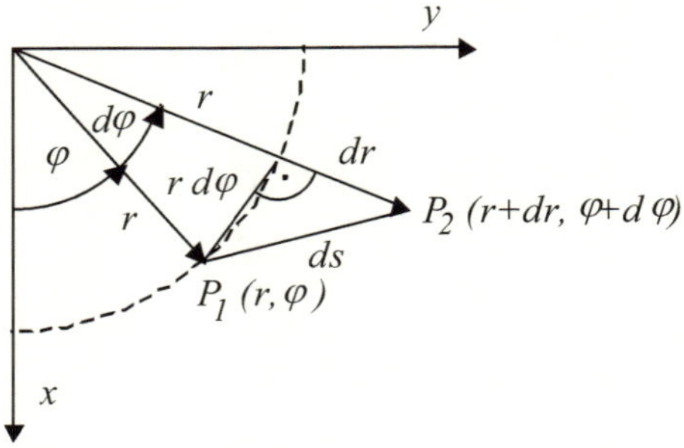

Fig. 16.7 The distance *ds* in other coordinates

For near points P_1, P_2 of the plane, the arc $rd\varphi$ and the chord of the circle differ negligibly little and, because of the theorem of PYTHAGORAS:

$$ds^2 = dr^2 + r^2\, d\varphi^2$$

As a proof by trial and error, one could make a spatial plaster model of Fig. 16.8, measure a series of point distances and compare them with (16.19). Of course, one can also calculate: The area is obtained by rotating the parabola

$$z = \sqrt{8\, M_S\, (\, x - 2\, M_S\,)},\; M_S = GM/c^2 \qquad (16.20)$$

around the *z*-axis. Here *x* becomes the plane distance *r* for all points on the surface of rotation. In Fig. 16.9 the parabola is drawn. As can be seen from the derivation

$$dz = \frac{4M_S}{\sqrt{8\, M_S\, (\, r - 2\, M_S\,)}}\; dr$$

Now calculate

$$\begin{aligned} ds^2 &= dz^2 + dr^2 \\ &= \frac{1}{A} dr^2 \end{aligned}$$

Thus we have the first term of the Schwarzschild metric (16.6), the second term in (16.6) is identical for our special case with

$$r^2\, d\varphi^2$$

and $rd\varphi$ is nothing other than the distance between two points on the same circle of latitude of the surface of rotation.

Here is a quote from WEYL [393.1]: "The geometry of the plane is therefore the same as it is in Euclidean space on the surface of revolution with the parabola $z = \sqrt{8M\,(r - 2M)}$." By plane is meant here the curved plane (two-dimensional surface) through the center of the sun.

The curvature of the surfaces for the case of Fig. 16.8 can be read quantitatively from the ratio of *ds*, the spatial distance between two near points, and the corresponding coordinate differences *dr* in the radial case and $rd\varphi$ in the tangential case. If *ds* is directed tangentially, *ds is* undistorted and agrees with $rd\varphi$, is *ds* placed radially, the corresponding *dr* decreases. If one considers *ds* as a scale, it is shortened (radial) or un-shortened (tangential) in a plane through the center of the sun, depending on the rotation.

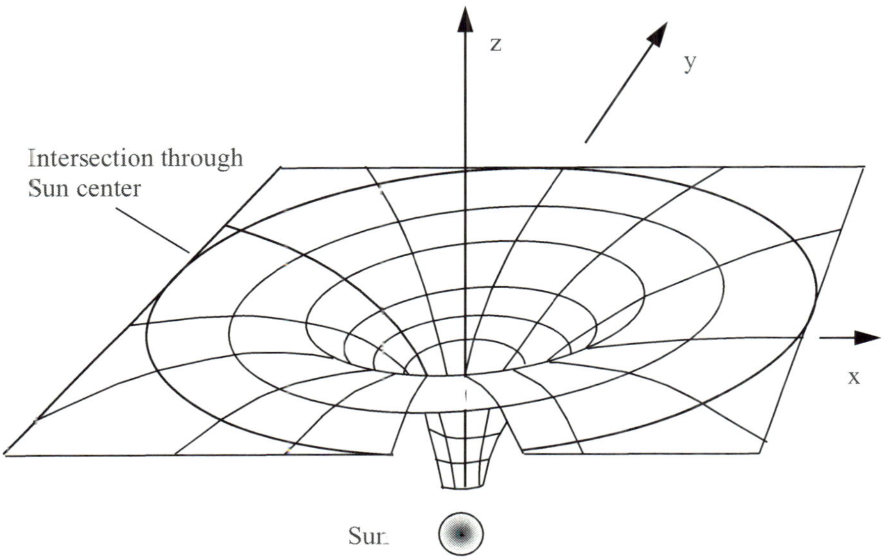

Fig. 16.8 Surfaces by rotating the parabola of the Fig. 16.9. [23], [338].

If one approaches the radius r = 2M radially to the radius $r = 2M_s$ then $dr = 0$, one moves downwards in the plane more and more steeply and does not reach the inside of the plane and the center. For large distances from the center of the sun, the surface becomes more and more like a normal plane. This shows that for large radii the Schwarzschild metric changes into normal (Euclidean) space, or in the general case into Minkowski space, which, however, is always Lorentz-interpretable.

One can also visualise the Schwarzschild metric for any other plane parallel to the midpoint plane at distance a. To do this, rotate the curve

$$z = \sqrt{8M_s(\sqrt{a^2 + x^2} - 2M_s)}$$

about the z - axis, see Fig. 16. 9. For $x = 0$ the following applies to the surface:

$$ds = dr$$

At this point a scale or the length of the spaceship is unchanged. This surface, however, only represents the scales that lie in the direction of the curve or the length of a spaceship, not its width, in true size. For the scales that are perpendicular to the curve, another surface is needed, which we will not show here. (Mathematically, the isometry problem shows up here, not every curved two-dimensional surface can be represented lengthwise in three-dimensional (Euclidean) space, s. ch. 18.5, but this does not affect the idea of shortening scales. Furthermore, although it is a reasonable model to illustrate scale contractions there is a difference to the geometric model of LI, ch. 20, for tangential distances.)

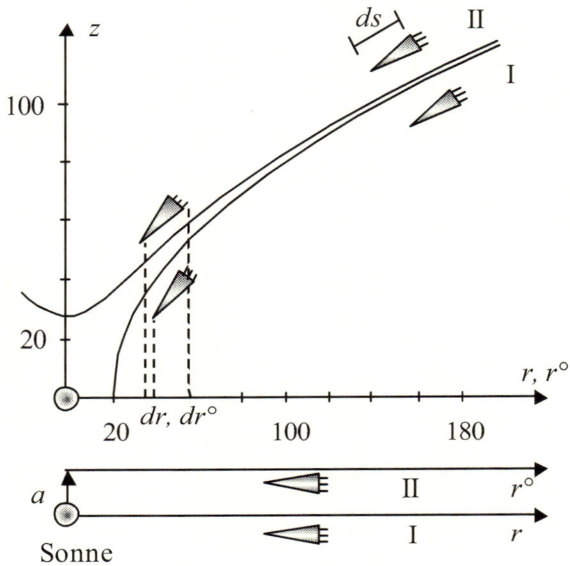

Fig. 16.9 Spacecraft in the gravitational field of the Schwarzschild metric

Curve I: The spaceship moves radially to the center of gravity and shrinks to zero. (parabola)

Curve II: The spaceship moves past the center of gravity at a distance *a*. For $r^\circ = 0$, $dr^\circ = ds$, since the spaceship is then a tangentially positioned scale. For large r°, the spaceship corresponds to a radially positioned scale and is shortened. (This geometric model differs from that of LI, ch. 20, for dr°).
ds: original length of the spaceship
dr, dr°: shrunken length of the spaceship in the gravitational field

16.5. Gaussian coordinate system for centrally symmetric gravitational fields

A Gaussian or curvilinear coordinate system has, like other coordinate systems, the purpose of assigning position parameters to points in space. In the model of the Schwarzschild metric, Fig. 16.8, this is clearly illustrated for the midpoint plane by the graphically depicted latitude circles and the parabolas intersecting them.

Mathematically, the latitude circles of the model are points with the same radial distance r from the z-axis, the parabolas differ only in their angle of rotation φ around the z-axis, so that each point of the midpoint plane is fixed by the coordinates r, φ. Physically, the parabolas in the gravitational field correspond to the fall lines of bodies with the initial velocity zero and for a circle of latitude all light sources resting on it experience the same red shift. Thus, longitude and latitude circles are physically determinable; the quantitative relationship between time dilation and radius r is known to us from formula (16.10).

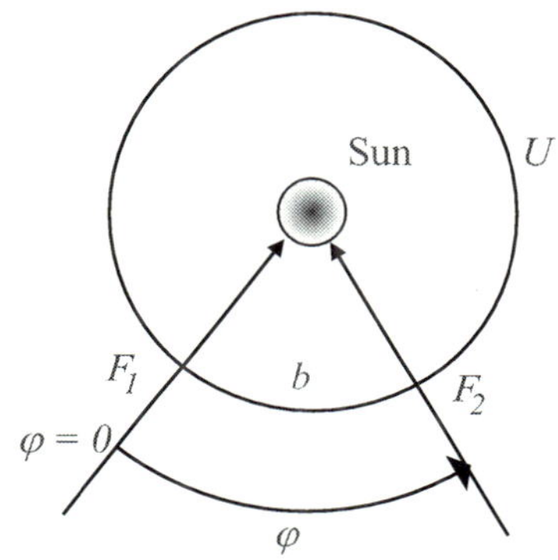

Fig. 16.10 The angle φ in of the Schwarzschild metric

U circumference of the circle of latitude
b Circular arc
F_1, F_2 Trajectories of freely falling bodies

Now a method is missing to determine the angle of rotation φ. Any desired fall line is first assigned the angle $\varphi = 0$, the angle of rotation of the remaining fall lines is then calculated from

$$\varphi = 360° \, b / U$$

with the arc length b and the circumference U of any latitude circle, as sketched in Fig. 16.10. For large distances from the center of gravity, the angle φ measured in this way is identical to that of normal (Euclidean) space, so that the angle φ of the fall lines at a sufficiently large distance can also be determined in the usual way with any desired accuracy. Neither is it necessary nor is it physically feasible to measure φ out of the center of the sun.

This construction can be transferred to the spatial, since the angle θ, like φ, can be measured in the usual way at a sufficiently large distance in any desired accuracy. If one does not want an approximation method, additional considerations analogous to those of chapter 18.4 are necessary. Suggestively the following: Spatially distributed, stationary light sources with the same redshift lie on a spherical surface, because they have the same radius r. All fall lines are located on this sphere. On this surface all the lines of free fall are perpendicular, just like the radii of spheres in normal (Euclidean) space. Midpoint planes intersect spherical surfaces of the same redshift in great circles, i.e. they divide the surface into two equal halves. One of the midpoint planes is marked and fixed to its φ as described. If one assigns $\theta = 90°$ to it (equatorial plane), all center planes perpendicular to it have constant φ and θ has all values from 0 to 180°. All constructions as on spherical surfaces in normal (Euclidean) space are permitted here because scales are not changed in the tangential direction and in this way the coordinates r, θ, φ can be assigned to each point. It does not matter whether the space of the Schwarzschild metric is curved in reality or is to be regarded as Euclidean, θ und φ does not differ.

Summarised in one sentence: The fall lines of bodies initially at rest have constant angles θ and φ, the values of which can be measured at great distances from the sun as in normal (Euclidean) space; the radial coordinate r can be calculated from the redshift of light sources at rest in the gravitational field.

16.6. The Einstein and Lorentz Interpretation of the Schwarzschild metric

There are two fundamentally different interpretations associated with the Schwarzschild metric, both of which are legitimate, and have been discussed by EINSTEIN [91] and others [338], [393.1].

Thesis I: The Schwarzschild metric describes a curved space or a curved space-time.

Thesis II: The Schwarzschild metric describes how scales and clocks change in gravitational fields, space itself remains flat.

Thesis I, analogous to the considerations of special relativity, is called the Einstein interpretation and Thesis II the Lorentz interpretation of general relativity.

Thesis I means:

Normal clocks and normal scales are to be regarded as ideal. The proper times of clocks at rest indicate the real times at their locations, while scales at rest measure the real distances between two points. Both are represented in the Schwarzschild metric by ds, i.e. reality is represented by the quantity ds, just as for a spherical surface the quantity ds represents the real distance between two points on it. The other quantities dr, $d\theta$, $d\varphi$ and dt are only the associated coordinate differences, they have no direct physical meaning.

The extent of the curvature of space can be illustrated with Fig. 16.8, just as for a spherical surface. Since the coordinates r, θ, φ, t, have no direct physical meaning, they cannot be measured with scales and they can be transformed into other coordinates at will without changing space or space-time itself. If, for example, one measures the length ds between two points near the sun as 1.5 m with a tape measure, then the 1.5 m is the real length ds, the real distance between the two points, the coordinate differences are purely calculated quantities.

The physical significance of coordinate systems for Thesis I lies in describing a given curved space in an application-oriented and elegant way. Thus, spherical coordinates in the Schwarzschild metric express the lines of constant curvature directly and vividly, in contrast to the x,y,z coordinates, which would also be usable.

At great distances from the center of gravity, the spacetime of special relativity applies approximately, i.e. the Minkowski space.

Thesis II means:

There are no ideal scales for gravitational fields, all normal clocks and all normal scales shrink as gravitational fields grow. The Schwarzschild metric describes in detail how. If, for example, a tape measure near the sun shows the amount *ds* of 1.5 m between two points, this is only a measured value, the real distance is *dr* if the tape measure is in radial direction and also in the other cases what the coordinate differences of the Schwarzschild metric show. In any case, this is the closest assumption, it will be modified later, see ch. 20.

The real spatial distances, the real time intervals are described by the coordinates r, θ, φ, t. However, they are not directly measurable, only *ds* values are measurable, which also include the proper time intervals $d\tau$ as a special case. The *ds* values must be converted into corresponding dr, $d\theta$, $d\varphi$, dt values using the Schwarzschild metric.

Spacetime is flat, as described with the Minkowski metric, but to be interpreted according to LI. All previous explanations of space and time of special relativity remain unaffected by general relativity for thesis II, apart from the changes of clocks and scales.

Finally, a few quotes on Theses I and II. They confirm that both theses are conceivable without contradiction within the general theory of relativity. Thus EINSTEIN states [91]: "The unit scale thus appears shortened with respect to the coordinate system in the amount found by the presence of the gravitational field when it is applied radially. ... When applied tangentially, the gravitational field of the point of mass has no influence on the length of the rod." WEYL [393.1] can be quoted with: "Admittedly, it is also possible to represent this state of affairs as follows: In truth, Euclidean geometry applies; but the gravitational field acts on the scales in such a way that a radially positioned scale, which is at the (true) distance r from the center of gravity, undergoes a contraction in the ratio

$$\sqrt{1 - 2m/r}:1$$

whereas on a scale placed tangentially, perpendicular to the radii, the gravitational field does not produce a length measurement." SEXL [338] states: "One can interpret this result in two ways: We have assumed that the intersection is a plane in which scales shrink. We have then determined the structure of the space axiomatically and seek to obtain information about the behaviour of scales in the experiment."

"However, another way of illustrating the results of our thought experiment is more convenient for many purposes. Since the shrinking of scales cannot be measured directly by bringing more scales closer, we can alternatively define that a scale always has the same length, regardless of where it is located."

"In this interpretation, the experiment does not provide information about the behaviour of scales, but about the structure of space."

AUDRETSCH [12]: "Indeed, alternative formulations of general relativity without "curvature" of space-time can also be given, which have significance for the unification of quantum mechanics and gravitation."

16.7. Experimentally verifiable conclusions from the Schwarzschild metric

The experimental proofs of general relativity presented in chapter 16.2 are conclusions from the Schwarzschild metric. For the clock experiment, this is quantitative, and for the others, it is easy to see qualitatively.

First, the quantitative consideration. For an infinitely distant clock, the coordinate r in the Schwarzschild metric is infinite, i.e. the factor A becomes one:

(16.25) $$ds^2 = - c^2\, dt^2$$

on the other hand:

(16.26) $$ds^2 = - c^2\, d\tau^2$$

and thus follows:

(16.27) $$dt^2 = d\tau^2$$

To explain (16.26): ds is supposed to be the spatiotemporal distance between two points, ds is supposed to express what can be measured in terms of lengths and times. This means:

a) If the scale is at rest

$$dt = 0$$

because the length measurement between two points happens simultaneously in a coordinate system and therefore ds indicates the measured length. The following applies: $ds^2 > 0$ if the points are different and $dt = 0$.

b) If the clock is at rest, the spatial distances for the clock are zero, since the clock always remains at the same point. ds should also express what is measurable in this case. However, this is only the proper time interval $d\tau$ the clock (multiplied by the factor c as in Minkowski space). This interpretation is appropriate because for large radii r the factor A tends towards one and only the measurement of a normal time remains. (The Greek letter τ is usually used instead of t for proper time). The following applies: $ds^2 < 0$, i.e. $ds^2 = - c^2\, d\tau^2$ is negative and $ds^2 = 0$ is explained according to formula (16.30).

It is unnecessary to think too much about why ds should be linked to length and time measurements in this form, it is a hypothetical assumption that is confirmed by experience. A metric is to be interpreted physically in this form. EINSTEIN says [99] that it took him a long time until he finally had the idea to regard ds as the actual measurable quantity and not the coordinate differences dx, dy, dz and dt.

Formula (16.27) is understandable with these explanations, it says nothing else than: The proper time of an infinitely distant clock at rest agrees with the coordinate time t. Or formulated differently, at a great distance from the center of gravity, the Schwarzschild metric transitions into that of Minkowski space and because of the Birkhoff theorem this also applies to the interior of a mass shell, see ch. 17.

If the clock is at rest at the position r_1, the following applies:

(16.28) $$ds^2 = - A\, c\, dt^2$$

and because of (16.26) and the relation for A one gets:

(16.29) $$d\tau_1 = dt_1\sqrt{1 - 2M_s/r_1}$$
$$M_s = GM/c^2$$

In words, if the time dt passes for the clock far away, the shorter time $d\tau_1$ passes for the clock at the location r_1. This is analogous to the moving clock in special relativity, now it is the factor A by which it goes slower. This factor decreases the smaller r becomes. Applies

$$r_1 < r_2$$

so is $$d\tau_1 < d\tau_2$$

This is exactly what has been claimed in Fig. 16.5.

Two of the four experiments were conducted with light signals. In these cases

(16.30) $$ds^2 = 0$$

From a physical point of view, this is very easy to see; first consider the limiting case with an infinitely large radius r, i.e. Minkowski space. $ds = 0$ is nothing other than a measurement of the speed of light along the path dx_1 in time dt and this must yield c, because if we rewrite the formula (16.2) for Minkowski space for this case, it reads:

$$0 = dx^2 + dy^2 + dz^2 - c^2\, dt^2$$

If you set dy and dz equal to zero, you have:

$$dx/dt = c$$

But dx/dt is a velocity v, its value is c, which is fulfilled for light signals.

From the Schwarzschild metric, the measurement of the speed of light - formula (16.30) should apply - results in the radial direction:

(16.31) $$\frac{dr}{dt} = \pm c\left(1 - \frac{2M_s}{r}\right) = c_{light,radial}$$

and in tangential direction, now $dr = 0$,

(16.32) $$\frac{rd\theta}{dt} = \pm c\sqrt{1 - 2M_s/r} = c_{light,tangential}$$

The speed of light in the gravitational field depends on the direction and on the location r.

The Saphiro effect measures the change in transit time of radar signals, i.e. light, over the entire path *l* Earth-Venus and back. If we think of *l* divided into *n* intervals *dl*, each dl has a speed of light that can be determined from (16.31) and (16.32), the sum of all *dl* results in a mean speed of light that differs from *c*, and the associated change in the transit time of the radar signals is measured.

The calculation of the curvature of light paths and the perihelion rotation of planetary orbits requires two further equations in addition to the Schwarzschild metric, which can be derived from the Schwarzschild metric with the help of the calculus of variations. [23], [357]. This is also demanding in detail, in contrast to the time dilation for clocks at rest in gravitational fields.

16.8. The Schwarzschild metric in other coordinates and the objection by WEYL

From WEYL [393.1] comes a noteworthy objection to Thesis II on the interpretation of the Schwarzschild metric. In order to understand it, it cannot be avoided to represent the Schwarzschild metric in other coordinates.

To do this, *r is* replaced by an *r''* defined by the following relationship:

(16.35) $$r = r''\left(1 + \frac{M_s}{2r''}\right)^2$$

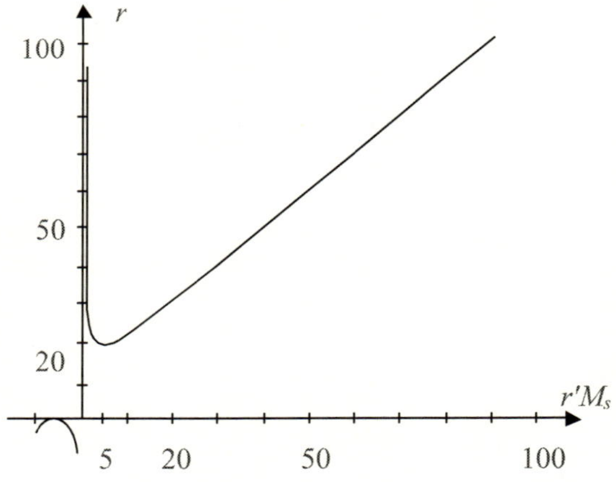

Fig. 16.11 Isotropic coordinates

If *r is* replaced by *r'' as* shown in the figure and θ, φ are taken over unchanged, the isotropic coordinates (r'', θ, φ) are obtained. For them, the size *ds* becomes direction-independent.

The relationship for *"r"* is shown graphically in Fig. 16.11. θ, φ are not changed. (r'', θ, φ) are called isotropic (same in all directions) coordinates because they describe *ds* independently of its direction, as will be shown. With them, the Schwarzschild metric is:

$$ds^2 = B_1^2\left(dr''^2 + r''^2(d\theta^2 + \sin^2\theta\ d\phi^2)\right) - B_2 c^2 dt^2 \qquad (16.36)$$

with

$$B_1 = \left(1 + {}^{M_s}/_{2r''}\right)^2$$

$$B_2 = \left(\frac{1 - {}^{M_s}/_{2r''}}{1 + {}^{M_s}/_{2r''}}\right)^2$$

Initially, only the case $dt = 0$ is of importance:

$$ds^2 = B_1^2\left(dr''^2 + r''^2(d\theta^2 + \sin^2\theta\ d\phi^2)\right) \qquad (16.37)$$

One can see the close relationship to spherical geometry, formula (16.3) is read:

$$ds^2 = dr^2 + r^2(d\theta^2 + \sin^2\theta\ d\phi^2) \qquad (16.38)$$

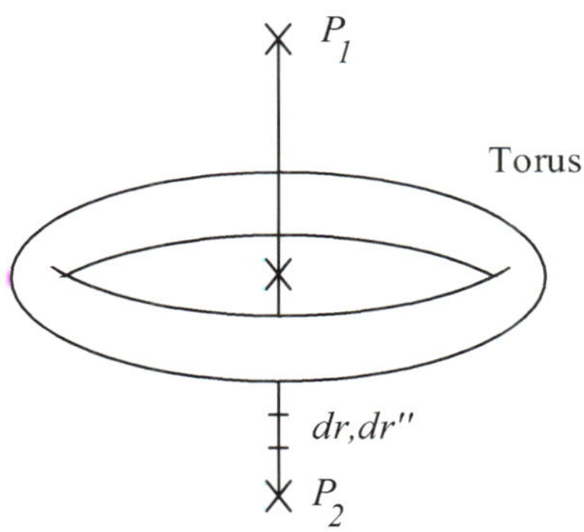

Fig. 16.12 Torus-Thought experiment

The coordinate differences *dr* or *dr''*, whose sum corresponds to the distance of the points P_1 P_2 in the absence of the torus, "save" normal (Euclidean) space.

and one can easily see what can be concluded from Thesis II for this representation of the Schwarzschild metric: Scales of length *ds* always shorten - radially or tangentially placed - by the same factor B_1. In contrast, for the original form of the Schwarzschild metric it is true that only a radially placed scale *ds* changes and beyond that by a different factor. The objection of WEYL [393.1] is: It is not decidable which length - *r* or *r''* - should be the true length. If the curved surface exists as real space, this has no consequence, but it does for the assertion that the surface is embedded in normal space.

Thus WEYL [393.1] states: "In the gravitational field we can, in accordance with our second normalisation (=formula (16.37)), also save Euclidean geometry by assuming that all scales, those placed tangentially as well as those placed radially, experience the contraction $(1+M/2r)^2$ at the distance r from the center of gravity. Thus there are here evidently an infinite number of equally possible and equally valid rules for the correction of the lengths read directly from the scales, each of which leads to Euclidean as the true geometry; no clue is there to select one of them as the only correct one in opposition to all the others."

As a counter objection to [393.1], let us assume that the thesis "*ds* shortens in real" is correct. Let us further assume that the original form of the Schwarzschild metric describes the shortening in accordance with reality. In this case, there is also the possibility of transforming r into r''. r'' is then not the real r, but it is a real physical quantity that can be calculated from r. It is then true for this r'' that it changes in the same way in gravitational fields, regardless of whether the scale ds is set radially or tangentially. The reason for this is ultimately the non-linear connection between r and r'' shown in Fig. 16.11. Even if r is real, there is the transformation to r''. We can therefore not agree with WEYL's objection, because the infinitely many transformations of r always exist. It only follows from Weyl's argument that the real connection between ds and r is not known and this does not prove that it does not exist.

Furthermore, one must consider: There are infinitely many coordinate transformations similar to (16.35) that are mathematically possible, but are already ruled out for physical reasons, namely all transformations that do not lead to $ds = dr$ for infinitely large distances. This condition is fulfilled by r''. As Fig. 16.11 shows, r'' does not differ from r for large values. How can one exclude that further restrictions follow from a more developed theory? If one agrees with thesis II, there will also be a theory that can say something about the causes of the length reductions. For such a theory, the curvature of space only illustrates the effects on scales and also explains why, despite scale changes, a generalised principle of relativity (covariance) applies, as it is assumed in the derivation of the Hilbert-Einstein field equations (Hilbert-Einstein equation [346]).

Physically and theoretically, it is not possible to decide between the two transformation formulas, but there is still the experiment: As in Fig. 16.12, place a massive ring (torus) between two points P_1 and P_2 that are far away from each other. P_1 and P_2 lie on the axis of the ring and the distance $P_1 P_2$ does not change in Euclidean terms. However, if it is measured with scales ds, a larger value results because of the scale reductions in gravitational fields. The sum of the corresponding dr must result in the Euclidean distance and this can only be true for one of the two formulae (16.5) or (16.36). (Formulas (16.5) and (16.36) must be converted for a torus, which is not trivial. A torus was chosen so that one does not enter areas within the Schwarzschild radius during the measurement).

A further objection is suggested: Imagine the spatial model of Fig. 16.8 in reality and constructed from solid rod elements with the sun at its center. The (wire) model is flat, there is no curvature to be seen. But if you remove the sun, the space becomes flat, the wire model can no longer exist as a plane and the curved model of Fig. 16.3 arises. The question is: Do not only the radial but also the tangential elements change? In LI the question can be answered by making an assumption about the "correct" coordinate system. If the *r,t* coordinate system of the Schwarzschild metric applies, only radially placed scales change; if the isotropic coordinate system applies, radially and tangentially placed ones change. Further considerations: If one leaves gaps in the wire model, they close differently, analogous to Fig. 9.3. For WEYL on the other hand, scale elements in curved spaces keep their original length, therefore it is not explainable at all where the forces come from that curve the surface. If one expands the plane into a spatial, spherical model and removes the sun, some rods must buckle or tear, in any case tensions arise without knowing which length element caused them. You can put it another way: If matter is removed from a curved space, the flat Minkowski space is created but what is the point-wise transition? If we consider, for example, the intersection points of loosely superimposed radial rods and tangential circular rings, their intersection points shift when the sun is removed. Are the points of the resulting flat space fixed on the radial or the tangential elements or somewhere in between? See ch. 10.5 and the idea of the inflationary cosmos, ch. 19.6. Here one can see the - so far neglected - physical consequence of the final mathematical consideration:

Apart from all previous arguments, there is a fundamental objection: The Schwarzschild metric describes only the inner properties of a surface, but the connection between *ds* and *dr* is an outer property of a surface. If the real space embedding the Schwarzschild metric should not exist, as the Einstein interpretation claims, this cannot be proven with the Schwarzschild metric alone, but only by additional considerations (see Chapter 18), quite analogous to the fact that the Lorentz interpretation cannot take the "correct" coordinate system from the Schwarzschild metric but must find it out by additional considerations or measurements.

16.9. The Singularity of the Schwarzschild Metric

Finally, a look at the possible limits of classical general relativity as revealed by the Schwarzschild metric.

The Schwarzschild metric has singularities for $r = 0$ and for $r = 2M_s$, i.e. individual abnormal points. At the point $r = 0$ the factor A becomes minus infinity and $1/A$ minus zero, at the point $r = 2M_s$ the factor A becomes zero and $1/A$ infinity. Since infinitely large values do not exist in nature, it is obvious to doubt the validity of the Schwarzschild metric here, especially since all experimental confirmations refer to radii that are large towards $r = 2M_s$.

According to prevailing opinion, general relativity does not eliminate the exceptional situation for $r = 0$, as it already exists in classical gravitational theory. For $r = 0$, the SM can no longer be valid. For $r = 2M_s$, the Schwarzschild radius, the most common view in the literature, and this corresponds to EI, is that this is a coordinate singularity. In other coordinates - e.g. in the isotropic coordinates, [357] - this singularity does not exist. It is not yet possible to prove or disprove this thesis here, that is done in chapter 21, but the special features for $r = 2M_s$ can be listed:

A finite ds at the point $r = 2M_s$ corresponds to $dr = 0$, i.e. for thesis II, at this point the scale ds disappears. It further means that the proper time $d\tau$ of a clock at rest there becomes zero.

Further consequences are: Consider an observer falling freely in the gravitational field. What is the spatial distance s from the start to $r = 2M_s$ and how long does it take to get there? The spatial distance s can be calculated, like the other quantities, and has a finite value; the proper time τ of the free-falling observer is also finite, whereas the coordinate time t has an infinitely large value - see Fig. 21.1. The physically sensible values for s and τ support the thesis of the EI that only the coordinates t and r are unfavourably chosen and should be replaced by others.

Due to the infinitely large coordinate time t, a paradoxical situation arises: "For a test particle (free-falling observer), the equations .... say that it needs the infinitely long time t to cover the finite distance s, but already arrives at its destination in finite proper time τ" [357]. Possible conclusion: "... the coordinate time t is physically not useful to describe the process" [357]. The Schwarzschild metric in the form (16.5) fails.

The coordinate time t is therefore not suitable for describing the process because one must assume that the test particle does not stop at $r = 2M_s$ but really falls down to $r = 0$. Under this condition, t in the vicinity of $r = 2M_s$ is no longer the time of a distant observer. On the other hand, one cannot rely only on the proper time τ, since it is not surprising that it remains finite, because at the decisive point $r = 2M_s$ it is zero, there the clock stops - that is how LI sees it. Moreover, t and τ have the same justification from a theoretical point of view. The introduction of the isotropic coordinates formally solves the problem in the sense of the EI, because they have finite values for $r = 2M_s$, but they are constructed to avoid the singularity, they are not a compelling consequence of the Hilbert-Einstein field equations (Hilbert-Einstein equation [346]). Many other coordinates can be constructed that do the same thing. This must be examined more closely, because the thesis that the free-falling observer enters another time τ, which is decoupled from our time t, is not comprehensible for LI. It would have to be proven: Not only time measurement changes but also time. As chapter 21 shows, LI and EI lead to differences that can be decided experimentally.

In contrast, for $r < 2M_s$ the Schwarzschild metric is regular again and suitable for making plausible statements about gravitational collapse and the formation of black holes - the end of massive stars when their nuclear fuel is depleted [357]. But here, too, the EI and LI differ; the different interpretations of $r < 2M_s$ are also discussed in ch. 21.

17. The Poincaré thesis and the empty, matter-free space of general relativity

According to general relativity, the curvature of space depends on the presence of matter; the smaller the accumulation of mass, the smaller the curvature of space becomes, and in the limiting case of vanishing masses, as is generally assumed and in agreement with the conclusions from the Schwarzschild metric, the flat Minkowski space of special relativity arises. But as long as matter exists, the demonstrably curved paths of light particles or other objects can always be interpreted as the effect of gravitational fields and not as the consequence of a space that is curved in itself. Strictly speaking, a curvature of space cannot be proven in experiments.

This is POINCARÉ'S thesis and JAMMER [173] describes it as follows: "Experiment can tell us nothing about the structure of space as such. It can only teach us about relations between material objects. Suppose, says POINCARÉ, that the triangulation carried out by GAUSS had resulted in a deviation from the sum of two rights, would that really have been a refutation of Euclidean geometry? Nothing would prevent us from continuing to use Euclidean geometry on the condition that light rays are curved."

Triangulation by K. F. GAUSS was the angular measurement of a triangle formed by mountain peaks about 100 km apart, carried out around 1810, which, although carried out very precisely, could not be accurate enough to be successful. The aim was to determine whether the angular sum of the triangle was equal to 180° (Euclidean geometry), less than (hyperbolic geometry) or greater than 180° (spherical geometry), with a deviation from 180° signifying a curvature of space.

The Poincaréan thesis is certainly correct and proven by the example of the Schwarzschild metric. Therefore, one is basically free to decide in favour of one of the two space-time theories. It would be possible, however, that both theses have different plausibility and are not equally elegant applicable to every observation in the same way. Such a situation would not be surprising especially if space is indeed non-Euclidean. In other words, an elegant, simple explanation of an observation with the help of a non-Euclidean geometry has a probative value in itself, and this especially when complicated theses have to be developed to save the Euclidean interpretation. The method chosen by GAUSS of measuring triangles had the chance of such an observation, because if there had been a deviation in all directions of space, without connection to masses of the surroundings, this would have been a convincing result to confirm a curved space. The experimental proof of light curvature in the vicinity of large masses does not have this probative force; here, the Poincaré objection applies in full, as will also be demonstrated in the following chapters.

What does the curvature of space look like in matter-free, empty space? It would be conceivable, after all, that measuring triangles with light beams in cosmic dimensions would still indicate a deviation from Euclidean geometry after deducting the gravitational influences. Such an experiment does not exist, of course; on the contrary, general relativity justifies itself precisely by predicting the curvature of light rays correctly as a function of the surrounding masses. But another possibility remains: What theoretical prediction does general relativity make for the limiting case of vanishing masses? In this limiting situation, is space curved or not on the basis of general relativistic considerations?

For the LI, as will be shown in Chapter 22, expanding closed and open spaces of the RWM (*k=0*,+1,-1) are a description for exploding dust stars, for which the underlying space is that of the SM. In the limiting case of vanishing masses, it therefore remains with the Minkowski space, i.e. for the LI with the three-dimensional Euclidean space - also in cosmic dimensions. For the EI there is both, the exploding dust star and the expanding universe and thus all spatial forms also for the limiting case of vanishing mass.

Does the situation change due to a cosmological constant that EINSTEIN [94] introduced into the Hilbert-Einstein field equations (Hilbert-Einstein equation [346]) and that has become the basis of many discussions? EINSTEIN later found the cosmological constant he introduced to be a mistake [95], [99], in the normal case it was regarded as zero by experts, so that for empty space it remained with the simplest thesis, it was regarded as Minkowski space or Euclidean. Recently, the constant has become current and positive again (see Chapter 19.7). This means the assumption of a far-reaching force accelerating the expansion of the universe, which thus counteracts gravity. Interpreted geometrically (relativistically), it contributes to a negative curvature of space (k = -1), for the LI this is again only a statement of how this force can be described. Since a cosmological constant includes a force and thus energy, space is material-free in the usual sense, but not empty since there is energy. Therefore, a cosmological constant does not answer the question of the curvature of empty space.

EINSTEIN took a startling view on this subject, different from these possibilities, and denied the existence of an empty space as a whole: "An empty space, i.e. a space without a (gravitational-) field, does not exist." [95] Elsewhere he says: "One can jokingly put it like this: If I let all things disappear from the world, then according to NEWTON the Galilean inertial space would remain according to my view, nothing would remain." [99]

To this it must be said: This is the most elegant way to solve a physical-philosophical problem, one proves that the problem does not exist at all. Thus, the problem of free will can also be eliminated by saying that man has no will at all. Since empty space does not exist, it is unnecessary to ask about its geometry.

There are various objections to this conception of empty space. Above all, the following one: From the general theory of relativity itself follows the existence of gravitational field-free and thus matter-free spaces. Thus, the interior of a hollow sphere is field-free. "The space inside a hollow sphere is field-free (flat), as in NEWTON'S theory of gravitation" [357]. Because of the much discussed so-called Birkhoff theorem [357], this can be regarded as proven and can be vividly imagined by analogy with a charged metal sphere, which, as a Faraday cage, also has no electric field inside it.

If a spherically symmetrical mass (sun) is placed in a hollow sphere of sufficient size, the gravitational field inside the hollow sphere corresponds to the Schwarzschild metric independently of the gravitational fields outside [357]. Also light rays do not propagate differently in the empty hollow sphere than in areas of the cosmos with vanishing mass density, so that one must also ask for these reasons why there should not be empty space but "nothing".

Although there is empty space because of the Birkhoff theorem in general relativity, this statement does not hold for two reasons because of quantum mechanics (QM). For QM, there is no emptiness, but only the 'non-empty vacuum' as the particle-free ground state of the quantum fields, see ch. 11.7. Moreover, for QM the negative gravitational forces of the cosmological constant have their origin there. This has two consequences that agree with LI. As the quantum ether requires, there is a particle-free state that can exert effects, as is now also proven by the cosmological constant and the quantum mechanical consequences of a cosmological constant do not fit at all to the expanding spaces of the RWM, s. ch. 24.13, but very well to a dust star expanding in space and time.

18. For the definition of curved space with the help of mathematics

18.1. Introduction

The Schwarzschild metric has shown us the part of general relativity that is best justified experimentally. If it holds, it could be shown how curved space in the vicinity of suns (and other spherical bodies) can be understood as flat, non-curved space together with scale changes due to the gravitational field.

More spectacular, but far more speculative, in general relativity is the idea of a finite, closed universes introduced into the discussion by EINSTEIN. The best way to understand this physical space is through the mathematical spaces that correspond to it. These include the spherical surfaces in three-dimensional and four-dimensional Euclidean space, whereby three-dimensional Euclidean space is the mathematical term for normal space.

Since everyone can imagine spherical surfaces and other arbitrarily shaped surfaces in normal space, everyone has a perfectly correct idea of curved spaces. It is only a question of specifying this idea in order to understand what a finite and closed universe is supposed to be.

18.2. Two-dimensional spherical surfaces

Let's start with the normal spherical surface. It is a two-dimensional curved space. It consists of all points of the normal space for which applies:

(18.1) $$x^2 + y^2 + z^2 = r^2$$

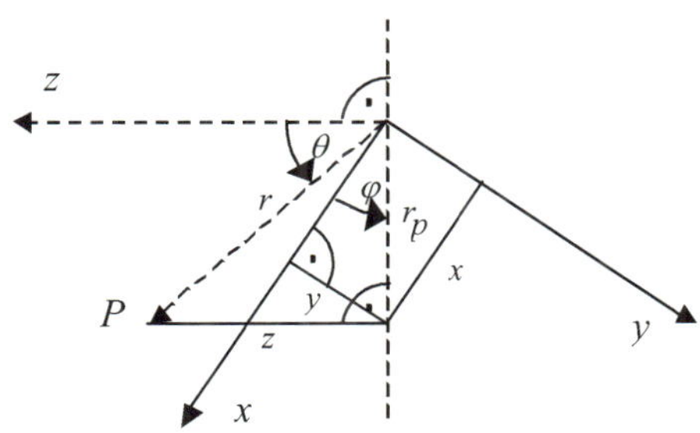

Fig. 18.1 The coordinates of point P from Fig. 16.6

The z-axis and P are folded into the x,y-plane by r_p

r = radius of the sphere with center at the origin of the coordinate system. x,y,z are the coordinates of the point $P = P(x,y,z)$ on the sphere. The formula (18.1) is read from the right-angled triangle of Fig. 18.1 by applying the PYTHAGOREAN THEOREM, first for the projection r_p from r, then for r. For a given r, each point P is defined by two coordinates - e.g. x and y - the third is calculated from (18.1), therefore the surface of the sphere is two-dimensional.

The spherical surface can also be represented in spherical coordinates - this corresponds to the longitude and latitude circles of the globe. When calculating, they have the advantage of not containing squares, but the disadvantage of using sine and cosine functions. The spherical coordinates are, see Fig. 16.6:

$$x = r\cos(\varphi)\sin(\theta)$$

(18.2) $$y = r\sin(\varphi)\sin(\theta)$$

$$z = r\cos(\theta)$$

with $$0° \leq \theta \leq 180°$$

$$0° \leq \varphi \leq 360°$$

Numerical example for pocket calculator (conversion of spherical coordinates):

$$\theta = 30°,\ \varphi = 60°,\ r = 1\text{ m}$$

$$x = 1\times 0.5\times 0.5 = 0.25\text{ m}$$

$$y = 1\times 0.866\times 0.5 = 0.433\text{ m}$$

$$z = 1\times 0.866 = 0.866\text{ m}$$

Check: x,y,z inserted into formula (18.1), gives $r = 1$ m.

The formulas (18.2) mean graphically that the angles θ and φ together with the constant radius r define all points of a spherical surface, i.e. $P = P(r,\theta,\varphi)$. The exact knowledge of the sine and cosine functions is in principle dispensable, because with given angles x,y,z are also geometrically constructible.

With these considerations it is easy to see why the spherical surface is called a two-dimensional space. It is a space because it consists of points and it is two-dimensional because the position of each point is determined by two independent coordinates θ and φ or x and y (or two others). The third coordinate is fixed (radius r) or can be calculated (z with formula (18.1)). Furthermore, a spherical surface is visually curved and thus a curved two-dimensional space, in contrast to the non-curved, two-dimensional plane.

How can the term curved be specified? One possibility is to say that a surface is curved if it cannot be completely placed in a plane. Consider the connecting line $P_1 P_2$ of any two points P_1 and P_2 of the surface in Fig. 18.2. If the line $P_1 P_2$ lies completely in the plane for all points P_1 and P_2, the surface is flat. In mathematics (differential geometry) this method is not used to define curvatures, because it does not provide a measure of the size of the curvature of a surface. The size of the curvature can be defined more simply by the reciprocals of the radii of circles that touch the surface in as high an order as possible (forging circles, [192], [199], [321]. If there are circles of contact with different radii, which is the rule, an arithmetic mean value is taken. For the spherical surface, the circles of contact are the great circles of the sphere itself, i.e. the meridians for the north and south poles. The curvature of a sphere is thus smaller the larger the radius of the sphere. A plane is graphically a sphere with an infinitely large radius, whose reciprocal value zero represents the curvature of the plane.

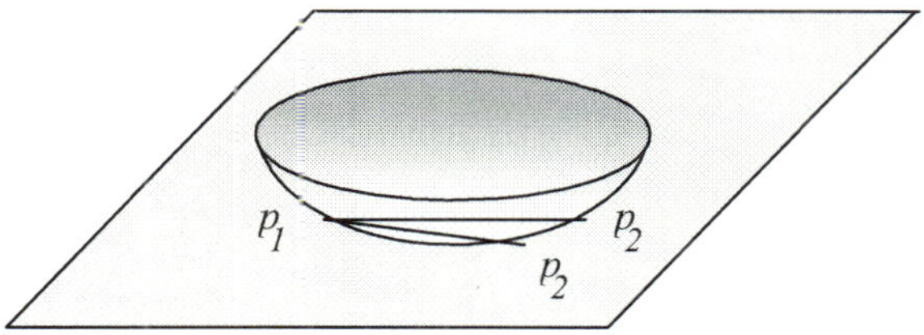

Fig. 18.2 A curved surface

cannot be placed completely in a plane. There are always points P_1, P_2, whose connecting line lies partly outside.

With this descriptive definition of curvature, both a sphere and a cylinder are curved. There is a second important and equally descriptive method of defining the curvature of a surface for which this no longer applies, but for which a sphere must be considered curved but a cylinder is not. This second definition is called the inner curvature of a surface in contrast to the above, outer curvature and uses the following considerations.

If you try to deform a spherical surface or a part of it, e.g. a spherical cap, into a flat circular disc, this is not possible without cracks or distortions. A circular cylinder, on the other hand, can be rolled into the plane without distortions. Thus, you cannot tell from a book page whether it has been printed in rotary printing or flat and a cylinder can be made from a sheet of paper without folding, whereas a sphere cannot, see Fig. 18.3.

This is easy to grasp mathematically. As is well known, the circumference of a circle is

(18.3) $$U = 2\pi r$$

If you draw a circle on the surface of the sphere - if the center is the north pole, you get the latitude circles - and measure the circumference and radius along the surface of the sphere with sufficiently small scales, you find:

(18.4) $$U < 2\pi r$$

It becomes extreme for the equator, here the circumference is exactly four times the radius. A circle on a surface is the totality of points that have the same distance from a given point, whereby all lengths on the surface are measured.

Accordingly, small squares with side length a on spherical surfaces have diagonals longer than $a\sqrt{2}$. On the other hand, there are surfaces - saddle surface of Fig. 18.3 - for which

(18.5) $$U > 2\pi r$$

becomes. These surfaces have a negative inner curvature, spheres have a positive inner curvature. If, on the other hand, you draw a circle on a cylindrical surface, there is no difference to the plane, formula (18.3) applies. It can also be expressed like this: Spherical geometry applies on the surface of a sphere, normal (Euclidean) geometry applies in the plane and on cylinders.

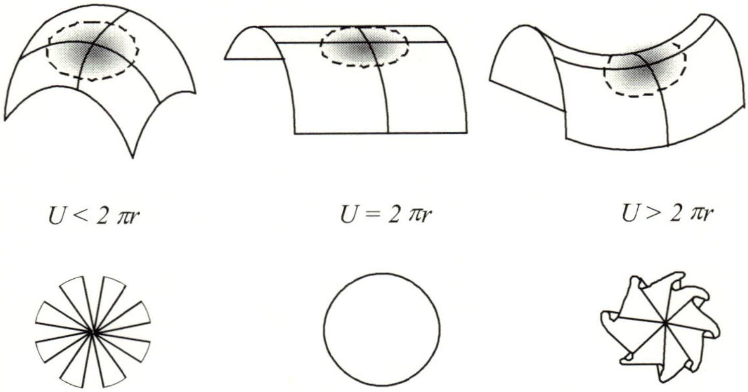

Fig. 18.3 Inner curvature

The spherical cap deformed into the plane, leads to cracks, the circular cut-out of the saddle surface is undulated in the plane, a cylindrical surface can be rolled off. The inner curvature of the sphere is positive, of the cylinder zero, of the saddle surface negative.

In the general case and thus valid for any figures, these inner properties of surfaces can be traced back to the distance between two points lying close together, described by the so-called metric. With the help of this concept, essential statements of the general theory of relativity about curved spaces become precisely comprehensible.

Consider two points of normal space that are close to each other. Their distance ds is:

(18.6) $$ds^2 = dx^2 + dy^2 + dz^2$$

if dx, dy, dz are the coordinate differences of the two points. This formula is called the metric of normal space. It expresses the simple fact that in normal space the Pythagorean theorem applies in particular to points that are close to each other.

For the spherical surface, see Fig. 18.4,

(18.7) $$ds^2 = r^2(d\theta^2 + \sin^2\theta \; d\phi^2)$$

Numerical example: On the earth's surface, at $\theta = 60°$, φ any, there are two places at a distance $d\theta = 1°$, $d\varphi = 1°$ - What is their distance in kilometers? $1° = 0.0175$ in radians, $r = 6370$ km, $\sin(60°) = 0.866$, $ds = 147$ km.

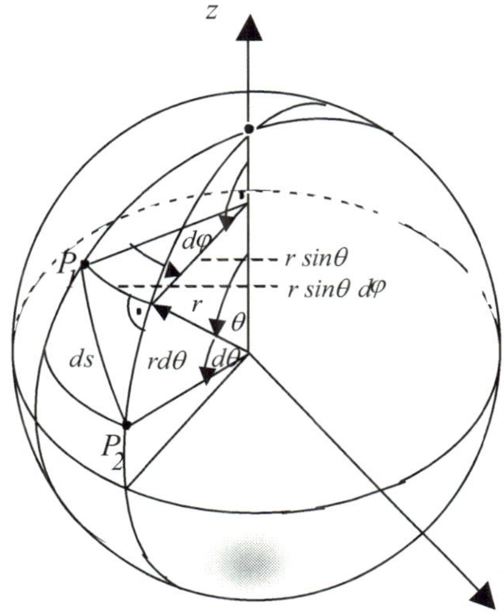

Fig. 18.4 Spherical coordinates of two points close to each other in Fig. 16 6

ds is calculated with the Pythagorean theorem and is sufficiently accurate if $d\theta$, $d\varphi$ are chosen to be sufficiently small.

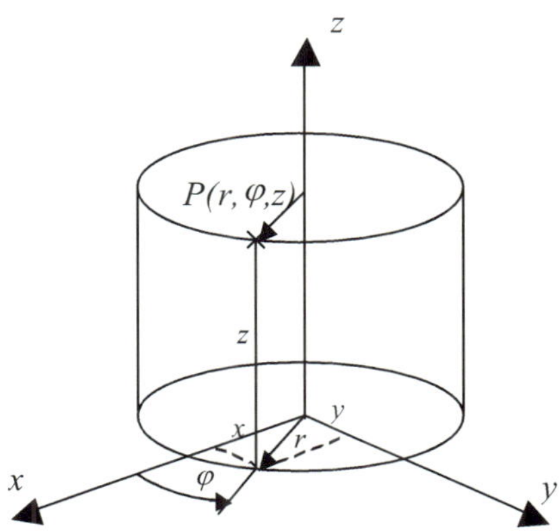

Fig. 18.5 Cylinder coordinates $P(r,\varphi,z) = P(x,y,z)$ with $x = r\cos(\varphi)$, $y = r\sin(\varphi)$, $z = z$.

The accuracy of the calculation can be increased as desired by dividing the great circle through both points several times and repeatedly applying formula (18.7), i.e. integrating. What is meant by close to each other is not absolutely defined and this is also not necessary, because with a metric like (18.7) the distance of any points on a surface can be determined as precisely as desired even without such a definition, normally by applying the above relation several times. All quantities r, θ, φ can be determined without leaving the spherical surface. The radius from the circumference of the sphere, the angle φ by uniform subdivision of the latitude circles, θ by corresponding subdivision of the longitude circles; we leave the details of the procedure undecided for the moment. If one assumes that the curved surface is the unknown space in which one lives, one has no other possibilities than to measure in the surface. The space in which the curved surface is embedded is not accessible, or such a space may not even exist.

Let us look at (18.7) in more detail, how does it differ from (18.6), or from

(18.8) $$ds^2 = dx^2 + dy^2$$

of the metric for the plane? As can be rigorously shown [192], no coordinates k_x or k_y can be found for the spherical surface in such a way that formula (18.8) applies to all points. For a certain point this is of course always possible by changing the unit of measurement, but with expressions for all points there are always some functions in front of the quadratic members. If the units of length are chosen so that $r = 1$, formula (18.7), for example, reads as follows

(18.9) $$ds^2 = d\theta^2 + \sin^2\theta \; d\phi^2$$

If sin(θ) were selectable for all θ equal to 1, then

(18.10) $$ds^2 = d\theta^2 + d\phi^2$$

for all points of the spherical surface. This would correspond to the metric (18.8) of the plane. However, there is no such thing. This is also plausible, because this would mean covering the spherical surface with a narrow net of small undistorted squares with the side lengths $d\varphi$ and $d\theta$, in other words, unrolling it into the plane. If this net is to be spherically curved, any squares must be deformed. With such considerations, the above assertion becomes plausible, a valid proof is much more elaborate [192].

The metric of a cylinder is obtained from the representation of the cylinder of Fig. 18.5 analogous to Fig. 18.4:

(18.11) $$ds^2 = r^2 d\phi^2 + dz^2$$

If one chooses the unit of length so that $r = 1$, one immediately has the metric of the plane, it applies to all points, the cylinder has the inner curvature zero.

Thus, for a two-dimensional space (a surface), we have three possibilities to determine whether it is curved. The first possibility is to embed it in the three-dimensional space and see if it forms a curved surface. According to this method, the spherical surface and the cylinder are curved (outer curvature). The second possibility is to take measurements in the surface itself and check whether, for example, the formula (18.4) applies to circles. Spherical surfaces have this inner curvature, cylinders do not. With the third possibility, the inner curvature of a surface can be examined with the help of its metric. If the metric can be brought to the form (18.8) by coordinate transformations, the surface has no inner curvature, but it can have an outer curvature, like a cylinder.

18.3. Three-dimensional spherical surfaces

Let us move on to three-dimensional curved spaces. Their curvature is not directly presentable, for this they would have to be embedded in a four-dimensional flat space that is also not directly presentable. A certain degree of clarity is made possible by analogies, sections and projections, the last two methods, because they lead to a reduction of the dimension. Essentially, the concepts of the previous chapter about two-dimensional spherical surfaces are generalised to three-dimensional spherical surfaces. One arrives at them via circle and sphere by analogy:

(18.12) $r^2 = x^2 + y^2$ (circle)

(18.13) $r^2 = x^2 + y^2 + z^2$ (spherical surface or sphere S_2)

(18.14) $r^2 = x_1^2 + x_2^2 + x_3^2 + x_4^2$

(three-dimensional spherical surface or sphere S_3)

If one chooses in (18.13) $z = z_0$, one obtains any number of circles

$$r^2 - z_0^2 = x^2 + y^2$$

These circles are non-point for different z_0 and lie in normal space in the planes for which $z = z_0$.

Correspondingly, in (18.14) $x_4 = x_{4,0}$ yields any number of non-point two-dimensional spherical surfaces

(18.16) $r^2 - x_{4,0}^2 = x_1^2 + x_2^2 + x_3^2$

They lie in three-dimensional spaces (x_1, x_2, x_3, $x_4 = x_{4,0}$), which have also no common points. In four-dimensional space, a point $P = P(x_1, x_2, x, x_{34})$ has four coordinates and $x_4 = x_{4,0}$ defines a three-dimensional space, just as $z = z_0$ defines a two-dimensional space (plane) in normal space. (We know: $y = y_0$ is a parallel to the x-axis in the plane.) Analogously there are infinitely many three-dimensional spaces in a four-dimensional space (infinitely many two-dimensional spaces in normal space) that are parallel to each other. "Being parallel" here means nothing other than "having no common points".

The four-dimensional space is created, see Chapter 11 and Fig. 11.1, by adding a fourth coordinate to the three coordinates of points in normal space and calculating in the same way as with three coordinates. There is no descriptive justification for this idea, i.e. one does not claim that four-dimensional space "exists" in the same way as normal space. Four-dimensional space is merely conceivable without contradiction. Phenomenologically seen, it exists as a thought space because its concept is free of contradictions, it does not exist as imagined space and also not as real, physical space, if one disregards time as a further coordinate.

The metric of the three-dimensional spherical surface can be obtained by considerations similar to formula (18.7). The result is [230]:

(18.17) $$ds^2 = r^2[d\alpha^2 + \sin^2(\alpha)(d\theta^2 + \sin^2(\theta)d\phi^2)]$$

with
$$0 \leq \alpha \leq 180°$$
$$0 \leq \theta \leq 180°$$
$$0 \leq \varphi \leq 360°$$

Numerical example: What is the distance between two points of a S_3 from the radius of the earth with the same coordinate $\alpha = 90°$, $\theta = 60°$, $d\theta = d\varphi = 1°$, $r = 6370$ km. Because sin(90°) = 1 and $d\alpha = 0$, the result is ds = 147 km, as in the previous numerical example.

Much more important than being able to derive the formula is to understand what it says. The quantity r is a constant, not a coordinate. It becomes a coordinate when the sphere S_3 is embedded in a four-dimensional space. On the sphere S_3 there are as coordinates the three angles α, θ, φ analogous to θ, φ for the S_2.

What are the properties of the spheres S_3 defined by formula (18.17)? A mathematician can first prove by calculation that formula (18.17) cannot be brought to the form (18.6), not even for other coordinates, and thus prove the non-zero inner curvature of spheres S_3. The same can be achieved with the second method of the previous chapter and examine for circles on the sphere S_3 radii and circumferences. Let $\alpha =$ 90° and $d\alpha = 0$, i.e. only the points with constant α are examined. (18.17) reduces to (18.7):

$$ds^2 = r^2(d\theta^2 + \sin^2\theta \; d\phi^2)$$

since sin(90°) has the value 1. The relationship (18.4) has already been proven for two-dimensional spherical surfaces. The same result also follows for $\theta = 90°$ and $d\theta = 0$, only one obtains a two-dimensional spherical surface in the coordinates α and φ. On the sphere S_3 there are thus infinitely many circles whose circumference and radius do not fulfil the relation (18.3). The sphere is therefore curved.

However, there is still a gap in the proof. Apply the same conclusions to the metric of normal space in spherical coordinates and choose in

(18.18) $$ds^2 = r^2(d\theta^2 + \sin^2\theta \; d\phi^2) + dr^2$$

r = constant and $dr = 0$. Here, too, one obtains the spherical surfaces (18.7) with the arbitrary, constant radius r

$$ds^2 = r^2(d\theta^2 + \sin^2\theta \; d\phi^2)$$

Of course, circles on them fulfil the relation (18.4), although normal space is not curved, e.g. because of (18.6). However, these circles all lie in a plane that intersects the spherical surface, and the normal relationship between radius and circumference applies to radii measured in this plane. Such a plane does not exist for spheres S_2 or S_3 unless S_2 is embedded in normal space or S_3 is embedded in four-dimensional space. Then instead of the spherical surface, one has the entire sphere and thus all the points inside. In addition, one has once again shown in a different way that the four-dimensional Euclidean space is flat.

To see the above consideration in more detail, choose in (18.17) α not equal to 0°, arbitrarily, e.g. for $\alpha = 60°$ the formula

$$ds^2 = r^2 \sin(60°)(d\theta^2 + \sin^2\theta \; d\phi^2)$$

For each of these α one obtains a spherical surface in the coordinates θ, φ with the radius r sin(α)) in a three-dimensional space. However, all of these three-dimensional spaces are non-point to each other; in contrast, for (18.18) the three-dimensional space is always the same. The non-point feature of these spaces is obvious: α always has another value, which means in four-dimensional space another point or another set of points; for the 60° example, such points whose α–coordinate is 60°. In normal space, for every constant θ in formula (18.7) there is a (latitude) circle and a plane containing it. All these planes are parallel to each other. Similarly, in four-dimensional space, for every constant α in formula (18.17), there is a two-dimensional spherical surface and a three-dimensional space containing it. All these three-dimensional spaces are parallel to each other, i.e. they are non-point to each other.

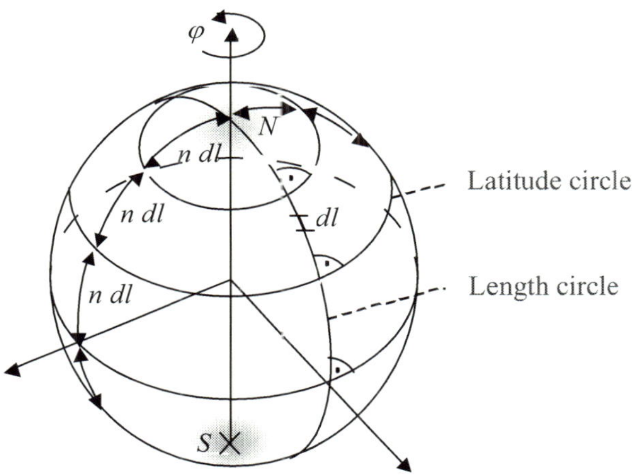

Fig. 18.6 To construct a coordinate system on the spherical surface S_2

Circles around N with radius $r = n\,dl$ are circles of latitude, circles of longitude intersect them perpendicularly. dl, φ, 90° angles are measured on the surface of the sphere, the surrounding space need not be known.

18.4. To construct a coordinate system for the three-dimensional spherical surface

Our following considerations on the construction of a coordinate system on a three-dimensional spherical surface are not intended to enrich the knowledge of a mathematician, they are not sufficiently abstract for that. They are intended to show, by way of example, how to recognise points, i.e. assign coordinates to them, in a three-dimensional curved space with known geometric properties using a compass and a ruler. If the universe is a three-dimensional sphere, the construction is transferable in principle. If it is feasible there in this or perhaps modified form, one also has a clue that the universe is in fact spherical.

Although the three-dimensional spherical surface S_3 with the metric (18.17) cannot be visualised, one can nevertheless construct the coordinate lines α, θ, φ analog to the normal spherical surface S_2 on it, see Figs. 18.6 and 18.8.

First, let a spherical surface S_2 and an ideal scale of length dl be given. Arbitrarily choose a point N on S_2 as the north pole. Latitude circles (θ = constant) are all n circles around N with the radius $r = n\,dl$ (the radius has all multiples of dl in succession). Longitude circles are the perpendiculars on the latitude circles through the north pole N, see Fig. 18.7. Subdivisions of an arc, right angles etc. are obtained by constructions analogous to normal geometry. In this construction, the meridians meet at one point, the south pole S, because of the spherical properties. The arc NS has the length π (180°) and dl can be calibrated in degrees.

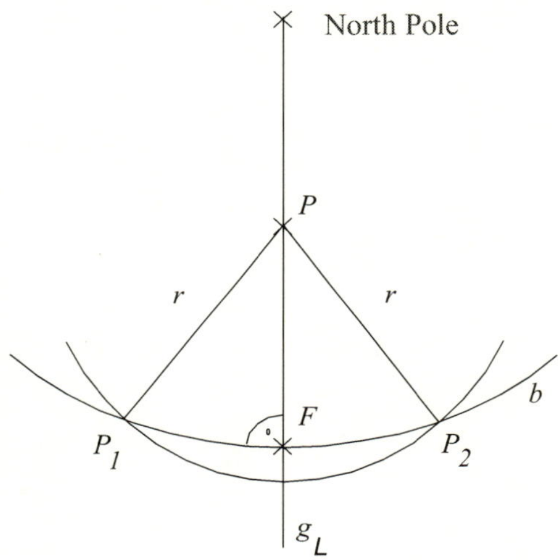

Fig. 18.7 Length circles on the spherical surface S_2

All points P which have the same distance from two given points P_1 and P_2 of the circle of latitude b form a meridian g_L with the base point F perpendicular to b. (This applies analogously to the perpendicular to a straight line in the normal plane).

Numerical example: One meridian minute corresponds to one nautical mile. What length does this mean? $dl\,/(2\pi\,6370\text{ km}) = (1/60°)/360°$. $dl = 1.85$ km.

With these hints, a coordinate system (θ, φ) can be constructed on S_2. If the meridians do not meet at one point, one is not in the S_2.

For the sphere S_3 one chooses any point N and constructs the n spherical surfaces around N with the radius $\alpha = n\,dl$. On each of these spherical surfaces, one defines the longitude and latitude circles as for S_2 as soon as one knows their north poles. To determine the north poles, a spherical radius is drawn. Its vertical continuation is the α line with $\theta = 0$, φ arbitrary. It meets all n spherical surfaces with $\alpha = n\,dl$ in their north pole.

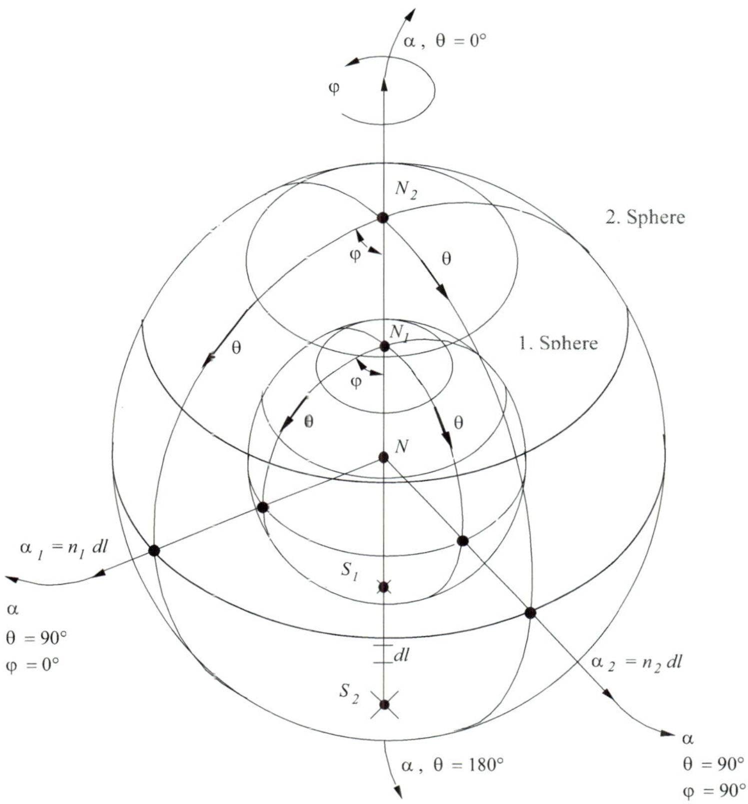

Fig. 18.8 For the construction of a coordinate system on the three-dimensional spherical surface S_3

$\alpha-$ lines are perpendicular to the S_2 spherical surfaces around N with the radius $\alpha = ndl$; an excellent α–line defines on the S_2 spherical surfaces their north poles N_1, N_2 ... and has there the coordinates θ with zero degrees and φ arbitrary. S_1 and S_2 designate south poles.

Vertical continuation means the construction of a perpendicular on a spherical surface. As in the plane, each point of a perpendicular has equal distances from all points on a circle around the foot point of the perpendicular.

All α lines must meet at a point S, the south pole of S_3. The arc NS is equal to π and thus dl can be calibrated in degrees for α. If this construction does not work, one is not in a S_3

The construction is only vividly imaginable at the beginning, it is not vividly imaginable how the α -lines over the south pole *S* lead back to *N* and at the same time the spheres α = constant become smaller again from 90°. If there is an S_3, however, one will notice that all α lines meet at a point *S*, have the length π and the radii of the spherical surfaces α = constant are given by $r\sin(\alpha)$. The two-dimensional spherical surfaces that arise when α or θ are chosen to be constant can be clearly represented.

The idea underlying this construction: All geodesics (= shortest connection, great circles on a sphere) starting from a point *N* must meet at a point *S* (analogous to the north and south poles). Spheres and perpendiculars serve only as aids for constructing the geodesics.

18.5. Interpretation of metrics as scale distortion

Consider the metric (18.7) and its spherical surface. This surface projected perpendicularly onto the *x,y*-plane results in a circular disc with radius *r*. The latitude circles remain undistorted, the longitude circles form the diameters of the circular disc, i.e. the element $r\sin(\theta)d\varphi$ retains its length as it is a segment of a latitude circle and $\rho d\theta$ is shortened to $r\cos(\theta)d\theta$ as it is an oblique straight line segment on a longitude circle. In the plane, the projection of a spherical surface therefore has the metric

(18.20) $$ds^2 = r^2(\cos^2(\theta)d\theta^2 + \sin^2(\theta)d\phi^2)$$

If we now assume that in the direction of the radii *r* the true distance *ds* is not given by $r\cos(\theta)\, d\theta = dr_p$ but given by

$$dr_p/\cos(\theta)$$

then there is no difference in principle to the metric of a spherical surface. This is illustrated by Fig. 18.9, the surveyor [370], progressing from the center to the edge of the circular disc, is constantly shrinking together with his meter stick. Once he has measured the radius *r*, the result is the same as the measurement on the spherical surface without a shrinking meter stick.

As explained in Chapter 16.6, a gravitational field can in principle be regarded as the cause of scale shrinkage.

Numerical example: At which position in Fig. 18.9 has the scale of the geometer shrunk to half? $ds = 1/2\, ds/\cos(\alpha)$; $\cos(\alpha) = 0.5$. For $\alpha = 60°$ the scale *ds* has shortened by half in the radial direction.

Analogous to the projection above one can project a three-dimensional spherical surface onto the sphere in normal space. From formula (18.17) it becomes

(18.21) $$ds^2 = r^2[\cos^2(\alpha)d\alpha^2 + \sin^2(\alpha)(d\theta^2 + \sin^2(\theta)d\phi^2)]$$

If we assume again that in the direction of the radii *r* the true distance *ds* is not given by $r\cos(\alpha)\, d\alpha = dr_p$ but by

$$dr_p/\cos(\theta)$$

then here too there is no difference in principle to the metric (18.17) of a sphere S_3

Figure 18.9 can be repeated; the geometer may now progress in all directions of space. Together with the shortening meter rule, an immediately clear representation of a three-dimensional curved surface is obtained.

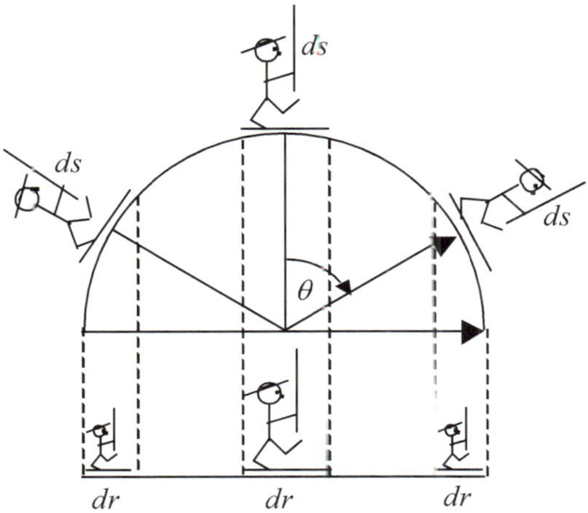

Fig. 18.9 The surveyor measures with his shrinking meter rule *dr* radially the same length as along a curved surface with the constant scale *ds* (145).

18.6. Other curved surfaces

To conclude this chapter, some remarks are essential, but dispensable for the understanding of a finite cosmological space of general relativity.

The difference between the outer and inner curvature of a space had emerged. The inner curvature can be derived from its metric. It is also what can be derived from the field equations of general relativity. General relativity therefore does not normally allow us to decide whether a curved space is really embedded in a flat space of a higher dimension, in other words, whether it also has an external curvature.

Fig. 18.10 A wavy cylindrical surface has a non-zero outer curvature.

The inner curvature is zero, as for a plane; also globally there is no difference to it.

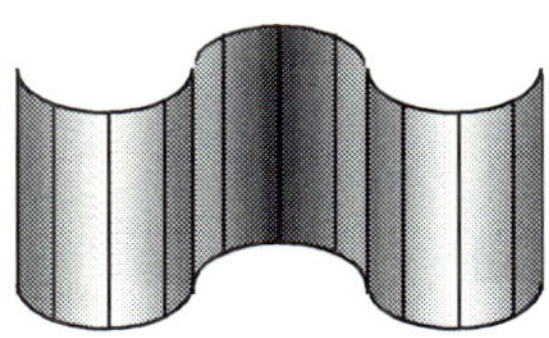

In some cases, however, statements can be made about the outer curvature of a space from its global properties. For example, a cylinder has the metric of the normal plane, but if you pass through it perpendicular to its height, you return to the starting point, so that you can distinguish between cylinder and plane by global measurements. For a surface curved like corrugated cardboard, see Fig. 18.10, no difference from the plane can be determined even by global examinations. In these cases, one needs the surrounding space embedding them.

Another point concerns the non-Euclidean geometries associated with metrics analogous to (18.17) [nic]. On the spherical surface, the so-called spherical geometry applies. If one combines the two radially opposite points on the spherical surface into one point, one obtains a hemisphere with a half equator. The metric does not change, you get a geometry similar to the spherical one, which is called elliptical [226]. Typical for these two geometries is: The sum of angles in the triangle is greater than 180°, to a given straight line (= great circle) there is no parallel straight line to a given straight line (= great circle) (but only parallel latitude circles; they are not a "shortest connection of two points").

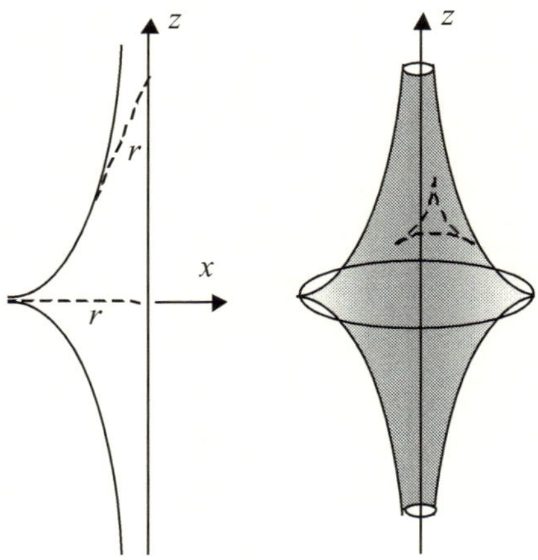

Fig. 18.11 Tractrix and pseudosphere

The pseudosphere has the same negative curvature at each of its points and is therefore the analogue of the sphere. Hyperbolic geometry applies to the pseudosphere.in which triangles have a sum of angles smaller than 180°. The pseudosphere is created by rotation of the tractrix.

.

Of much greater importance is the metric in which the sine function for α of (18.17) is replaced by the hyperbolic angle function sinh(α). The metric is:

(18.30) $$ds^2 = r^2\left[d\alpha^2 + \sinh^2(\alpha)\left(d\theta^2 + \sin^2(\theta)d\varphi^2\right)\right]$$

with α any

$$0 \leq \theta \leq 180°$$
$$0 \leq \varphi \leq 360°$$

It defines hyperbolic geometry (of GAUSS and LOBACHEVSKY) [86]. It can be represented in normal space by the so-called pseudosphere, Fig. 18.11, but only partially, just as a cylinder represents only part of the normal plane. The real distance between two points on the pseudosphere, as it can be calculated from (18.6), agrees with that from formula (18.30). The (so-called stereographic) projection onto the normal plane yields a circular disc in which the hyperbolic straight lines (= shortest connection of two points in hyperbolic space) become circles perpendicular to the edge of the circle, or diameters (which are also perpendicular to the edge of the circle and are circles with an infinitely large radius). This corresponds to the famous Poincaré's model of hyperbolic geometry, see Fig. 18.12.

The most important properties are: The sum of angles in the triangle is less than 180°, as can be seen immediately if you draw triangles on the pseudosphere. For every point *P* outside a straight-line *g*, there are infinitely many straight lines through *P* that do not intersect *g*, i.e. are parallel. Hyperbolic and normal (Euclidean) geometry differ from each other only in one axiom, the parallel axiom. The elliptical and spherical geometry differ from them in other axioms, too.

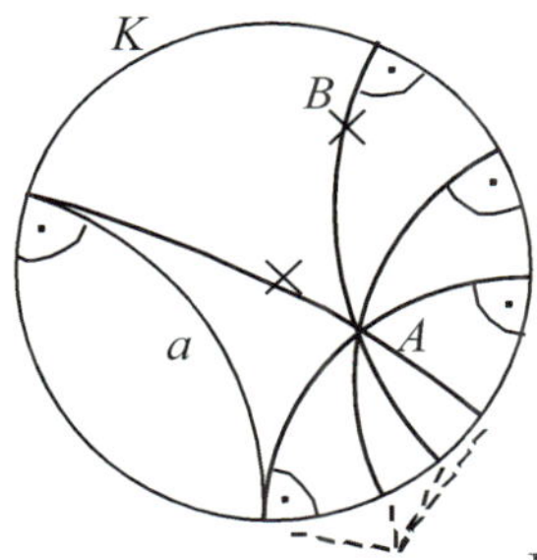

Parallels to a through A

Fig. 18.12 Poincaré model of hyperbolic geometry in the unit circle

All diameters of the circle *K* and all circular lines perpendicular to the circumference of *K* are hyperbolic straight lines. To the straight line *a* there are infinitely many parallels through the point *A*, i.e. straight lines that do not intersect the straight line *a*. The Euclidean axiom of parallels is no longer fulfilled, all other axioms of Euclidean geometry apply, so exactly one straight line is determined by two points *A* and *B*

The metric (18.30) of hyperbolic geometry has the same meaning for cosmology (infinitely expanding universe) as the metric (18.17) (finite universe). As the Poincaré model shows, it too can be interpreted with changes in scale as part of the normal plane. Here is a quotation: *"The occasional refusal in textbooks to use the Poincaré model as early as possible to illustrate the facts of hyperbolic geometry is apparently caused by a desire to provide psychological support for the idea that hyperbolic geometry is a possible geometry of real physical space."* SCHREIBER [325].

The importance of this chapter has decreased since the 3rd edition. It still serves the understanding of what the EI means by an expanding universe. However, expanding spaces require cosmological simultaneity, which must be rejected for two physical reasons, see ch. 19.8.4 and 19.8.5. The spaces of the RWM reduce to those of the SM, as will be shown. The SM does not require three-dimensional spherical surfaces for either the EI or the LI.

"Without doubt, the world is not created in time, but with time.
... Before the world, time could not be … .
Rather, in One with time also the world was created."
AUGUSTINE, *God's State*

19. Cosmology and Robertson-Walker metric (RWM)

19.1. Introduction

The experimental confirmation of the general theory of relativity lies in the statements of the Schwarzschild metric. However, the most spectacular theses on curved space, which have received the most public attention, concern cosmology, the doctrine of the shape and origin of the cosmos.

Cosmology is speculation. But although it was not known for a long time whether the expansion of the universe is infinite or finite, whether the observed expansion of the universe will continue indefinitely or reverse itself after a finite time, and only the associated, theoretically possible space models were known, today (2021), footed on observation, an infinite, forever expanding universe with the cosmological constant $\Lambda > 0$ is favoured. With the previous chapter on the mathematical concept of curved space, one has a mathematical idea of the connection between metric and coordinate system for two- and three-dimensional spherical surfaces (more abstractly called spherical spaces). These mathematical concepts, interpreted physically, turn out to be possible cosmological spaces.

In the foreground of the following considerations is the Robertson-Walker metric (RWM), applied to the case of the finite closed space because this corresponds best to the unconventional space-time philosophy of the EI and has special theoretical significance, even though in the meantime the majority regard our universe as non-finite and non-closed. For more details, see ch. 19.7.

ch. 19.3 - 19.7 can be skipped - except for a look at ch. 19.5 - if the RWM is known from textbooks. ch. 19.8 shows that the RWM is also suitable for collapsing and exploding dust stars (supernovae) without using terms such as expanding or closed space. The following chapters problematise the notion of 'cosmological' simultaneity and expanding space- Chapters 19.8.4 and 19.8.5 are of particular importance. This fulfils the prerequisites for one's own judgement on EI and LI of general relativity. In addition, one has a clear basis for all further considerations on RWM, especially ch. 22.

19.2. The finite curved cosmological space from the point of view of the Schwarzschild metric

In order to understand the thesis of a finite universe as a possible consequence of the RWM, the question of the finiteness of the universe should first be addressed on the basis of previous findings. If one takes a matter-free space as a basis, there were three opinions: Nothing exists, empty space exists (see chapter 17) or the non-empty vacuum of QM exists. If we take a spherical mass as a basis, the Schwarzschild metric yielded the following results:
a) an infinite space, since the radius r is allowed to grow indefinitely
b) in infinity, this space is a Minkowski space, i.e. it is flat there, not curved, in other words Euclidean.

Roughly speaking, a universe is formed by adding to a spherical mass many spherical masses, as many as the universe contains suns, and distributes them in space. Then one may assume:
a) Space does not thereby become smaller, i.e. it remains infinite,
b) near each spherical mass, it has the shape of the Schwarzschild metric,
c) at a great distance from all of them, it is flat, i.e. not curved.

The main idea is: the universe should not become finite in this process, but remain infinite. In other words, if you assume the universe consists of only one spherical mass, it is infinite, as the Schwarzschild metric states; if you add many more masses, as many as the universe comprises, it should not shrink overall.

Why the universe is nevertheless considered to be possibly finite is shown by the discussion of the so-called cosmological principle and the associated Robertson-Walker metric - which, along with the Schwarzschild metric, forms the basis of classical general relativity. The subsequent discussion of the special relativistic limiting case, chapter 19.8.4, shows why the above considerations on infinite space can remain.

19.3. The cosmological principle and the co-moving coordinate system

The universe is partly hierarchical:
a) planets with moons orbit around suns and form solar systems
b) Solar systems move around a common center and form a galaxy.
c) Galaxies form a galaxy system and orbit their common center of gravity.

Thus our Milky Way system forms a galaxy and, together with the Andromeda Nebula and other galaxies form the local group, a galaxy system [317]. There is some evidence that galaxy systems are also structured in superclusters and orbit a common center but no other structures are suspected beyond this [357], [246]. Galaxies in the far distance appear as discrete points in the starry sky, the starry sky is neither uniformly bright nor dark and the idea of a homogeneous universe can at best be justified on average.

Fig. 19.1 Cosmological principle: "The world, my child, nothing but cages and those in a big one."

(Dust grains, galaxies or birdcages - evenly distributed in space they form an open or closed expanding universe for the EI). [357]

It is precisely this assumption that no place in the universe is distinguished from the others, that all directions are of equal value, that the world appears in the same way in all of them (isotropy) and in all spatial areas, regardless of where these partial areas lie, the world has the same shape (homogeneity), this assumption is made and called the cosmological principle. {Actually isotropy follows from homogeneity, nevertheless both terms are always mentioned together}.

The cosmological principle, Fig. 19.1, can be regarded as fulfilled for the galaxy clusters of the universe. They are imagined as grains of dust that have certain distances from each other. Different dust grains have different coordinates; the billions of suns with their planetary systems within a galaxy system have the same position in cosmological observations.

In terms of thought, a number of objections to the cosmological principle are possible. It is the success that justifies it, since a number of observations are explained surprisingly well with this simplest cosmic model [357].

World models with galaxy clusters as dust grains can be developed for any coordinate system. So-called co-moving coordinates have proved to be particularly useful. One starts from spherical coordinates α, θ, φ (ch. 18), each galaxy cluster can be localised in the universe by the three coordinates. θ and φ are angles, in normal space α is a distance in the radial direction, in three-dimensional curved space it is also an angle, but radial distances can also be given in degrees if you calibrate accordingly. Thus, a nautical mile is a minute of arc on the surface of the earth. The universe is described in co-moving coordinates if each galaxy retains for all times exactly the coordinates it was given at the beginning. To do this, consider the balloon in Fig. 19.2 with coins attached to it. Their position is determined by θ and

φ. The radius changes the more the balloon is inflated, but it is the same for all coins. As it turns out, the corresponding situation applies to the universe, it changes solely under the influence of gravitational forces. As a result, all clusters of galaxies keep their position among themselves unchanged like coins on a balloon. What changes in time is a quantity that is analogously called the world radius. Reduced to the balloon example, there are only two questions:

(a) what kind of curved space does the universe form at any given time,
(b) how does the radius of the universe change as a function of time?

The second question is discussed here only as far as it contributes to the understanding of the first. For the cosmological constant, see ch. 17 and 24.13.

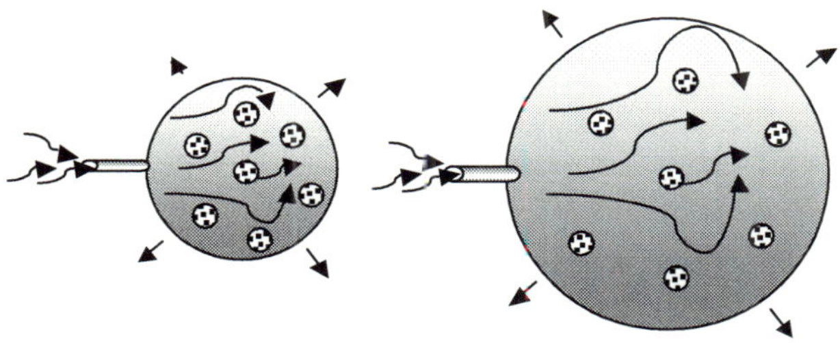

Fig. 19.2 A balloon is inflated. The coins on its surface ...

move away from each other faster the further they are from each other. Each coin can consider itself as the center of the expansion [230]. The coordinates of the coins on the balloon do not change, only its radius.

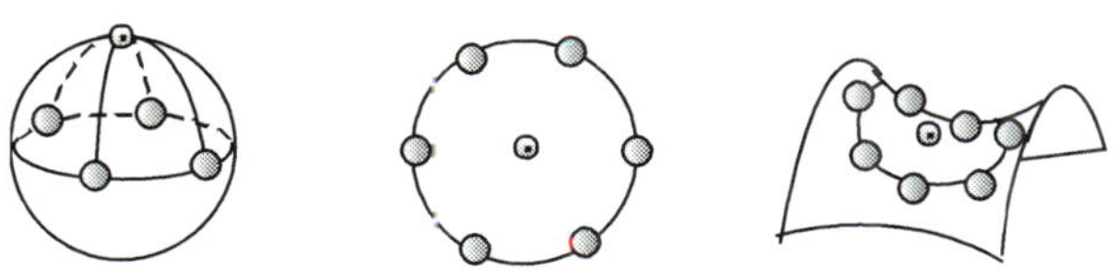

Fig. 19.3 Galaxy distribution in the universe

At the same distance from the observer and with the same density, there are fewer galaxies on a sphere than in the plane. On saddle surface there are more

The simultaneity of two events can be elegantly defined for this model. The synchronisation of clocks with light signals, as in special relativity, fails not only because of distances, but also because galaxies are accelerated relative to each other and clocks at rest relative to each other could run at different speeds because of different gravitational forces. But from each galaxy, the instantaneous world radius (like the circumference of the balloon) can be observed and the same clock time exists for all galaxies if the world radius has the same value for them. This theoretically beautiful definition of simultaneity is problematic, as the special-relativistic limiting case, Section 19.8.4, shows. In agreement with the majority, we accept it for the time being because it is indispensable for the relativistic definition of a finite, closed universe.

19.4. The Robertson-Walker Metric (RWM)

Just as the Schwarzschild metric can be derived from a centrally symmetrical mass distribution and this metric describes the space belonging to it, the Robertson-Walker metric applies accordingly, if one may assume the cosmological principle. In its most general form, it includes all three types of constant space (spherical, normal, i.e. Euclidean, hyperbolic). If we restrict ourselves to the theoretically most important case of a finite space, the spherical form of the Robertson-Walker metric applies:

(19.1) $$ds^2 = R^2[d\alpha^2 + \sin^2(\alpha)(d\theta^2 + \sin^2(\theta))d\phi^2] - c^2dt^2$$

(For the other two cases sin(α) is to be replaced by α (normal space) and sinh(α) (hyperbolic space).so in normal space r is described by the product Rα).

It means:

α, θ, φ the spherical coordinates as for the sphere S_3 in formula (18.17)

$d\alpha$, $d\theta$, $d\varphi$ are the differences of these coordinates for two near points,
dt is the time difference for two closely succeeding events, expressed in the coordinate time t .
ds is the spatiotemporal distance between two close points, as for the Schwarzschild metric, and describes what is measured with ideal scales and clocks.

Since t simultaneously represents the proper time of the clocks moving with it, it would be more understandable to use τ instead of t. R is called the world radius. R is a function of time and can be calculated with the general theory of relativity under additional assumptions (see Fig. 19.7 and ch. 19.7). More precisely, $R(t)$ or $R(\tau)$ is a scaling function and only the relative radius of the universe. This because for k = 0, -1 the universe is infinite but $R(\tau)$is always finite. (k is contained in another form of the RWM, formula (22.1), and is discussed in ch. 19.7).

Every galaxy cluster has the constant coordinates α, θ, φ, during the entire development of the universe, i.e. there is a co-moving coordinate system. The origin of the coordinate system can be any point of the finite universe, since all points of the universe should be equivalent because of the cosmological principle.

The discussion of the Robertson-Walker metric that now follows is essentially a physical interpretation of the mathematical exposition of constant curved spaces of ch. 18 and is similar in many ways to the study of the Schwarzschild metric.

The simplest way to understand formula (19.1) is the last element $c^2\, dt^2$. t describes the coordinate time. In contrast to the Schwarzschild metric, it agrees with the proper time of all clocks at rest in the moving coordinate system, because if $d\alpha$, $d\theta$, $d\varphi$ are zero then :

(19.2) $$ds^2 = -c^2dt^2 \quad \text{better} \quad ds^2 = -c^2d\tau^2$$

ds is the spatiotemporal distance between two events; in this special case, since the spatial coordinates do not change, *ds* describes the proper time interval of clocks at rest in the co-moving coordinate system. In comparison, the corresponding part of the Schwarzschild metric (16.5) still has a factor *A*, which describes the difference between the proper time and coordinate time and thus expresses the lag of clocks in gravitational fields. Clocks at rest in the co-moving coordinate system are in fact free-falling clocks. By what factor they go slower than clocks far away outside the field is not answered by (19.2). Such a question fits poorly or not at all with the cosmological principle for which all points are equivalent and an outside does not exist (more follows in ch. 19.8.4 and ch. 22).

If we set *dt* equal to zero in formula (19.1), we obtain the spatial part of the Robertson-Walker metric:

(19.3) $$ds^2 = R^2[d\alpha^2 + \sin^2(\alpha)(d\theta^2 + \sin^2(\theta))d\phi^2]$$

ds is now the spatial distance between two near points. The instantaneous value of the world radius *R* is irrelevant for the discussion of space curvature.

Formula (19.3.) can be compared directly with formula (18.17). They are identical, the spatial part of the Robertson-Walker metric represents a three-dimensional spherical surface. This is the nearest interpretation. Another possibility would be - as for the normal (Euclidean) plane the cylindrical surface - the elliptical space to which half of the three-dimensional sphere corresponds. All explanations about mathematical curved spaces in chapter 18 are transferable to the cosmos. With them one has a clear idea about three-dimensional curved spaces, their projection into normal space and the introduction of a curved coordinate system.

Because of its great importance, the mathematical consideration for finite spaces shall be repeated once again for cosmological contexts: A finite and closed universe is spoken of because the Robertson-Walker metric corresponds to the metric of a three-dimensional spherical surface. General relativity does not offer any other weighty arguments for this thesis, at least not those that have become part of textbooks. The actual physical achievement lies in the derivation of the metric from the Hilbert-Einstein equations (see [346] for the name), the geometric interpretation is the easier step.

Once a finite universe has been theoretically derived, the obvious but fundamental question arises: What possibilities are there to recognise and prove the existence of a finite, closed universe? This will be discussed below and, analogous to the Schwarzschild metric, the Einstein and Lorentz interpretation of the RWM will also be discussed.

In order to deepen the understanding of a finite, curved space, first some considerations on the questions: Where is the edge of the universe?, what is the position at which the big bang (see below) happened?, can we leave the universe? Such considerations are sensible or senseless, depending on the prerequisite.

If one assumes that there is only the spherical finite space of the universe and nothing beyond it, these questions are meaningless. First of all, all mathematical spaces are conceivable as real space without contradiction, especially a three-dimensional spherical space. Outside of it there is supposed to be nothing, then there is also no edge, this concept presupposes an outside. The position of the big bang does not (no longer) exist in today's universe, this applies quite analogously to the balloon example. Where is the place on the balloon where the inflation began? Where on the balloon is the place to which it contracts again when the air is released?

One can only express it in this way: At every point in time there is a different, initially growing spherical space, all galaxies have a certain position in it, once a larger space has continuously formed, the previous one no longer exists. This can be demonstrated and therefore considered proven for two-dimensional spherical surfaces; it can be proven for three-dimensional spherical surfaces by looking at two-dimensional sections that form two-dimensional spherical surfaces. A finite universe cannot be left, because if a position α, θ, φ has admissible values, then it lies somewhere in the universe and not outside, or they are inadmissible, then this position does not exist. For example, start somewhere with a spaceship. You can choose any three mutually perpendicular directions and move "straight", i.e. on the shortest path in one direction with the result: As on every spherical surface, you finally return to the starting point. If one does not move "straight" but uniform, the same happens even faster, because one moves on a circle with an even smaller radius.

Another new point of view concerns the expansion of the universe, in the Einstein interpretation it does not mean a radial velocity of galaxies in space away from a starting point, but an expansion of space between the galaxies, the galaxies do not move in the process. In the balloon example, the coins do not move on the surface and the surface only because, contrary to the premise, the balloon envelope is embedded in a space of higher dimension. In this sense it said in [20] "... that the escape velocity defined in this way is due to the expansion of space and not to the galaxy's own motion in space." This is true for all k. Space is expanding and not: the galaxies are moving through existing space.

The above questions only make sense if one imagines the three-dimensional sphere embedded in a four-dimensional Euclidean space. By analogy, the balloon is somewhere in normal space. By inflating, the surface of the balloon forms a continuously changing, always new spherical surface in space. The starting point does not exist on it, but it exists. It is the center of the sphere in the surrounding space, a certain point in the surrounding space. In the same way, there is also the edge, more generally: points outside the spherical surface. In this case, is it possible to leave the universe with a spaceship? That depends on the physical assumptions. The surrounding space is supposed to exist, then in principle it is also reachable, but from the above formula for the metric it cannot be inferred how this could happen. From the metric it cannot be inferred whether the surrounding space exists, it fits the two assumptions: there is the surrounding space and there is not.

In cosmology, when we speak of a finite universe, we mean a universe that is *not* embedded in a comprehensive space and to ask the previous questions under this condition is rightly considered pointless. It is mathematically and physically legitimate to describe the motion of galaxies relative to their proper times, and the result, an expanding universe that is closed ($k = 1$), is impressive. For the LI, however, this is not all, see ch. 22.

19.5. The Lorentz interpretation of the Robertson-Walker metric as scale changes

Just as the Schwarzschild metric could be understood as the formulaic expression for the change of scales in a centrally symmetric gravitational field, the Robertson-Walker metric can also be understood as the formulaic expression for the possible changes of scales in fields for which the cosmological principle is fulfilled. Since there are many ways to map the Robertson-Walker metric onto normal space, it is not obvious which of these mappings should be called the LI of the RWM. In the simplest case, the three-dimensional sphere is projected onto a sphere in normal space. This has been done in ch.18 with formula (18.21). The Robertson-Walker metric takes the form:

(19.5) $$ds^2 = Adr^2 + r^2(d\theta^2 + \sin(\theta)d\phi^2)$$

with $A = \dfrac{1}{\cos^2(\alpha)}$ or as a function of r: $A = \dfrac{1}{1 - r^2 / R^2}$

Where: $0 \leq r \leq R$, $r = R\sin(\alpha)$, $dr = R\cos(\alpha)d\alpha$

The universe is now a normal (Euclidean) space, its points have the spherical coordinates r, θ, φ. If the distance between two points is measured with normal scales ds, ds gives the true length only if ds is set tangentially (perpendicular to r). If ds is set radially, the true length is $ds\cos(\alpha)$. In Fig. 19.4 this is shown graphically for the case $\theta = 90°$, the equatorial plane. This is analogous to Fig. 16. 8 of the Schwarzschild metric and corresponds mathematically to Fig. 18.9.

(However, this model has a theoretical meaning for the LI, only. As chapters 21 and 22 show, the LI regards the RWM as a special case of a time-dependent SM and thus it is clear that a mapping of the SM should apply but with a function of t, see Fig.22.1. Nevertheless, the following arguments remain correct.)

The coordinate origin of the SM forms the center of the universe and is also the location of the big bang. (see below). Galaxies exist for $k = 1$ only within the sphere with the radius R. The expansion of the universe is reduced to the expansion of a metagalaxy and is also described here by the time dependence of R.

In this model, a spaceship can leave the spherical part of space. It depends on the magnitude of the gravitational forces at distance R how much energy is needed to do this, but in principle it is possible. Light always escapes, because the gravitational redshift can make the frequency arbitrarily small but not zero.

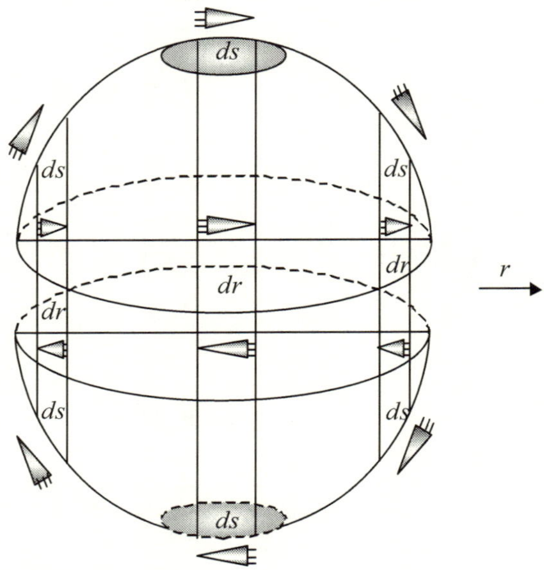

Fig. 19.4 Robertson-Walker metric

The Robertson-Walker metric can be visualised as a surface of rotation like the Schwarzschild metric.

For $\theta = 90°$, the result is a spherical surface. Light or a space traveller moves in curved space along a great circle (Einstein interpretation) or in a normal (Euclidean) plane along a straight line, whereby the scale *dr* shrinks with increasing distance from the center of gravity of the universe and no longer indicates the real distance *ds*. This could fit the LI, in fact it looks different again, s. ch. 22.

19.6. Observations in the universe

The Robertson-Walker metric contains a factor *R(t)* in both the spherical and the non-spherical form. *R(t)* does not need to be known in order to interpret the possible cosmic spaces. This changes when it is to be verified by observation which of these spaces really exists.

First of all, today's universe, as it is observable, is discontinuous, non-homogeneous and anisotropic if one thinks of the accumulations of matter in the form of suns and planets and the large free space between them. The same applies to galaxies, only clusters of galaxies are approximately evenly distributed across the universe. No direction in the universe is then excellent (isotropic) and at every location the density of the galaxy clusters is approximately the same (homogeneous). Only in these size ranges do observation and the Robertson-Walker metric fit together.

A first essential observation in the universe is the discovery of the radial escape velocity of all galaxies relative to any point of observation (e.g. the Earth). For this motion, the famous Hubble's law applies:

$$v_{galaxy} = H(t)\ d_{galaxy} \tag{19.6}$$

v is the escape velocity of a galaxy at distance d from the observer, $H(t)$ *is* the Hubble constant. The further away a galaxy is from the observer, the greater the speed at which it moves away from the observer [338], [317], [246], [379].

This law can only be observed from Earth, but it is easy to show that it must then be the same for every other observation location, see Fig. 19.5. This means, however, that every galaxy in space can regard itself as the center of the universe. For further explanation, consider the balloon example, Fig. 19.2. The balloon is inflated, the coins on it move further and further away from each other, and the further away they are from each other, the faster they move. This is relative to each of the coins. Additionally, one can imagine that the balloon expands to infinity, but also that when a maximum size is reached, it releases air again and shrinks. Both ideas apply to the development of the universe.

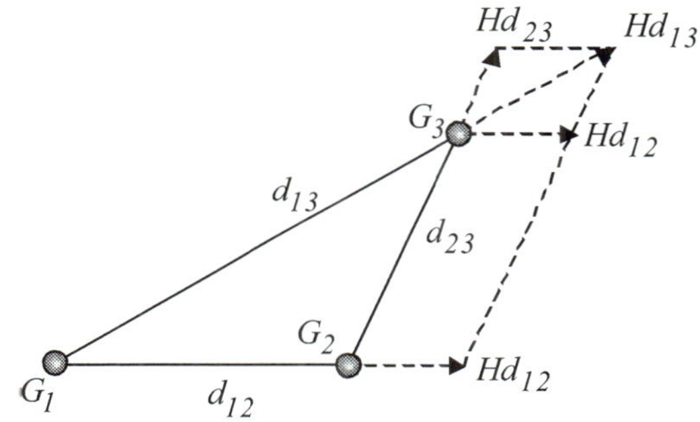

Fig. 19.5 If Hubble's law $v = H\,d$ applies relative to the galaxy G_1, then because of the similarity of the triangles of sides d_{12}, d_{13}, d_{23} and $H\,d_{12}$, $H\,d_{13}$, $H\,d_{23}$ also relative to G_2. It follows: G_3 moves away relative to G_2 with the speed $v_{23} = H\,d_{23}$.

The expansion of the universe cannot be observed in terms of any earthly dimensions. Through complicated calculations it can be shown that the galaxies themselves, like the pieces of money on the balloon, do not change their dimensions. (If one regards the flight of the galaxies quite simply as the flying apart of fragments after an explosion, then it goes without saying that the fragments do not change their size in themselves). The flight of the galaxies has been demonstrated by the redshift of the light they emit. The higher the escape velocity the lower the frequency of light signals received on Earth, they turn from blue to red.

The radial escape velocity of the galaxies is considered to be the result of an explosion that occurred about 15 billion years ago, the Big Bang, and it is natural to ask where in the universe it could have taken place. Because of Hubble's law, every galaxy can consider itself to be the center, at most one can be. If one interprets the Robertson-Walker metric according to the LI, the location of the big bang is the center of gravity of the universe at this point lies the center of the sphere onto which the space of the Robertson-Walker metric is projected. If one views the universe relativistically as a three-dimensional sphere, then there is no center point. The three-dimensional sphere changes its volume, it can shrink to zero. Like the balloon, it has no distinguished point and the starting point is not on the balloon or the sphere, or somewhat less precisely: all points were the starting point of the Big Bang.

Another important cosmological observation is background radiation (CMB). These are microwaves from space that hit the Earth from all sides almost isotropically. It is assumed to be left over from the beginnings of the universe - according to the following model: During the Big Bang, hot plasma first spread out in all directions, and the galaxies with the suns then formed from this plasma. The first atoms are an intermediate stage. The light waves they emitted are still visible today if they originate from gas remnants that were at a greater distance from the Earth. Since all matter - gas remnants as well as galaxies - move away almost at the speed of light if they are only far enough away (Hubble's law), bright light becomes a microwave due to the red shift.

One property of today's universe that does not contradict the Standard Model but cannot be explained with its help, lies in the relatively small density fluctuations of the universe. This results in an even more homogeneous universe in its early days, since initially small fluctuations in density (would have to) intensify in the course of evolution. What the Standard Model lacks is a "smoothing mechanism" [113] that could explain the homogeneity of the universe today. (This applies to the EI, for the LI density fluctuations are less serious, see below).

FLIESSBACH [113] describes this phenomenon as follows: "The cosmological standard model provides ... no explanation for this extraordinary flatness (note: of the universe). One can also argue the other way round in terms of time: Small curvatures ... at an early time must have led to large curvatures today. From this point of view, today's (relative) flatness is a mystery. (Note: $k = 0$, s. ch. 19.7)

"A related problem is the density fluctuation problem: Initially very small density fluctuations intensify under the influence of gravity. The present-day relative homogeneity of the distribution of matter on a large scale, but in particular the isotropy of the background radiation requires in the Standard Model an improbably homogeneous universe in its early days."

In the technical literature, the so-called inflationary model is introduced for the above problem and a particularly rapid expansion (inflation) is assumed for the initial time of the universe [113], [196]. The quantum mechanical considerations associated with this cannot be reproduced here, as they are not a topic for space-time concepts of special and general relativity.

(For LI, gravitational forces do not exist from the beginning, but only when particles with non-zero rest mass have emerged, as ch. 21.3 shows. An inflationary model at the beginning is therefore very obvious).

Another observed possible intermediate stage on the evolution to a present-day galaxy are the X-ray emitting quasars (probably active galactic nuclei, AGN's). Galaxies and their stars, quasars together with less luminous objects, and the gas clouds, from which the background radiation are emitted are located at different distances from Earth, so that one can draw Figure 19.6 [246]. If you look at distant objects, you look back into the past at the same time, because the objects become visible as they looked when the light was emitted.

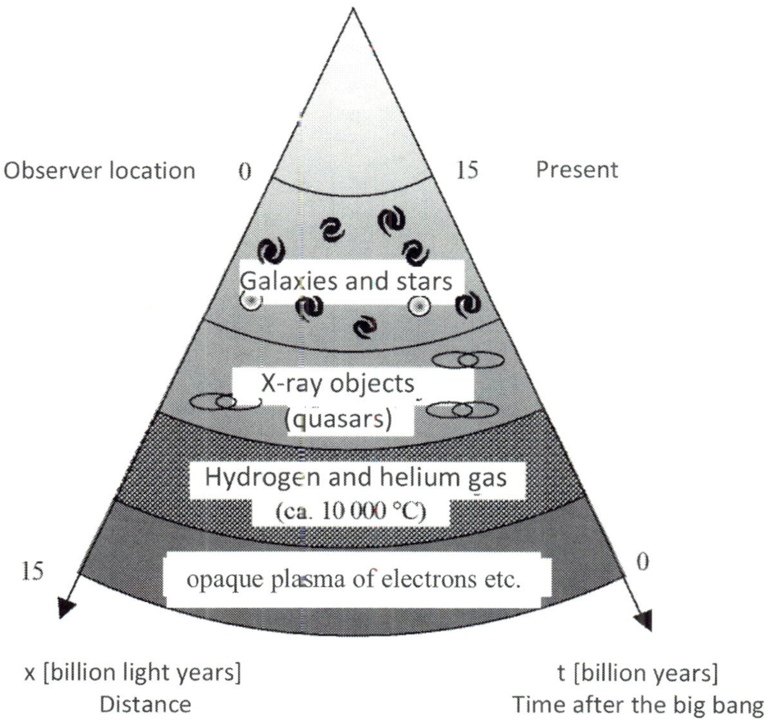

Fig. 19.6 The view into space. Objects at great distances are seen as they looked a long time ago. [246] The distances are mean values.

For the background radiation in particular, the following question arises: It is said to originate from objects that are about 13.5 billion light years away. It was therefore emitted about 13.5 billion years ago. At that time, the universe still had the small expansion of approx. 300000 light years, the gas cannot have been at all at the distance of approx. 13.5 billion light years from where the background radiation is said to come today. The explanation for this seemingly contradictory fact lies in the co-moving coordinate system. Consider the background radiation that originates in a gas cloud at a certain position $(\alpha_0, \theta_0, \varphi_0)$. The gas cloud retains this position as long as it exists and is taken over by the galaxy that evolves from the gas cloud. The origin of the background radiation is considered to be the position $(\alpha_0, \theta_0, \varphi_0)$ which today has a distance of about 13.5 billion light years. Today, a galaxy is also located at this position, as can only be seen at a distance of more than 13.5 billion light years.

In the same way, one can explain why the background radiation hits the Earth from all directions, although the universe only had the size of a galaxy about 13.5 billion years ago and one could conclude that the background radiation should only arrive from this direction. If the background radiation, when it came into being, moved uniformly from all directions towards the position that the Earth has today, the same is true today for the radiation that has not yet reached the Earth. Despite the initially small distances, such radiation exists because the speed of light was significantly smaller than c at the beginning of the expansion due to the high gravitational forces. (Details depend on the assumed geometry, i.e. on k.)

Hubble's law, which was proven by the red shift and the discovery of the background radiation are the main supports for the thesis of a big bang, the so-called cosmic standard model. It is not accepted by all astrophysicists - with today's (2021) observational results a rather hopeless position. The discovery of the cosmological constant, see the following chapter, clarifies the cosmological standard model.

19.7. The cosmological alternatives: The infinite Euclidean and infinite hyperbolic universe.

The Robertson-Walker metric resulted from the cosmological principle. Chapter 19.4 dealt with the spherical part of the Robertson-Walker metric, which refers to a finite universe. The other two alternatives refer to a Euclidean infinite or hyperbolic infinite universe. All three differ from each other by a factor k, if one chooses a different representation for the Robertson-Walker metric shown by formula (22.1). [[all] except [89], [91]]

$k = 1$, spherical space, the kinetic energy of the universe is low, the present expansion of the universe will reverse.
$k = 0$, normal (Euclidean) space, the expansion of the universe ends at a very great distance, at infinity; kinetic energy and gravitational energy balance each other out.
$k = -1$, the kinetic energy of the universe exceeds the gravitational forces; expansion continues even at any (infinite) distance.

All three cases can be distinguished by the time dependence of the world radius R. For the physically simplest case of the dust model - one assumes that the galaxies behave like dust grains, they do not form a gas that has an internal pressure - $R(t)$ can be calculated, the result is shown in Fig. 19.7. What they all have in common is the thesis of a big bang. These simplest but meaningful cosmological models are called the Friedmann cosmos after its discoverer.

Of course, it would suffice to observe the world radius R as a function of time t to enable a decision between the three models. The observation periods are not sufficient for this. However, the following observation possibilities remain:

a) The cases $k = 1, 0, -1$ differ in their kinetic energy or in the magnitude of the existing gravitational forces. These depend on the total mass of the universe and, for a given volume, on the density of matter in the universe. If it is above a critical value, the universe is finite and the expansion will reverse. The density of the universe can be determined by counting galaxies and other observations. Since there is so-called dark matter in the universe, which is not directly observable, doubts are allowed, but "the currently observed mass density is below the critical density... we would have to live in an open world" STEPHANI [357].. Today (2010) $k=0$ applies, see below.

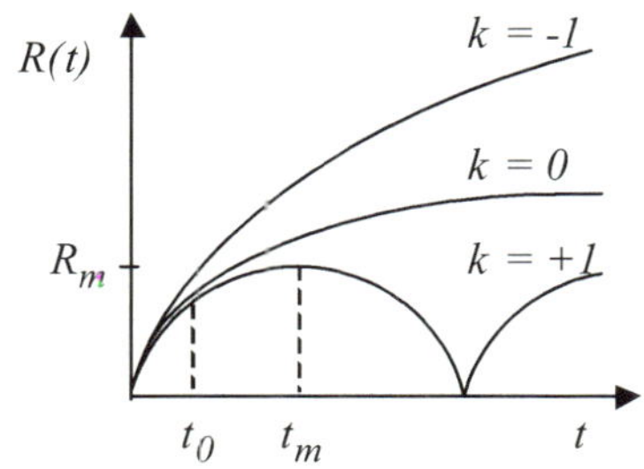

Fig. 19.7 The relative radius $R(t)$ of the universe as a function of time for the Friedmann cosmos.

$k = -1$ infinite hyperbolic space
$k = 0$ infinite normal (Euclidean) space
$k = +1$ finite spherical space (for numerical values for $k = +1$ see appendix)
R_m : maximum radius, t_0: present time, t_m: time of greatest expansion

b) If light is emitted in a finite universe in any direction, it arrives back at the starting point after a finite time. It cannot leave the universe, its always "straight" path is circular. The time required for this can be calculated; it is exactly as long as the universe exists. If the light signal is emitted shortly after the big bang it has not yet completed its round trip when the universe collapses. WHEELER [230].

c) The mathematical relationships for circumferences of circles, sums of angles in triangles, volumes of spheres, the fact that there are no parallels - all the properties of spherical geometry are fulfilled in a finite universe, but cannot be verified with usual scales because of the large distances required.

d) If one assumes that the galaxies in the universe are equally distributed and counts the galaxies as a function of distance, a different result is to be expected for the different forms of space, see Fig. 19.3 (before ch. 19.4). For a finite universe, the number of galaxies is smaller than for normal space (and largest for the hyperbolic case), because the corresponding circumferences or spherical areas are smallest for finite spaces. Corresponding estimates have been carried out, but have no significant result [338], [357].

Finally, a few quotes on the question of whether the universe is demonstrably finite:

"The question of the structure of the universe at large is ... unsolved. Finite or infinite universe - a decision is not possible at present." SEXL [338]

"As the best value compatible with all observational data, we have assumed (note: for the mean density, the value) 0.1. According to this, we live in an open, continuously expanding universe of decreasing density." UNSÖLD, BASCHEK [379] The - rarer - counterthesis is held by BERRY [23] and, modified, by a Bonn-Potsdam research group [20].

Today (2010), due to extensive astronomical observations [158], [295.1], [192.1], [62.1], [260.1], the so-called cosmological standard model is valid. It remains with an expanding, open space (k = 0), to which a positive cosmological constant Λ is added. For numerical values, see [189.1] and ch. 25. A positive Λ results because the observed Hubble constant H_0 gives a smaller value for more distant supernova events than for closer events, [192.1]. H_0 must always give the same value even for distant galaxies, but an equal H_0 is obtained only with the assumption of a positive Λ. If at all, a negative Λ was to be expected, since only gravitational forces were known, which mean a negative acceleration. By a positive Λ, properties of the particle-free ground state of the quantum fields come into question [22]. This would be another property proving the existence of an ether, ch. 11.7.

In the meantime (2021), further observations (different values of H_0) prove that the isotropy and homogeneity of the universe does not exist as expected, even a rotation of the visible universe seems to be detectable, which both are in favor of the LI, see ch. 24.11.

19.8. Deepening the Einstein and Lorentz interpretation of the RWM

19.8.1. Overview

In chapters 18 and 19.5, Figs. 18.9 and 19.4, an obvious approach to the LI of the RWM was discussed. This was based on a closed, expanding space. A number of arguments follow as to why this variant of the LI shall become modified. The reasons do not lie in the mathematical statement of the RWM, the function $R = R(t)$ or its derivation from the Einstein field equation (Hilbert-Einstein equation [346]), but, especially for k = 1, in its interpretation as a closed and expanding space. There are a number of objections to this, two of which are serious:

(a) The RWM describes not only the expansion of the universe but also the explosion and collapse of dust stars. Here the idea of a closed space makes no sense, ch. 19.8.3.

(b) The cosmological simultaneity underlying the RWM does not fit the special-relativistic limiting case, Section 19.8.4.

Let us first take the EI and LI of Chapters 18 to 19.7 as a basis: Whether the universe is finite or infinite can only be decided by observation; the general theory of relativity allows both variants. Independent of this and also only to be decided by observation is the question of whether the space of the universe is really curved (Einstein interpretation) or whether there are scales and clocks influenced by gravitational fields, as corresponds to the Lorentz interpretation. These alternative interpretations have been explained in detail for the finite case and also apply quite analogously to a hyperbolic infinite universe, which will not be discussed further.

For a distinction between the two alternatives - which, due to Poincaré's thesis, can at most turn out to be differently plausible - the following arguments can be put forward for a distinction between the two alternatives. They are largely based on the differences between inner and outer curvature, and between local and global surface properties, which have been discussed mathematically in ch. 18.

a) For the Robertson-Walker metric, not only the three-dimensional spherical surface fits as spaces, but also the elliptic space, which is also closed, as well as open models, namely rotational surfaces of elliptic integrals [192]. The decision for the three-dimensional sphere is justified by the fact that it represents the most general and simplest solution. However, this is not sufficient, if the space is curved, one may choose any of these curved spaces as long as no further distinguishing criteria become known.

b) Globally, there is a difference between the spaces of the two interpretations. For example, if one has the metric $ds^2 = dx^2 + dy^2$, which is valid both for the normal plane and for a cylinder, one can distinguish between the two with light signals or by means of space travel. Light signals emitted in the direction of the cylinder axis disappear at infinity, perpendicular to it they arrive at the starting point after finite time from the opposite direction. The situation is similar with the finite three-dimensional sphere: light rays or spaceships can never leave it. All α, θ, φ positions can be reached, but these points always lie on the three-dimensional sphere. Light rays emitted from the point α, θ, φ reach their opposite pole 180° - α, $180° - \theta, \varphi$ and fly straight on there and return to the starting point "from behind".

These predictions are not compatible with the scale interpretation; in the global domain, it differs from the Einstein interpretation: If a radially emitted light signal reaches the edge of the universe (θ, φ arbitrary, $r = R(t)$), it leaves it (like light leaves a collapsing star) and continues to move in the now galaxy-free space for which the Schwarzschild metric applies. That is, the light beam disappears into infinity and suffers a gravitational redshift in the process. Something else applies to a spaceship. It too leaves the universe and moves in the space of the Schwarzschild metric. In the process, it constantly loses speed because of gravitational forces. If it becomes zero at a finite distance, the spaceship turns around, re-enters the universe, returns to the starting point and thereby comes from the same direction in which it left the starting point.

Of course, there are many reasons why such an endeavour fails, but it illustrates a difference in principle in both models. As explained before, however, a round trip of light particles, for example, takes far too long to be considered as an experiment.

c) Since the general theory of relativity must contain the Newtonian limiting case if the gravitational forces are small, which is the case in cosmic dimensions, one may expect the same for the associated space model: If the kinetic energy of the galaxies is not sufficient for a hyperbolic or Euclidean infinite space, by NEWTON'S law of gravity a universe arises that collapses again. But light waves and spaceships can leave it, the universe has an edge. It is not conceivable in the Newtonian limiting case that a radially moving object, when it has reached the edge of the world, jumps there to the opposite side of the universe in order to then return radially inwards to the starting point from the opposite direction. In a finite closed universe, however, this is expected. In the following chapter, it will be shown how well the Friedmann model also fits the classical ideas.

d) The decisive arguments lie in the application of the Robertson-Walker metric to collapsing stars or to the formation of galaxies regarding the versions of cosmological simultaneity. The Robertson-Walker metric is also fulfilled for these cases, as will be shown in the concluding chapter.

With these arguments, the Lorentz interpretation, in whatever variant, gains weight and the interpretation as a change of scale comes to the fore, but the mathematical and physical methods developed with the idea of a finite, curved world model will retain their importance and the relativistic theses on space and time will always remain a deepening of the classical position and vice versa.

19.8.2. Newton's law of gravity and the big bang

The most important cosmological results of general relativity are the thesis of the expansion of the universe, which is expressed in Hubble's law, the thesis of the big bang and the associated idea of a spherical, Euclidean or hyperbolic space. The exact same predictions also result from the application of Newton's theory of gravitation. Here, the curved space-time is linked to the explosion energy of some massive object and, as with every explosion in space, the three cases are also possible for the big bang:
$k = +1$. The explosion energy of some massive object is smaller than the gravitational energy, i.e. the explosion pieces come to a standstill at a finite distance and, because of the existing gravitational forces, move back to the starting point where they collapse.
$k = 0$. The explosion energy is as great as the gravitational energy, and the explosion pieces have zero velocity at infinity.
$k = -1$. The explosion energy is greater than the gravitational energy, i.e. the explosion pieces escape to infinity.

The same predictions also apply to other details. For example, the so-called cosmological constant in Newtonian cosmology correspond to repulsive forces that only become effective at great distances [23]. Likewise O. HECKMANN [147]. The investigations in [23], [147] specify chapter 19.2.

It is not surprising that the Newtonian theory of gravitation agrees with the general theory of relativity in cosmic areas, since small gravitational forces prevail between the galaxies due to the large distances and this leads to almost identical equations of motion in both theories. It is surprising that different space models should nevertheless apply. On the one hand, the infinite space of the SM and, on the other hand, the expanding and, in the case of $k=+1$, finite closed space of the RWM. This difficulty does not exist for the LI, see ch. 21, 22.

19.8.3. The Robertson-Walker metric and the gravitational collapse of stars

In cosmology, the Robertson-Walker metric leads to the idea of the universe as spherical, Euclidean or hyperbolic space or alternatively to the idea of an infinite normal (Euclidean) space with contraction of scales and retardation of clocks in the presence of matter. Important was the case of a finite space. The essential argument for its existence was the interpretation of the Robertson-Walker metric with as few assumptions as possible. Since we are in the interior of the expanding universe and the idea of looking at the universe from outside is ill-suited to the concept of the universe, the assumption of a finite spherical or elliptical space is undoubtedly elegant. However, it contradicts the different application of the Robertson-Walker metric to the gravitational collapse of stars and the formation of galaxies and stars.

The following quotations on this subject: LANDAU-LIFSCHITZ [190, p. 383] "It should be pointed out that this metric is identical with the Friedmann solution for a cosmos homogeneously filled with dusty matter." In STEPHANI [357] the corresponding equations (19.1) of the RWM - here it is called TOLMAN solution - and the function *R(t)* of Fig. 19.7 can be found.

The RWM thus also describes the gravitational collapse of a star. If the nuclear fuel in a star is consumed, there is no longer an equilibrium between the thermal pressure due to nuclear fusion and the gravitational pressure. The star cools down and collapses [113]. How does such a process take place? A plausible model for the collapse discussed in the literature is: At the end of nuclear fusion, the star consists of nothing but dust particles, because the cooled stellar matter no longer forms a gas that has an internal pressure. In this state, only gravitational forces act, and the dust particles collapse into a neutron star a few kilometers in size or, with greater mass and thus greater gravitational forces, into a black hole (or the borderline case that comes close to a black hole).

These are, as was said, the prerequisites for the applicability of the Robertson-Walker metric or the Tolman solution. The only existing force is gravity, dust particles are the atoms in the star, the galaxy clusters in the expanding universe. Only one difference exists: The star collapse is observable from outside, its gravitational field is described outside the star's surface by the Schwarzschild metric, and the mass of the star is measurable. Correspondingly, the dust model is applied to the formation of stars from nebular fields in galaxies. In all cases it makes no sense to say that the collapsing or forming star forms a finite, closed, spherical or elliptical space and that outside this space there is nothing. It is not clear why anything else should apply to cosmic space. In addition, the following quotation: "An expanding "dust star", i.e. a section of a Friedmann cosmos, which is surrounded on the outside by a static

Schwarzschild metric (cf. the model of the collapsing star ...), is perhaps also a quite good world model." STEPHANI [357]

"What did God do before he created heaven and earth?
He prepared hell for all those who ask such questions."
AUGUSTINE

19.8.4. Expanding space and cosmological simultaneity for the special relativistic limiting case

The special-relativistic limiting case of an expanding universe problematises the idea of an expanding space and the usual cosmological definition of simultaneity. For this purpose, the cosmological principle is made more precise and applied to the case of an expanding universe with vanishing density. For this limiting case, both cosmological simultaneity and expanding space are demonstrably artificial and superfluous conceptualisations. This justifies imagining our universe, as it is visible, as a metagalaxy and a dust star (with the galaxies as dust grains).

1. the special-relativistic limiting case

Fig. 19.8 illustrates the evolution of the universe in the case of vanishing acceleration, i.e. $q = 0$. The cosmological principle is assumed, as in Fig. 19.9. All galaxies G_i are spatially isotropic and homogeneously arranged and the following applies

$$v_i = r_i/t$$

in the direction of r_i. By comparison with Hubble's law $v_i = H(t)r_i$, the Hubble constant results in $H(t) = 1 / t$. We consider G_0 to be at rest, but it could also be any other galaxy. The masses of the galaxies are small and their distances large, so that gravitational forces can be neglected.

If we calculate backwards, then at the time $t = 0$ all G_i are at one and the same place, namely G_0. This corresponds to the big bang if one imagines that at this time all G_i have acquired their velocity v_i and that they were previously at rest. The usual scaling factor $R(t)$ is calculated from $R(t) = \text{const} \cdot t$ and is something different from the world radius $r_{\max} = v_{\max} t_0 < ct_0$, although $R(t)$ is often called that. The Hubble constant is also given correctly for fig. 19.9 by the usual definition: $H(t) = \dot{R}(t)/R(t) = 1/t$. The curvature of space is zero, since Minkowski space is involved.

On the other hand, this limiting case must belong to one of the three possible universes $k = +1, 0, -1$. For the corresponding formulae in the literature, e.g. [20], [147], only the case $k = -1$ for $\rho_{density} \to 0$ has a suitable solution with $q = 0$ and $H(t) = 1 / t$. This is also shown by problem 19.17 in Lightman [198]. $k = -1$, however, means a hyperbolic and not a Euclidean space as is valid for fig. 19.9. The explanation for this is simpler than expected (and problematic for the Einstein interpretation). In the special-relativistic limiting case, there is simultaneity because of Einstein's clock synchronisation, in this case relative to

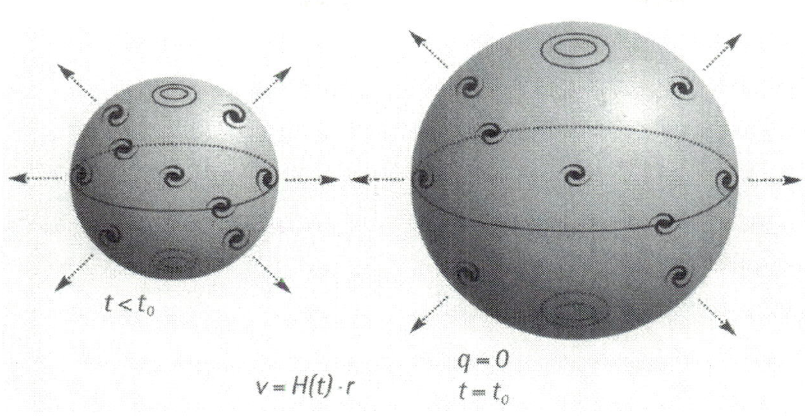

Fig. 19.8 The expanding universe.
In the special-relativistic limiting case, gravitational forces are neglected ($q = 0$).

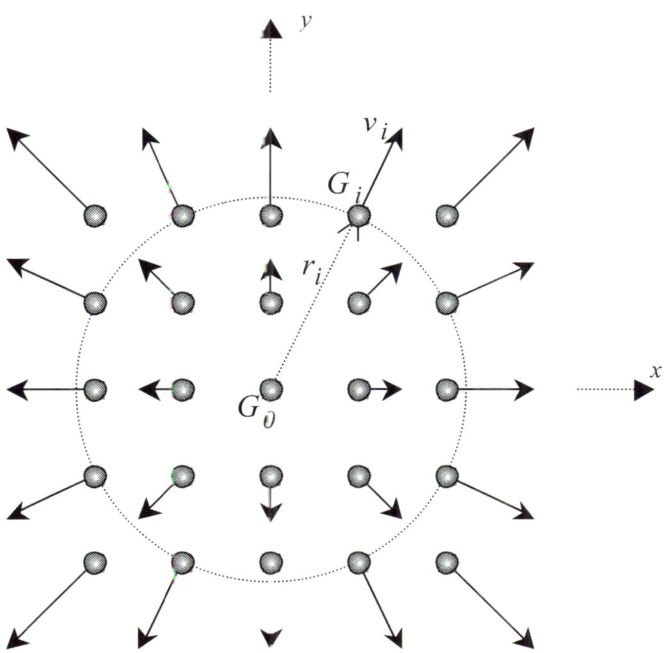

Fig. 19.9 The cosmological principle for galaxies

For all galaxies G_i $v_i = r_i / t_i$. The galaxies G_i are uniformly distributed. This gives a model for a homogeneous isotropic universe with the Hubble constant $H(t) = 1/t$ and the acceleration $q = 0$, since gravitational forces are to be neglected.

to the inertial system G_0. Deviating from this, however, cosmological simultaneity is taken as a basis in the expanding universe and has to be applied in fig. 19.9, too. This means: all galaxies G_i have equal rights and simultaneity exists if all clocks moving with the galaxies show the same time. Certainly, this definition of simultaneity is inadmissible for our limiting case since the clocks are not at rest relative to each other (special relativity becomes false, the principle of the constancy of the speed of light is violated if cosmological simultaneity is introduced), but doing so in the limiting case of fig.19.9 one sees: space is becoming hyperbolic.

To prove this, measure the radius and circumference of the circle of Fig. 19.9 when the clock in G_0 shows the time t_0.

(a) Regarding Einstein simultaneity the circumference at G_i has to be measured at the proper time

$$t_i = t_0\sqrt{1-v_i^2/c^2}$$

of clocks at rest on G_i (time dilation of the clock at rest in G_i) because then coordinate time is t_0.

(b) Regarding cosmological simultaneity time dilation demands measurement of the circumference at G_i (and at the other galaxies on the circumference) at proper time t_0 of the clock at rest in G_i. This is coordinate time

$$t_i = t_0 / \sqrt{1-v_i^2/c^2}$$

(note the division sign, because only then does the proper time of G_i have the value t_0). This is special-relativistic later than t_0, the circumference is measured larger:

$U > 2\pi r_i$ i.e. you have a hyperbolic space. More precisely: The circumference of the circle at time

$$t_i = t_0 / \sqrt{1-v_i^2/c^2}$$

is

$$\begin{aligned} U_i &= 2\pi v_i t_i \\ &= 2\pi v_i t_0 / \sqrt{1-v_i^2/c^2} \end{aligned}$$

U_i is the circumference determined by G_i at cosmological time t_0. The radius of G_i determined by G_0 at cosmological time t_0 is $r_0 = v_i t_0$, the corresponding circumference is $U_0 = 2\pi r_0 = 2\pi v_i t_0$. It is $U_i > U_0$, so $k = -1$.

All that actually remains is to adapt the simultaneity definitions of the cosmological models to the special-relativistic limiting case. This is also appropriate by a more detailed discussion:

First of all, one could object that in Fig. 19.9 the expansion of the space is shown. It is not the G_i that are moving away from each other, but the distance between the G_i is increasing as a function of time, i.e. space is expanding. Since the G_i are at rest, the speed of the clocks of the G_i does not change and cosmological simultaneity exists. This is possible at least. For a better understanding, imagine an independent observer in Fig. 19.9. He can decide by observation what the situation is. If there was an explosion, he knows which of the G_i was the location of the big bang. In addition, after a finite time, he himself will be inside the dust star. But if space expands, then so does the space to the observer and the observer will always be outside the expanding space of the G_i. This would be measurable in principle and thus the expansion of space is an operational, i.e. a philosophical concept.

But there is no such or equivalent observation. Despite the lack of experimental confirmation, the belief in an expanding space is tremendous: "The *Doppler formula* should not obscure the fact that the escape velocity defined in this way is caused due to the expansion of space and not to the galaxies' own motion." [20]. Other quotations can be cited. Where is the timidity in allowing the cosmos to become wrongly hyperbolic or spherical? The Lorentzian alternative is not even referred to.

Of course, these cosmological spaces have merits through their simplicity and form special mathematical-physical spaces, but why should there be more? The expansion of space has not been proven and (until proven otherwise) only makes sense in mathematical-physical spaces. The answer to the other questions also remains open: How is the expansion of space to be distinguished experimentally from a relative velocity of two galaxies? The Robertson-Walker metric applies to exploding dust stars as well as to an expanding universe. It has been known since GALILEO "that those physical processes which determine terrestrial matter also exist in principle on extraterrestrial bodies." [368]. What should be different in this case?

If one takes into account the definition of simultaneity that is compatible with the special-relativistic limiting case all three cosmological standard models have the value $k = -1$, because they only represent a variation of the Schwarzschild metric as will be shown in detail in Chapter 22. $k = 0$ exists only approximately and $k = +1$ not at all. The simplest explanation of the observations is: Our universe, as far as it is visible, is a metagalaxy, which can be described as an exploding dust star with galaxies as grains. This justifies the heuristic considerations of chapter 19.2.

2. the special-relativistic limiting case as a consequence of cosmological simultaneity and expansion of space

It is to be shown that the hypothesis of cosmological simultaneity, together with the hypothesis of an expanding space, yields equivalent physical formulae as before. First of all, it is obvious to reject the idea of an expanding space because the principle of the constancy of the speed of light seems to be violated. If a light signal is emitted from the origin G_0, it reaches G_i at the coordinate time t_i and in cosmological time (= proper time of G_i)

$$t_i' = t_i\sqrt{1-v^2/c^2}$$

If the speed of light is calculated using the cosmological time then t'_i applies

$$v_{Licht} = s_i / t'_i > c\,.$$

But this objection comes to nothing, according to the Einstein interpretation, space expands during the propagation of light. Only when the light arrives at G_i is the distance from G_0 equal to r_i. On the return, it is exactly the other way round.

More precisely: Since the proper time interval of light is zero, the Robertson-Walker metric yields $cdt = R(t)d\alpha = dr$, i.e. to each time t belongs a spatial interval dr, which is traversed in the time dt by light with the speed c. *The sum of all dt gives the proper time t' dα and the sum of all dr is the total distance travelled ≠ r dα.* The sum of all dt gives the proper time t'_i and the sum of all dr is the total distance travelled $\neq r_i$ The speed of light was always c for each interval dt, so it is also c in total. This applies to all other formulas accordingly, both views are not mutually exclusive because of different formulas. Thus, in the LI, the redshift is calculated for light from a galaxy G_i in the special-relativistic limiting case according to the Doppler formula (19.9) and the cause of this effect is the relative velocity v. For the Einstein interpretation, expanding space causes the redshift z, the galaxies are at rest and the redshift increases as the light propagates with increasing 'radius' $R(t)$ according to formula (19.7). Both are the same:

$$1 + z = \frac{R(t_0)}{R(t'_i)} \tag{19.7}$$

$$= \frac{t_0}{t'_i} \qquad \left(\text{wegen } R(t) = \text{ const } \cdot t\right)$$

$$= \frac{t_0}{t_i\sqrt{1 - v^2/c^2}}$$

$$c(t_i - t_0) = vt_i$$

$$= s_i$$

$$\frac{t_0}{t_i} = \frac{v}{c} + 1 \tag{19.8}$$

In (19.7) inserted:

$$1 + z = \frac{1 + v/c}{\sqrt{1 - v^2/c^2}} \qquad \text{(Doppler formula)} \tag{19.9}$$

$$z = \frac{\lambda_{beob} - \lambda}{\lambda} \text{ : Redshift}$$

t_0 Start of the light signal at G

t_i Arrival time of the light signal at G_i

t'_i Proper time of G_i belonging to t_i.

Formula (19.7) and (19.9) from [20].

Numerical example: The microwave background is a radiation that had its maximum at a wavelength of $\lambda \approx 10^{-6}$ m (3000°K) when it was created; today the maximum is observed at $\lambda_{beob} \approx 10^{-3}$ m (2.7°K). By what factor has the universe expanded radially since then?
(By the factor $1+z \approx 1000$ or $1+v/c \approx 2$ in the case of cosmological or special-relativistic simultaneity).

Mathematically, both interpretations are equivalent, because one may describe any movement relative to t or τ. In their physical premises, both interpretations are not equivalent:
a) If one says that the redshift of the galaxies proves the expansion of space, one postulates a new physical effect.
b) If one says that the redshift is a Doppler effect due to the explosion of a metagalaxy, one refers to known physical effects.

Where does one get the right to this new hypothesis? Should mathematical beauty suffice? There are further counter-arguments. Quintessence models, for example, fit fit well with thesis b) [vor20] and the smallness of the cosmological constant- "one of the most outstanding challenges in modern theoretical physics" can also be explained well with thesis b), see ch. 24.13. So, what reasons are there not to at least tolerate the Lorentz interpretation for almost 100 years?

19.8.5. The expansion of the universe as a function of t and τ

For the expanding spaces $k = 0, -1$ the following physically contradictory facts exist: At the time $\tau = 0$ is $R(\tau) = 0$ and from this follows $r_{Welt} = 0$. For $\tau = \varepsilon$, where ε is an arbitrarily small positive number, $R(\tau) > 0$ and thus $r_{Welt} = \rho R(\tau)$ is arbitrarily large, because ρ can be arbitrarily large, in particular $r_{Welt} >> c\tau$. This is physically contradictory, because for a point-like initial state $r_{Welt} = 0$, the world remains finite for finite times, $r_{Welt} < c\tau$, since matter propagates with $v < c$.

The EI explanation: Nothing moves here, the space between the galaxies expands. If a finite piece of space of the size ε moves between an infinite number of particles, we have an infinite space from now to afterwards. Logical, but do we have to believe that? If all particles move away from a center, something like an explosion has taken place, this alternative has to be refuted and if that succeeded, one would first of all have an unsolved problem and without further reasons one does not have the right to postulate something like the expansion of space. The significance of the EI may lie in something else: see end of ch. 22.3.

The LI explanation for the special-relativistic limiting case: Because of

$$t = \frac{\tau}{\sqrt{1 - v^2 / c^2}}$$

at any time point $\tau = \varepsilon$ with ε arbitrarily small, and for each particle with $v \approx c$ the value of t is arbitrarily large. Now cosmological simultaneity is assumed i.e. everything with the same τ exists simultaneously. This means that the world is infinitely extended from now to later. The 'trick': one does not ask for the future $t > t_0$ and the present ($t = t_0$) but for the same proper time ($\tau = \tau_0$). For LI, it follows that the spaces k = 0, -1 are mathematical spaces with physical meaning, but not space and time in which we live.

For spaces with k = +1 something else happens, even for $r_{Welt} = r_{max}$ the entire universe is in a black hole, s. ch. 22.3.

"To be a blameless member of a flock of sheep, one must first of all be a sheep."
EINSTEIN 1935

20. The Lorentz Interpretation (LI) of general relativity

20.1. Approach to an exact definition

The Lorentz transformations (LT) can be derived from the exact definition of the LI of special relativity in ch. 6. There it was also shown that with the assumptions

(20.1)
$$\Delta x' = \gamma \Delta x$$
$$\Delta y' = \Delta y$$
$$\Delta z' = \Delta z$$
$$\Delta t' = \Delta t \,/\, \gamma$$

the derivation of the LT is also possible and in chapter 4.5 that with

(20.2)
$$m(v) = \gamma m(0)$$

the conservation of energy and momentum can also be introduced. With both together, the core of special relativity is available. The advantage of this approach is that the assumptions (20.1) and (20.2) are experimentally confirmed, which is only conditionally true for a general principle.

In accordance with this approach, the LI of general relativity is to be extended in relation to EI by the assumptions (20.3) - (20.6). How can these assumptions be justified?

(a) They can be derived from a general principle, as will be indicated below.

(b) They are experimentally confirmed or open to it, see below and ch. 20.2.

(c) They have a theoretical gain, analogous to the inertial system requirement, ch. 10.8.6. For example, they have a law of conservation of energy for the special case of freely falling objects, which is not part of classical general relativity, see ch. 21.3.

The approach for an exact definition of the LI of general relativity is: As in classical general relativity apply:

Postulate 1: Principle of equivalence - in local inertial systems the formulas of special relativity apply

Postulate 2: Covariance principle - The laws in gravitational fields can be represented as vector and tensor equations in 4-dimensional pseudo-Riemannian spaces of index 1 which are valid for any coordinate system, or similar constructions (pseudotensors).

LI: The covariant laws thus derived describe what is measurable from and in gravitational fields.

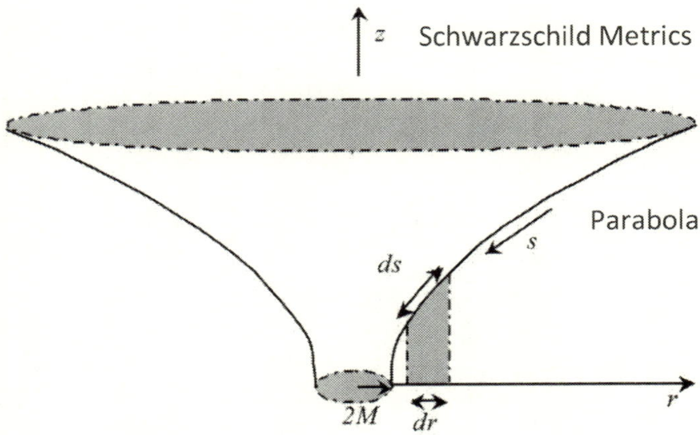

Fig. 20.1 Radial scale changes, assumption (20.3). They correspond to the representation of WEYL, Fig. 16.8 and 16.9.

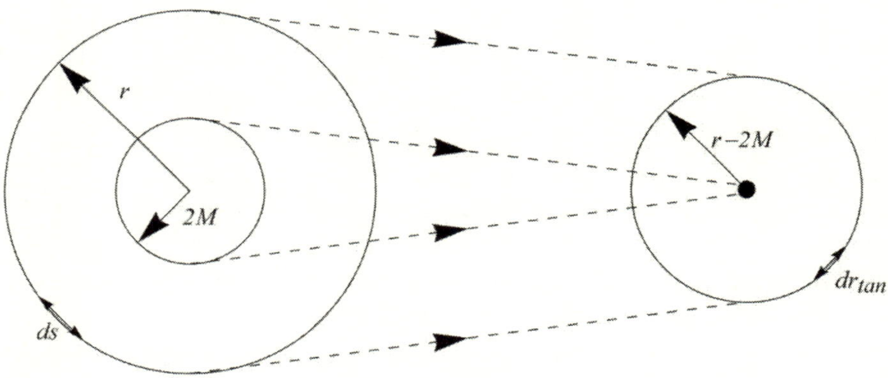

Fig. 20.2 Tangential scale changes, assumption (20.4).

The tangential scale changes occur in addition to Fig. 16.8 and 16.9. The interior of the Schwarzschild sphere becomes point-like, black holes become a theoretical limiting case, because for spherical objects:

$$\frac{1}{2\pi}\int dr_{\tan} = r - 2M$$

(Furthermore, for EI the general relativistic vector and tensor equations are laws of nature that directly express what the real physical properties of physical objects are. The respective Riemannian spaces are the real, 4-dimensional spacetime in which we live. The following postulate 3 (and 3') is not needed by the EI.)

The following postulate 3 (and 3') is needed by the LI to describe the mapping of the respective Riemannian space onto space and time and is the basis of a physical extension of the EI.

Postulate 3. For each general relativistic metric (Kerr metric, SM, RWM, etc) there is a set of mapping rules that corrects for the influence of the gravitational field on the gauges.

Example: The time measured in the gravitational field must be converted into the real time, the time outside the gravitational field. For the SM this is equation (20.5).

In the following we consider the postulate 3 at the example of the SM. Thereby we assume additionally that the center of gravity rests in an IS. (Whether this IS is the excellent IS of the special relativity theory or moves relative to it, can be left open, because the measured values do not differ on account of the RP).

Postulate 3'. For the Schwarzschild metric (SM) in the (r, t, θ, φ) coordinate system, formula (16.5), the mapping rules (20.3) - (20.6) apply.

(20.3) $$dr_{rad} = ds(1 - 2M/r)^{1/2}$$

(20.4) $$dr_{\tan} = ds(1 - 2M/r)$$

(20.5) $$dt = d\tau(1 - 2M/r)^{-1/2}$$

(20.6) $$dm_r = dm(1 - 2M/r)^{1/2}$$

$2M$: Schwarzschild radius.

On the right are the measured quantities, on the left the real quantities calculated with them. These assumptions are not arbitrary but can be read from the general rela tivistic formulae and can be verified experimentally. (20.3) - (20.6) show how the gravitational field of the SM changes radially and tangentially placed (measuring) rods, clocks and masses quantitatively, see Figs. 20.1 and 20.2. They apply to the SM and RWM, for other metrics (Kerr metric ...) they are not examined but should be similar for this purpose. Without these assumptions, space and time are to be regarded as 4-dimensional and curved, with them as 3-dimensional and 1-dimensional and as Euclidean. The loss of philosophical meaning causes a gain in physics, whereby (20.6) has special meaning, s. ch. 21.

Example: The LI distinguishes between measured and real values of physical quantities. Is dr a quantity that describes measured values or real values? What is true for r? dr_{rad}, see (20.3), and the integral

$$\frac{1}{2\pi}\int dr_{\tan} = r - 2M$$

are real physical quantities, but not $r = U/2\pi$, U: measured circumference along a circle of latitude. What lies on the funnel of the SM is for LI and EI measurable. An important application is shown by the formulae (22.14) and (22.15).

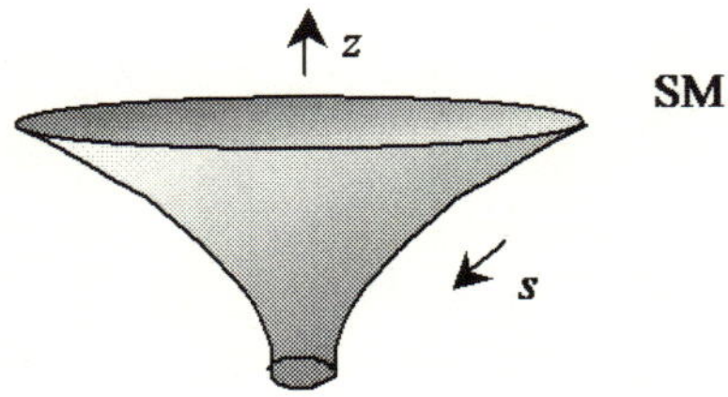

20.2. Plausibility and verifiability of the additional assumptions

Equations (20.1) - (20.2) and (20.3) - (20.6) refer to different areas of physics but have a similar structure. This is no coincidence, they follow a universal principle, as it also applies to QM and QF: Measuring object and measuring device together shape the measuring result. The influence of the measuring device can no longer be neglected, as in classical physics.

This deepening of the general relativity by the LI means no loss elsewhere: The general theory of relativity remains - with minor changes - because of postulates 1 and 2. This is necessary because the assumptions (20.3) - (20.6)) can only be justified with general relativity, since the metric must be known.

Assumption (20.5) is experimentally confirmed, [142], [143], the important relation (20.6) can be derived from general relativistic equations (for the special case of particles at rest in the gravitational field, s. ch. 21), (20.3) and (20.4) are in principle experimentally verifiable for objects to which the RWM is applicable, s. ch. 22.3.

Does the LI extend classical general relativity or restrict it? In the main, the LI does not change anything, because the formulae remain the same. The LI is a restriction, because it makes a choice where general relativity offers several alternatives. The RWM, for example, allows both the exploding dust star expanding to infinity and the expanding space of the universe. The LI opts for the former, see ch. 19. In the SM there are two fall times, the finite one for the proper time and the infinite one for the associated coordinate time. The LI is an extension, because classical general relativity declares the second fall time to be meaningless, for the LI both have physical meaning - with essential consequences, a new law of conservation of energy applies, s. ch. 21.3, and black holes become a theoretical limiting case, s. ch. 21.5.

All of the following chapters, especially Chapters 21 and 22, aim to prove that the modified view of black holes fits all the metrics of general relativity and is experimentally testable. Unintentionally, this solves "one of the most oustanding challenges in modern theoretical physics" [22] to the smallness of the cosmological constant since this problem exists only for classical general relativity, s. ch. 24.13. These are all additional reasons to consider LI as an extension of EI.

21. Black holes and conservation of energy of the SM

21.1. Conservation of energy for light

EI: For physicists, it is basic knowledge that light experiences a so-called gravitational redshift when it rises in the gravitational field because photons, considered as particles, have to 'fight' against the gravitational forces. The formula developed by POUND and REPKA [279] experimentally confirmed to 1%.

$$\frac{\Delta\lambda}{\lambda} = \frac{gh}{c^2} \tag{21.1}$$

g = acceleration due to gravity, h = height of fall.

In this sense, STEPHANI states [357]: "When light propagates in a gravitational field, not only the direction of the light rays changes, but also the frequency of the light." In RINDLER [299] "... as light climbs up a gravitational gradient, its frequency decreases ..." but also "... standard clocks fixed in a stationary gravitational field at low potential go slower than clock fixed at higher potential." Both together contradict each other, because one cannot prove the decreasing frequency of light with a clock that runs slower by the same factor. The effects would cancel each other out. Nevertheless, both statements mean something right.

Finally, PETKOV states [261]: "Therefore, in the gravitational redshift, it is the velocity and wavelength of a photon that changes whereas its frequency does not change." A contradiction to the other two quotations. If the convinced relativist PETKOV is right, the above contradiction is eliminated, only the clocks run slower, but it means that the equivalence principle cannot be literally true. For that, the proper times of free-falling clocks would have to show the real time. If they run slower, the speed of light measured in the local inertial frame is equal to c, but after conversion to the real speed it is not equal to c. This contradicts the principle of the constancy of the speed of light. The real time can be determined because the frequency of light is supposed to remain constant.

Conclusion: Within the EI, contradictory views on gravitational redshift are held with good reason. only the most important - the judgement on the measurement result for the redshift - is unanimous. Thus, the equivalence principle can remain, but only in a weakened form, as only valid for the measurement results. The question arises: Has someone not thought properly here and can these contradictions be remedied or is the EI incomplete on this point?

No problems arise for the LI: For them, (20.5) applies, i.e. clocks run slower in gravitational fields. Light waves are generated with a lower frequency from the beginning and retain this frequency when ascending or descending in the gravitational field, because the atom generating the light wave is a clock and must therefore oscillate more slowly. The frequency remains constant because the gravitational fields act like optically dense media. In them, the wavelength and speed of light changes, but not its frequency - because of $E = h\nu$ and the conservation of energy, this must be the case. Anyone who disputes this is once again reducing the amount of redshift and contradicting the experiment.

Quantitative: For the LI apply

(21.2) $$\nu = \nu_0(1 - 2M/r)^{1/2}$$

and the formulae (16.31) and (16.32) for the real speed of light in radial and tangential directions.

Example: The D_1 line of sodium has the wavelength $\lambda_{D1} = 589{,}6$ nm. The real frequency ν_{real} of the D_1 line emitted from the solar surface has the value $\nu_{real} = \nu_{D1}(1 - 2M_{Sonne}/R_{Sonne})^{1/2} = \nu_{D1} \times 0.9999979$. ν_{real} is numerically the value measured at infinity, it has been experimentally confirmed to be 5%. [21.3]

The equivalence principle changes its meaning for the LI, it applies to *measurement results* just as the relativistic formulae predict measurement results. Example: The *measured* frequency of the sodium D_1 line has its theoretical value, regardless of where it is generated in the gravitational field, if it is measured with clocks that are at rest at the point of origin. The speed of light *measured* there has the value c, and the same applies to theoretical values of the energy e.g. of decay processes, especially for the rest mass. Everything as required by the equivalence principle. But the *real* values are something different than the *measured ones*. Both are needed in chapter 21.2.

"What EINSTEIN says is not so stupid."
W. PAULI as a student

21.2. Fall laws

For the following fall laws, the EI denies the validity for $r = f(t)$, formula (21.9), in contrast to $r = g(\tau)$, formula (21.10). But the relativistic laws of motion of the SM are experimentally well confirmed and thus it is also the special case of free fall. For LI, therefore, $r = f(t)$ is not objectionable and, as it turns out, has special, hitherto unnoticed physical properties. The considerations of this and the following chapter deepen chapters 16.8 and 16.9 on the SM.

For the SM, the law of conservation of energy results for radial free fall

(21.3) $$\left(1 - \frac{2M}{r}\right)\frac{dct}{dc\tau} = E_G$$ [60], [327.1], [234.1]

(21.4) $$1 = \left(1 - \frac{2M}{r}\right)\left(\frac{dct}{dc\tau}\right)^2 - \left(1 - \frac{2M}{r}\right)^{-1}\left(\frac{dr}{dc\tau}\right)^2$$ follows directly from the SM, [357]

From (21.4) together with (21.3) one gets:

(21.5) $$\left(\frac{dr}{dc\tau}\right)^2 = E_G^2 - \left(1 - \frac{2M}{r}\right)$$

From (21.3) - (21.5) we get $\tau = g(r)$ and $t = f(r)$, Fig. 21.1, formulae (21.9) and (21.10) as well as

(21.6) $$E_G = \left(1 - \frac{2M}{r_{v=o}}\right)^{\frac{1}{2}}$$ with $v = 0$ (reversal point)

(21.7) $$E_G = 1$$ with $v_\infty = 0$

(21.8) $$E_G = \frac{1}{\left(1 - \frac{v_\infty{}^2}{c^2}\right)^{\frac{1}{2}}}$$ with $v_\infty > 0$

The energy E_G in formulae (21.6) - (21.8) is the energy per unit mass times c^2 and for a given particle to be multiplied by mc^2. m: Rest mass of the freely falling particle.

According to D'INVERNO [60] is $t = f(r)$:

(21.9) $$t - t_0 = -\frac{2}{3\sqrt{2M_s}}(r^{3/2} - r_0{}^{3/2} + 6M_s r^{1/2} - 6M_s r_0{}^{1/2}) + 2M_s \ln \frac{\left[r^{1/2} + (2M_s)^{1/2}\right]\left[r_0{}^{1/2} - (2M_s)^{1/2}\right]}{\left[r_0{}^{1/2} + (2M_s)^{1/2}\right]\left[r^{1/2} - (2M_s)^{1/2}\right]}$$

And $\tau = g(r)$ for proper time τ

(21.10) $$\tau - \tau_0 = \frac{2}{3\sqrt{2M_s}}(r_0{}^{3/2} - r^{3/2})$$

i.e. τ has a finite value at the point $r = 2M_s$. Although it takes an infinitely long time t to reach $2M_s$, as formula (21.9) shows, the proper time τ required for this remains finit, as formula (21.10) shows.

r_0: starting point of the free fall at time t_0.

1) From $\tau = g(r)$, its inverse function $r = g^{-1}(\tau)$ and their derivatives, one sees: The kinetic energy increases steadily and reaches its maximum at $r = 0$, as one would expect. Relative to a local inertial system at rest at r, the following applies

(21.11) $$v = \frac{ds}{d\tau} = c\sqrt{\frac{2M}{r}}$$ as follows from (21.3), (21.5), assumptions (20.3) + (20.5). (Note the difference between the proper time of the free-falling particle and the proper time in the local inertial frame, both are denoted by τ.) I.e. the total energy in the local inertial frame at the point $r = 2M_s$ is infinite, because the particle has reached the speed of light.

(21.12) $$v = \frac{dr}{dt} \to 0 \quad \text{for} \quad t \to \infty,$$ $2M_s$ will not be exceeded.

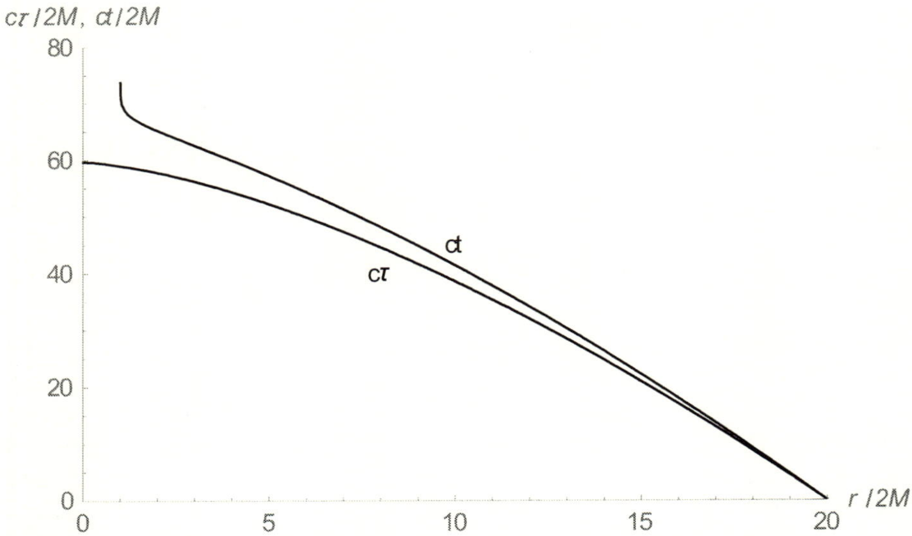

Fig. 21.1 The functions $\tau = g(r)$ and $t = f(r)$, more precisely $c\tau = g(r)$ and $ct = f(r)$

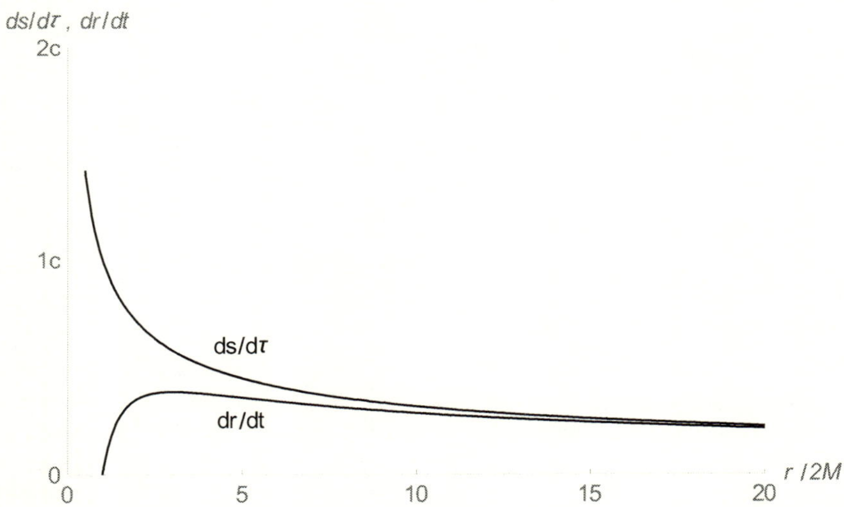

Fig. 21.2 dr/dt , the derivative of the inverse function of $t = f(r)$ and $ds/d\tau$, formula (21.11), as a function of r.

The velocity dr/dt is the derivative of the inverse function of $t = f(r)$ and is the fall velocity measured in the r,t reference frame, $ds/d\tau$ is formula (21.11) and is the fall velocity measured in the local inertial frame at rest at r. For $r = 2M_s$ is $dr/dt = 0$ and $ds/d\tau = c$.

2) For $r = f^{-1}(t)$, inverse function of (21.9) holds:

(21.13) $$b = \frac{d^2 r}{dt^2}$$ Initially the acceleration is $b > 0$, then $b < 0$.

(21.14) $$\lim_{r \to r_{SM}} \left(\frac{d\tau}{dt} = E_G^{-1} \left(1 - \frac{2M}{r} \right) \right) = 0$$, a free-falling clock runs slower and slower and stops when $r = 2M_s$. (from (21.3)).

The two curves $r = g^{-1}(\tau)$ and $r = f^{-1}(t)$ differ in one essential point. For $r = g^{-1}(\tau)$, a free-falling object reaches the center $r = 0$ after finite proper time and is then in a black hole, according to $r = f^{-1}(t)$, the object never reaches $r = 2M_s$, it does not fall into a black hole because it does not reach a value $r < 2M_s$. The EI decides for $r = g^{-1}(\tau)$, the LI for $r = f^{-1}(t)$ and $r = g^{-1}(\tau)$.

The arguments of the EI: In ch. 16.9, quotations from STEPHANI [357] are reproduced. Further quotations: "It is thus clear that the Schwarzschild time coordinate t is unsuitable for the description of this motion". [60] "It follows that there is a great difference between the description of the motion in terms of proper time τ of a clock on the falling particle (r goes all the way from $r = R$ down to $r = 0$ ...) and a description in terms of the Schwarzschild-coordinate time t appropriate for the faraway observer (... r is only brought asymptotically down to $r \sim 2M_s$). Thus the second description of the motion leaves out, and has no alternative to leave out, the whole range of r values from r=$2M_s$ down to zero: perfectly good physics, and physics that the falling particle is going to see and explore, but physics that a faraway observer never will see and never can see." [230]

The LI **arguments:**

(1) On MISNER, THORNE, WHEELER [230]: You can believe it, but you don't know it, whether you lose good physics if you favor $r = g^{-1}(\tau)$. What 'good physics' does the inside of a point have? Until the experimental proof of black holes (BH), one may disagree.

(2) The laws of falling are a special case of the equations of motion, which are experimentally well confirmed. t is not only a coordinate time, it indicates the time that elapses outside the gravitational field.

(3) The following easily imaginable experiment could disprove the claim that the free-falling particle reaches the interior of the black hole (BH): Position an observer at certain distances at all locations r up to $2M_S$. If $r = f^{-1}(t)$ applies, there are places after any length of time t that the free-falling particle has not yet passed. Its proper time always remains smaller than the proper time $\tau_{2M} = g^{-1}(2M_s)$ belonging to $r = 2M_S$. This is unproblematic within LI - a free-falling clock always runs slower and would stop if $2M_s$ were reached.

Above all, (3) is a clear, experimentally verifiable difference between EI and LI.

21.3. Law of conservation of energy for freely falling particles

In Chapter 3.3, the following contradiction was described for a stone falling freely in the gravitational field of a star: On the one hand, the total energy of the stone increases (at least at the beginning, see Fig. 21.2), on the other hand, the total energy of the star does not change and is eliminated as an energy supplier [299].

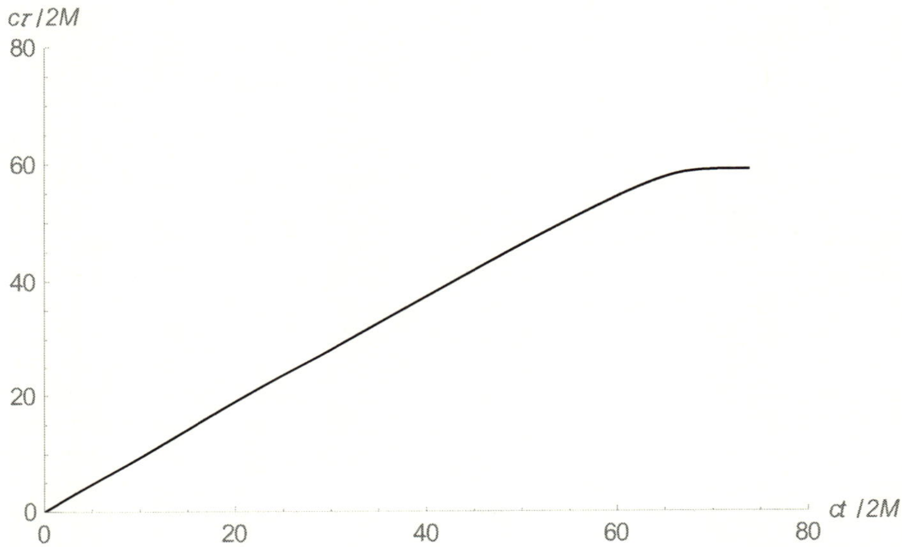

Fig. 21.3 The function $\tau = h(t)$, more precisely $c\tau = h(ct)$

In the following, the equations $r = f^{-1}(t)$ and $r = g^{-1}(\tau)$, formulae (21.9) and (21.10), of the SM are assumed to be valid. In principle, they are experimentally verifiable and to restrict their validity means to deny the validity of the general theory of relativity. In particular, both formulae yield a relationship $\tau = h(t)$ (see Fig. 21.3) between the proper time of the stone and our time t outside the star, which would be testable, as ch. 20.2 has shown. It follows from this: As long as t is finite, the stone has not reached the black hole and therefore it never reaches it - either one agrees here or one denies a relativistic formula and its prediction.

The following applies to the EI: The total energy of the star increases with the fall time because the kinetic energy increases with the fall time (see Fig. 21.2) and the rest mass remains constant. So the total energy of the star must decrease so that $E_{Stone} + E_{Star} = \text{const}$ is fulfilled. This requires the conservation of energy. How the energy transport from the star to the stone takes place would not need to be known. The contradiction for the EI arises because, as RINDLER states. [299, p. 265], the star does not transfer any energy to the stone and, moreover, in the law of falling bodies (21.3) the total energy E_G is a constant.

The LI solves the problem with claim (20.6), the rest mass of the stone decreases during free fall. Thus the kinetic energy can increase without changing the total energy of the free falling stone, E_G in equation (21.3).

This leads to two further, but solvable, problems for the LI:

1) If the stone exceeds the turning point of $r = f^{-1}(t)$, see Fig. 21.2, the velocity and therefore the kinetic energy decreases, as does the rest mass, where does the now surplus energy remain?

2) According to general opinion, light experiences a red shift during ascent and a blue shift during descent. Light has no rest mass that could change. Where does the energy go and where does it come from?

Point 2.) was solved in chapter 21.1. Point 1.) requires an additional hypothesis. Quantum mechanics (QM) and quantum field theory (QF) could help. The wave and particle picture of QM and QF allow a qualitative interpretation of the motion sequence during free fall. Initially, the particle picture applies you have physical bodies whose kinetic energy increases as the velocity of the bodies increases, then contrary to expectation the velocity decreases and approaches zero. The same is true for waves in increasingly dense medium, there they can keep their energy and frequency, but at the same time their propagation speed can decrease. Consider assumption (20.6). What becomes of a particle when it reaches the Schwarzschild radius? The rest mass is zero, so it is light? No, that contradicts the conservation of baryon number. What remains? The particle approaches the Schwarzschild radius, but does not reach it, the rest mass remains finite and the baryon number does not change. Decreasing rest mass could mean an increase in wave properties, which in turn could mean an increase in wave energy. In any case, this fits with the decreasing speed, because the denser the optical medium, the lower the speed of light or other waves, without changing their frequency and thus their energy. Altogether an effect of quantum gravity? Yes, if the hypothesis is correct. This is in principle experimentally testable. No, not acceptable? The - perhaps only - alternative is then to call into question the relativistic law of falling bodies and thus general relativity, and the author rejects this until the opposite is experimentally proven. Thus, one has the following hypothetical theorem of conservation of energy:

$$E_G = E_{Ruhe} + E_{kin} + E_{Welle}$$

(21.15)
$$= m_0 c^2 \left(1 - \frac{2M}{r}\right)^{\frac{1}{2}} + m_0 c^2 \left(1 - \frac{2M}{r}\right)^{\frac{1}{2}} \left[\frac{1}{\sqrt{1 - v^2 / c^2}} - 1\right] + E_{Welle}$$

with $v = \frac{dr}{dt} = \frac{dr / d\tau}{dt / d\tau}$, quotient of the formulae (21.3) and (21.5). Fig. 21.4 and 21.5 show the total energy E_G graphically.

These formulae fit the considerations about the Newtonian gravitational potential, ch. 24.8, formulae (24.7) - (24.9) and Einstein's law of conservation of energy, ch. 24.9.

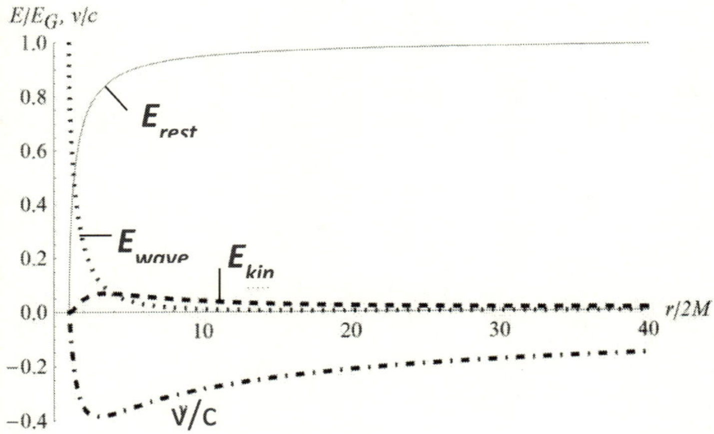

Fig. 21.4 The proportions of the total energy E_G and the velocity $v = dr/dt$ for radial free fall

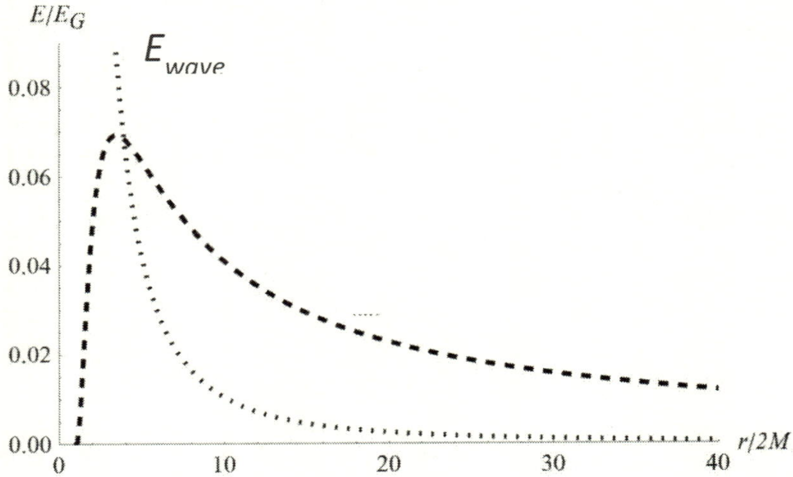

Fig. 21.5 The proportions of the total energy E_G for radial free fall - section of Fig. 21.4

The EI also has a law of conservation of energy with the formulae (21.3), (21.5), but one does not understand what it means, one does not understand why $E_G = \text{konst}$ should apply although $v = dr/dt$ is growing, one does not understand the boundary transition $v = dr/dt \to 0$ for $r \to 2M$, again although $E_G = \text{konst}$. In addition, there is the contradiction that in the local inertial frame at the Schwarzschild radius, the particle energy for the EI becomes infinitely large, because an observer at rest there measures the velocity $ds/d\tau = c$ for a free-falling particle ($E_G = 1$).

The gravitational force can be understood for LI with the above hypotheses as the effort of gravitational fields to reduce the rest mass of a particle. No gravitational forces then act on particles with zero rest mass. For such particles, gravitational fields are optically dense media that influence the speed and trajectory of the particles. No new theory is necessary for this, it already exists, see LANDAU-LIFSCHITZ [190] and STEPHANI [357], see ch. 24.6. A logical consequence: Visible matter does not react to gravitational forces already at the beginning of the big bang, but only when particles got rest mass (Higgs mechanism). Without rest mass, there were no gravitational forces at the time of the Big Bang, and an inflationary phase is almost a given. Here, too, one can see the great importance of the CERN experiment to prove the Higgs particle [21.4], [22]. See also ch. 24.8.

The author does not want to overestimate these hypotheses. He is impressed by the fact that the equivalence principle and the formulae of classical general relativity derived from it by EINSTEIN could be valid from the classical particle to a wave model of quantum gravity ... and once again the assumption goes further that this also applies to dark matter.

"To get to the source, you have to swim against the current".
Stanislaw J. Lec, writer, Poland

21.4. Energy of a particle at rest in the gravitational field

Contradictory statements on total energy

The elementary question of this chapter is: What is the total energy E_G of a particle at rest in the r,t reference frame of the Schwarzschild metric (SM)?

From the formulae for the radial free fall (21.6) we get:

(21.20) $$E_G = mc^2\left(1 - \frac{2GM}{c^2 r}\right)^{1/2}$$

This is at least qualitatively correct, because energy must be expended to remove the particle from the gravitational field. Only then does the particle have the total energy mc^2 and therefore the total energy must be smaller in the gravitational field.

On the other hand, the equivalence principle applies. A particle at rest in the local inertial frame (i.e. the free-falling particle) has a total energy equal to its rest mass:

(21.21) $$E_G = mc^2$$

Formula (21.20) and (21.21) contradict each other, as will be shown.

First of all, they refer to different reference frames (to the r,t reference frame of the Schwarzschild metric (SM) and to the local inertial frame, respectively) and the latter is even accelerated. Relative to different reference systems, a particle may have different energies, that is not a contradiction. But: At the time $t = 0$, the starting time of the free fall, the free-falling particle is also a particle at rest in the r,t reference frame, because, although the acceleration is $b \neq 0$, its velocity is $v = 0$. The free-falling particle is at rest for $t = 0$, but formula (21.21) already applies because of the equivalence principles. Since the special theory of relativity is applicable and there the total energy of a free particle depends only on v and not on b, the free-falling particle at the time $t = 0$ and the particle always at rest at the same place thus have the same total energy (21.21). This statement is initially only valid relative to the local (accelerated) inertial frame.

The following consideration - it is similar to section: Change of measuring instruments in the gravitational field - shows that the above statement is also valid for particles at rest relative to the r,t reference frame. At the time $t = 0$, bring to the particle its antiparticle and measure the annihilation frequency. It does not matter whether one measures the annihilation frequency and thus the total energy of the particle with clocks at rest in the r,t reference frame or in the local inertial frame. For $t = 0$, the result agrees because the clocks only differ in their acceleration state, their relative velocity is zero and the special theory of relativity is applicable. In an inertial system, a clock only changes its frequency when $v > 0$ applies and not already when $b \neq 0$. This is also confirmed by: E. REBHAN [293.1], page 245: "Pairs of similar measuring instruments at rest relative to each other, one of which falls freely and the second is fixed in the gravitational field, display the same lengths and times." Applied here to clocks. Formula (21.21) thus also applies relative to the r,t reference frame of the Schwarzschild metric. It follows that: **Formulas (21.20) and (21.21) are a contradiction**.

What does classical general relativity (EI) say about this? For it, both formulas are experimentally verifiable. Formula (21.20) applies if one measures the annihilation frequency with clocks that are synchronous to clocks outside the gravitational field. Formula (21.21) applies when using standard clocks, i.e. normal clocks transported into the gravitational field. Standard clocks correctly reproduce the passage of time in space-time curved by masses. Therefore, formula (21.21) is the correct energy. Furthermore, formula (21.21) cannot be derived from the equivalence principle, but formula (21.20) can - another argument in favour of (21.21).

The EI cannot explain why formula (21.20) represents the conservation of energy in free fall. Furthermore, formula (21.21) does not explain the Newtonian limiting case.

What does the LI say about the contradiction? LI solves it with the following arguments. Formula (21.20) is correct because it represents the conservation of energy in free fall and because it satisfies the Newtonian limiting case. The latter can be shown more precisely. To do this, develop (21.20) into a series:

(21.23) $$E_G = mc^2 - \frac{GmM}{r} \pm ...$$

The 2nd member is the negative Newtonian gravitational energy. Approximately, formula (21.20) is the rest mass minus the Newtonian gravitational energy. Formula (21.20) thus fulfils the Newtonian limiting case, formula (21.21) does not.

The LI still has a clear argument why formula (21.21) is wrong. Clocks run slower in gravitational fields, formula (20.5). Therefore, the measurement of energy with standard clocks needs to be corrected, the measurement error due to slower running clocks has to be taken out, see below. The only remaining problem is the equivalence principle, which now leads to a wrong result with formula (21.21). The solution for the LI is to modify the equivalence principle.

Modified equivalence principle

The equivalence principle is so proven and convincing that it cannot be called wrong Therefore, formula (21.21) must also be correct.

Let us slightly reformulate the equivalence principle: For *measurements* in the gravitational field, the *measurement results* in the local inertial frame can be predicted with the help of special relativity.

For our application case, the following then applies: The *measurement* of the total energy E_G with measuring instruments at rest in the gravitational field yields the value that formula (21.21) predicts. This is no longer a contradiction to (21.20), since the LI assumes that measuring instruments change in the gravitational field.

Change of measuring instruments in the gravitational field

Let us consider the change in the measuring instruments when measuring E_G. We choose a mentally simple measuring procedure. Bring the antiparticle to the particle at rest and measure the annihilation frequency of the two photons produced. Then applies:

(21.24) $$\begin{aligned} E_{G,measured} &= mc^2 \\ &= h\nu_{\tau,measured} \end{aligned}$$

$\nu_{\tau,measured}$: Annihilation frequency measured with a clock at rest in the gravitational field.

τ: proper time of the clock at rest in the gravitational field (τ-clock)

On the other hand, it follows from the SM (and this is also confirmed experimentally):

(21.25) $$\tau = t\left(1 - \frac{2GM}{c^2 r}\right)^{1/2}$$

In words. If time passes for the t-clock at rest outside the gravitational field $t = 1\text{s}$, time passes for the τ -clock by the factor

(21.26) $$\left(1-\frac{2GM}{c^2r}\right)^{1/2}$$

smaller time τ. I.e. for the measured frequencies applies:

(21.27) $$\nu_{\tau,\text{measured}} = \nu_t\left(1-\frac{2GM}{c^2r}\right)^{-1/2}$$

(21.27) is inserted into (21.24):

(21.28) $$\begin{aligned} E_{G,\text{measured}} &= h\nu_t\left(1-\frac{2GM}{c^2r}\right)^{-1/2} \\ &= E_G\left(1-\frac{2GM}{c^2r}\right)^{-1/2} \end{aligned}$$

or

(21.29) $$E_G = mc^2\left(1-\frac{2GM}{c^2r}\right)^{1/2}$$

thus formula (21.20).

In words: If one takes into account the change of the measuring instruments in the gravitational field, in this case the clock running slower in the gravitational field, the total energy of a particle at rest in the gravitational field can also be derived from the equivalence principle.

Formula (21.20) can be formally derived very simply by writing down the scalar product of the covariant, four-dimensional energy-momentum vector for a particle at rest. Since the particle is at rest in the gravitational field, of the four-dimensional energy-momentum vector the three components $p_1\ p_2\ p_3$ are zero and the square of the four-dimensional energy-momentum vector

(21.30) $$p^2 = p_\mu p_\nu g^{\mu\nu}$$

provides on the one hand $p^2 = m^2c^2$ and on the other hand

(21.31) $$p^2 = \left(1-\frac{2GM}{c^2r}\right)^{-1}\frac{E_G^2}{c^2}$$

by taking into account that the four-dimensional energy-momentum vector is a covector. Formula (21.20) also follows from this.

Discussions

The above comments on the conflicting formulae (21.20) and (21.21) have great significance for a decision for or against the LI, so the following are some arguments from various discussions, [vor29], [vor30].

1) A well-known gravitational physicist from Cologne writes: "There is no problem with the proton in the gravitational field according to the general theory of relativity." "The proton at rest therefore clearly has a mass (energy) that is smaller than the rest mass." Formula (21.20) is correct, no comment was made on the contradiction with formula (21.21).

2.) V. PERLICK, Bremen. "Your formulae (2) [(21.20)] and (3) [(21.21)] both give the energy as measured by an observer moving on t lines. But the clocks used are different: equation (2) [(21.20)] is correct if the clock, shows coordinate time t; equation (3) [(21.21)] is correct if the clock shows proper time \tau (as one assumes when applying the equivalence principle)." "Both are correct. It is just that in one case it is the energy with respect to one clock, in the other case with respect to the other clock. Both clocks are well-defined and well-understood, and the conversion formula, as you correctly write, is also well known. There can therefore be no question of an 'incompleteness' of the theory." Both formulae are therefore correct. Elsewhere, formula (3) [(21.21)] is preferred: "But, of course, the GRT distinguishes clocks, namely the so-called 'standard clocks', which display proper time."

3) ANNETTE ZIPPELIUS, Göttingen and H.-R. TREBIN, Stuttgart, Ombudspersons of the DPG (2013), write: "A very simple argumentation based on the classical limiting case shows that there is no contradiction between the formulae 2 [(21.20)] and 3 [(21.21)] on particle energy." "So the acceleration b plays the decisive role. That is the flaw in your argument ..." Regarding the quote from E. REBHAN: "With this quotation taken out of context ... we do not [regard] the arguments already presented [as] refuted ..." The author's conclusion: Since the quotation is *not* taken out of context, the contradictory formulae (21.20) and (21.21) are valid in one and the same reference frame, the r,t reference frame of the Schwarzschild metric.

21.5. Conservation of Energy in the Inner Schwarzschild Metric.

Conservation of energy in the inner SM is another use case of the formulae (20.3) - (20.6). It shows that the new formulas of the LI bring about a deeper understanding of those of the EI and do not replace them. For the inner SM, the total energy of a star can be calculated in terms of its gravitational mass. Equation [230] serves this purpose

(21.35) $$m_{grav} = \int_0^r 4\pi r^2 \rho dr$$

It follows from relativistic theory and is therefore adopted by the LI. In the EI, this formula, although correct, is not to be understood. The EI would have to require

(21.36) $$m_{EI} = \int \ 4\pi r^2 \rho_{\text{measured}} ds$$

with $$ds = \frac{dr}{\left(1 - \frac{2M(r)}{r}\right)^{\frac{1}{2}}}$$

because then the following applies: The integral over measured volume elements $4\pi r^2 ds$ times the density measured therein ρ_{measured} is the mass available for the EI. But then $m_{EI} \neq m_{grav}$.

For the LI, the formula (21.36) is comprehensible.

(*gemessen* the same as *measured*)

(21.37)
$$m_{grav} = \int_0^r 4\pi r^2 \frac{dm_{gemessen}}{dV_{gemessen}} dr$$
$$= \int_0^r 4\pi r^2 \frac{dm_{gemessen}}{4\pi r^2 ds} dr$$
$$= \int_0^r dm_{gemessen} \left(1 - \frac{2M(r)}{r}\right)^{\frac{1}{2}}$$
$$= \int_0^r dm_{real}$$
$$= m_{real}$$

Nothing more has been done than to convert measured quantities into real quantities with the help of the assumptions (20.4) - (20.6) - and the actual achievement is the formula (21.35) supplied by general relativity.

"He who digs a pit for others ..."
Proverb

21.6. Black holes of the SM and their observability

Black holes (BH) are a prediction of the EI from properties of the Schwarzschild metric (SM) and are also predicted by the EI in other places, such as in the inner Schwarzschild metric (inner SM) and in the Tolman-Oppenheimer-Volkhoff equation (TOV). For the LI, BH are only a theoretical limiting case and following this chapter, it will be shown with what modifications by the LI this is equally true for the inner SM and the TOV. The differences between the two interpretations can in principle be verified by astronomical observation, e.g. by a doubtless proof of the BH taking into account the objections of the LI.

The SM describes the external gravitational field of a centrally symmetric star and is given by (16.5):

(21.38)
$$ds^2 = \frac{1}{A} dr^2 + r^2 (d\theta^2 + \sin^2 \theta d\varphi^2) - Ac^2 dt^2$$

with $A = 1 - \frac{2GM}{c^2 r}$ resp. $A = 1 - \frac{2M_s}{r}$ mit $2M_s = \frac{2GM}{c^2}$

($2M_s$ is thus just another notation for the Schwarzschild radius.) For the EI, the prediction of BH's is a direct consequence of the SM. For $r = 2M_s$ this equation is singular, because $1/A$ becomes infinite. This singularity is considered to be recoverable. A freely falling particle exceeds the spherical surface $r = 2M_s$ and thus falls into the BH after finite proper time, see formula (21.10), the tearing forces occurring in the process (tidal forces) are finite and, for sufficiently massive and thus large BHs, without consequences even for a space traveller. The journey into an BH, however, is a journey of no return, within an BH even infinitely large counterforces are not sufficient for a reversal of motion. "An observer can cross the horizon in finite proper time and (hopefully) convince himself inside of the correctness of the theory presented here; but he can never report to the outside." [357]. Some of the difficulties with BH for the EI are: infinitely long fall time as well as infinitely high energy, see previous chapter.

One might think that the EI has not yet registered the latter, questionable fact of an infinitely high energy. This is not the case, where it is helpful, it is pointed out, as here with BH's: Because the energies for $r = 2M_S$ in the local inertial system and thus also the gravitational forces become infinite, the collapse to the BH is compelling and doubts about BH's are rejected, see above. ch. 21.2, ... and the theorem of conservation of energy is not observed.

For the LI, BH only exists as a theoretical limiting case. First, the LI must show that this theoretical result is compatible with the relativistic formulae, which has already been done. Thus, the above violation of the conservation of energy does not exist for the LI and also the equivalence principle EP is not violated. Because of assumption (20.6)

$$dm_r = dm\left(1 - 2M_s/r\right)^{1/2}$$

the real masses and rest masses are smaller than the measured ones. However, the EP applies to the masses *measured* in the local inertial frame, which can assume arbitrarily high values. The LI may therefore remain the same for the SM: The state $r = 2M_S$ cannot occur because of the infinite fall time in equation (21.9) for $r \to 2M_S$, BH's cannot arise.

At this point, some, perhaps observable differences for BH's in the EI and supermassive stellar objects (SMO) of the LI. As both Figs. 21.6 and 21.7 show, the (almost) point-like object of the LI is also observed as a sphere with $r \geq 2M_S$. A circumference measurement for BH's of the EI and the corresponding stellar objects of the LI has approximately the same results. But the almost BH's of the LI are degenerate objects and perhaps additionally hot and can radiate. The radiant energy can be produced by nuclear fusion as in normal stars, by accretion, but also from gravitational energy produced by contraction of the supermassive stellar object. As the considerations of RINDLER, ch. 24.8, almost the entire rest mass of each particle can be converted into energy during free fall, and because of (21.9) this can take an arbitrarily long time, so that such energy is continuously produced. If one has BH's of the EI, only accretion of matter remains.

Fig. 21.6 Black hole with accretion disk and angular momentum. With a co-rotating charge, it also has a magnetic field.

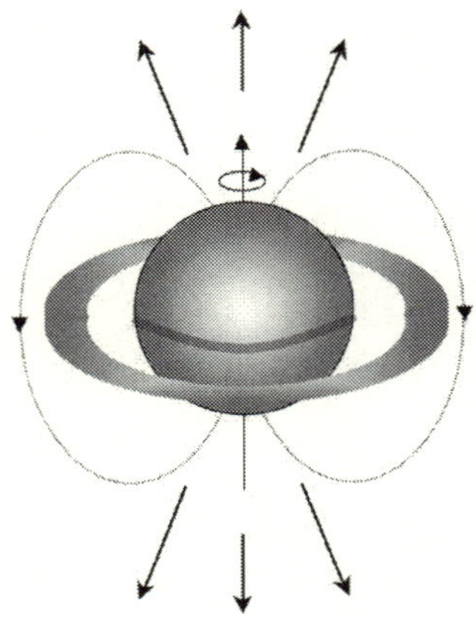

Fig. 21.7 Supermassive stellar object.

In addition to a black hole of Fig. 21.6, it can radiate, build up very strong magnetic fields, generate jets and have a turbulent boundary layer.

The differences for objects without angular momentum, without charge and without accretion are: EI: No radiation, no magnetic fields, LI: Radiation emission, albeit strongly redshifted, magnetic fields, flares, particles emission. The differences for objects with angular momentum (spin), no charge Q and no accretion: EI: No changes. LI: above effects are amplified. The differences for objects with angular momentum, with charge and no accretion: EI: Coulomb electric fields, magnetic dipole fields. LI: The electric and magnetic fields of the EI occur in addition.

If one also has accretion of matter, which is the normal case for galactic centers, an accretion disk can form and there are no more fundamental differences, BHs can also shine for the EI and emit particles. Nevertheless, there are great differences in the efficiency of these phenomena, because Q may be assumed to be practically zero (magnetic fields do not apply to the EI, although to the accretion disc), while the spin can have large values, which favors particle emission and flares in the sense of the LI (see magnetic reconnection [mag]). For further details see [230, §33]. A very important difference: SMO's can create a turbulent boundary layer from falling material, which can become very hot and radiate, see Fig. 21.8. For current observations, see also ch. 24.11.

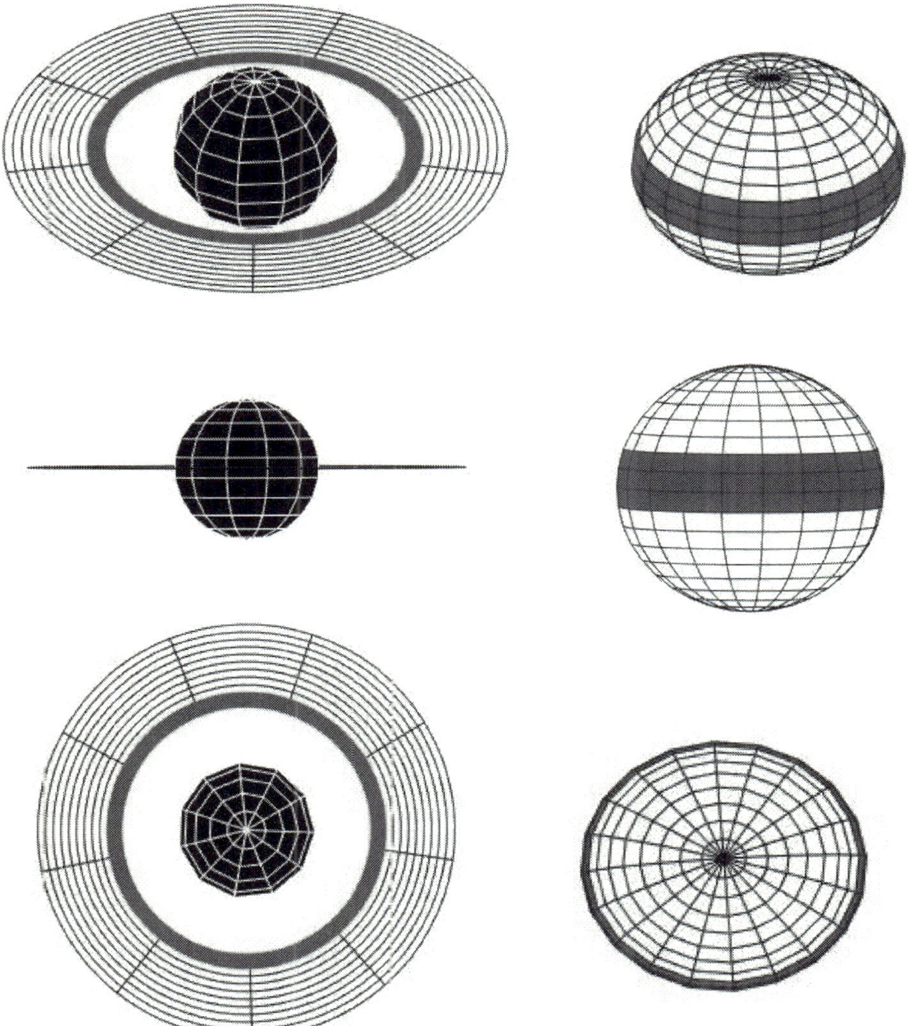

Fig. 21.8. The 6 figures show some observable differences between BH's and SMO's.

Figures 1 and 2 are the 3-dimensional sketch of an BH with accretion disk and an SMO with turbulent boundary layer. If one looks at the edge of the accretion disk, Figure 3, the surroundings of the BH appear dark, whereas the SMO is luminous because the turbulent boundary layer has a certain width. Figures 5 and 6 illustrate that a frontal view of the accretion disk does not reveal any differences between an BH and an SMO. In both cases, one sees a ring, on the one hand the luminous, inner edge of the accretion disk or on the other hand the luminous turbulent boundary layer.

Fig. 21.9 fits to both cases, Figures 5 and 6. The ALMA observation of Sgr A* fits better to Figure 4 than to Figure 3. [163.1], [66.1], [jb.12, Fig.1].

It is not known which of the mentioned differences between BHs and SMOs will become observable. The author considers the different properties of the innermost edge of the accretion disk in BHs and the turbulent boundary layer in SMOs to be particularly promising at present (2021).) in SMO's.

Unfortunately, the innermost edge of the accretion disk arbitrarily approaches the turbulent boundary layer for a sufficiently high angular momentum of the BH, so that one does not know which of the two one observes, M. THÜRAUF, [366.3], [jb.14, Fig. 3]. Both can assume arbitrarily high temperatures and make the galactic centers bright. To distinguish between them, one therefore needs an additional measurement of the spin of the BH. If one is lucky, the measured spin is low or differences are observable even without knowledge of the spin, see Fig. 21.8, Figures 3 and 4. In particular, Figure 4 fits well with the ALMA observation of Sgr A* and could be confirmed by the EHT observation of Sgr A*. [163.1], [66.1], [jb.12, Fig.1]. In other words: If the turbulent boundary layer of Sgr A* glows as shown in Fig. 21.8, Figure 4, then the photo of Sgr A* has a bright center, whereas in the case of a black hole one expects a dark center, as in Fig. 21.9. About the turbulent boundary layer and Fig. 21.9 in a moment.

According to the majority of physicists, BH's are experimentally confirmed. Each of us knows the picture, Fig. 21.9.

It shows the "first photo" of a black hole and is considered the ultimate proof that black holes exist. **Nobel laureate GENZEL** now publicly questions this. Here is a quote from Spiegelgespräch, 10. 10. 2020, [122.1], [jb.16, Fig. 1]

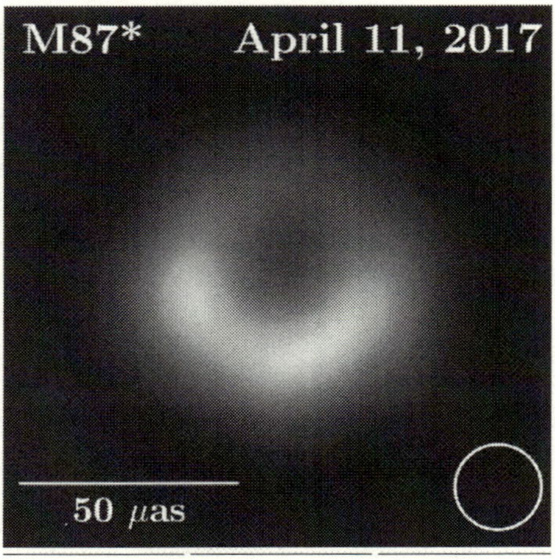

Fig. 21.9 Inner edge of the accretion disk of an BH, turbulent boundary layer of an SMO or outer wall of a jet - it is not yet decided beyond doubt whether this is the "first glimpse of an BH". Taken from [105.1].

Spiegel: "Wherever black holes are reported, you see this picture. And you're telling us now that we don't even know what it actually shows?"

Genzel: "That's how it is. It could be that we are seeing the shadow of the black hole as it is commonly depicted. But it could also be that it is the outer wall of a jet coming straight at us at the speed of light. . .."

An expert in the newspaper SZ and others in television reports express similar restraint. In other words, the existence of black holes has not yet been proven, even though this was presented as such at several press conferences of the EHT group. So far, SMO's of the LI are not being considered.

Further information on the turbulent boundary layer: If there are SMOs, one may assume - analogous to accreting neutron stars - that a turbulent boundary layer forms on the surface of the SMO by matter falling from the accretion disk or from the wider environment. The turbulent boundary layer radiates the energy released by accretion and makes the SMO's bright. In the case of BHs of the EI, it is instead the inner edge of the accretion disk that heats up and emits radiant energy. This effect is additionally possible for SMO's.

The turbulent boundary layer has several functions, it releases accreted material to the SMO and increases its mass, provided that the angular momentum is not too high. Matter with too high angular momentum is ejected from the boundary layer, becomes part of the jets and thus leaves the turbulent boundary layer. In accordance with GENZEL, this could form the outer wall of a jet. The angular momentum of the jets should be higher than that of the SMO and from its measurement an upper limit for the spin of the BH could become determinable.

Without excluding BH's, one can see the scope for SMO's. Updated observational results can be found on the author's home page, [jb.12], [jb.15]. The influence of magnetic fields is not yet taken into account.

Further details on possible BH's follow from the question: In the context of BH's, are there other facts, corresponding to relativistic principles and physically credible, which speak for or against the LI or EI? In the following chapter it will be shown how simply SMO's for the equation of state $p = 1/3\, \rho c^2$ can be derived with the help of TOV. This speaks in favor of the LI. Subsequently, the inner SM and the TOV for other equations of state and then the RWM are examined in more detail. With sufficient mass, an infinite gravitational force could arise, higher than any other counterforce following from the equation of state, so that in the case of a gravitational collapse for the LI, the unphysical state, $r_{real} = 0, \rho = \infty$ would have to arise. It will be shown that this state does not occur for stellar objects described by the inner SM and the TOV because, at the latest, the counterforces are greater than the gravitational forces for degenerate objects. The same applies to the RWM, ch. 22.

21.7. Calculation of supermassive objects using TOV

The Tolman-Oppenheimer-Volkoff equation (TOV) is:

$$\frac{dp(r)}{dr} = \frac{G}{c^2}\left[\rho(r) + \frac{p(r)}{c^2}\right]\left[m(r) + 4\pi r^2 \frac{p(r)}{c^2}\right]\left[1 - \frac{2Gm(r)}{c^2 r}\right]^{-1} \quad (21.40)$$

Furthermore:

$$\frac{dm(r)}{dr} = 4\pi r^2 \rho \quad (21.41)$$

r is the radial coordinate of the stellar object to be calculated. $m(r)$, $\rho(r)$ and $\boldsymbol{p}(r)$ are, respectively, the mass, density and pressure of the stellar object as a function of r. The TOV is the elementary tool to calculate the mass, density and pressure inside the stellar object as a function of r. Another relationship necessary for this is the equation of state, see glossary. To find it out, one must ask: What happens when an ideal, relativistic Fermi gas is compressed and its volume reaches almost zero ($r \approx 0$)? The answer, it changes its equation of state from a low relativistic to a high relativistic one, namely to $p = 1/3\,\rho c^2$. At least this is the case for LIGHTMAN [198]: "A relativistic zero-temperature Fermi gas has the equation of state $p = 1/3\,\rho c^2$ ". This is a very sympathetic equation of state, because for it the TOV is analytically solvable. For the proof, one sets the solution of LIGHTMAN [198], problem 14.5,

$$m(r) = (3/14)\left(\frac{G}{c^2}\right)^{-1} r \quad (21.42)$$

$$\rho(r) = (3/14)\left(\frac{G}{c^2}\right)^{-1}\left(4\pi r^2\right)^{-1} \quad (21.43)$$

$$p(r) = (1/14)\left(\frac{G}{c^2}\right)^{-1}\left(4\pi r^2\right)^{-1} c^2 \quad (21.44)$$

into the TOV and confirms by simple calculation the equality of both sides of formulae (21.40) and (21.41). This solution describes a stellar object of *infinite* mass and extent, because the larger r the larger m in formula (21.42). The maximum allowed r must fulfil the condition $p(r_{max}) = 0$, because the surface pressure of a star must disappear, otherwise the star expands and is unstable. This condition is only fulfilled when r is infinite. This solution is not exactly practical, but it is the starting point for calculating stellar objects with *finite* mass and radius. To do this, one calculates - as is usual in the calculation of neutron stars - a stellar object composed of several, piecewise different equations of state. With appropriately chosen equations of state, one arrives at an object that in the interior has the equation of state $p = 1/3\,\rho c^2$ and in the outer shell with lower pressure the equation of state $p = const\,\rho^{4/3} c^2$. The latter is well known and is regularly used in the calculation of neutron stars. The LI thus assumes that collapsing matter reaches a highly relativistic state $p = 1/3\,\rho c^2$ and then the application of TOV for objects composed as above yields a supermassive object of arbitrary mass.

Any relativistic ideal Fermi gas can reach the highly relativistic state $p = 1/3\,\rho c^2$ as long as it remains stable at high pressure. It is a question for nuclear and particle physics how far this is fulfilled for a fermi gas of neutrons. If not, it is plausible to assume that a quark-gluon plasma forms from the neutrons. This means that instead of a massive neutron star you get a quark star. Of course, other types of particles could also form the center of massive stellar objects, but these are also questions for nuclear and particle physics. TOV shows that supermassive objects with masses millions of times that of our Sun are possible. The main difference here is the correspondingly larger radius of the innermost region with state $p = 1/3\,\rho c^2$. The following two sections each give an example of this.

Supermassive objects (SMO) with $p = 1/3\,\rho c^2$ using the TOV

Normally, the TOV cannot be solved analytically, it must be solved numerically There are mathematical algorithms for this, because for the mathematician the TOV is nothing more than a differential equation that is to be solved numerically. The corresponding MATLAB implementation of the TOV is applied to $p = 1/3\,\rho c^2$ [jb.10], [jb.11]. This is a test of the MATLAB implementation, because the result is known, it must agree with the analytical solution above. In Fig. 21.10, Figure 1, you can see the result for the mass, formula (21.43). It is a linear function (straight line) and has a mass of $m_{\max} =\sim 4.1 \times 10^6 M_{Sun}$ for $r_{\max} = 2.8 \times 10^{12} cm$, it agrees - within the reading accuracy - with the analytical solution. From Fig. 21.10, Figure 2, it can be seen that the radius r is always larger than the Schwarzschild radius r_{sm}. The curves for p and ρ , which are not reproduced, also correspond to the analytical solution. The algorithm was terminated as soon as the total mass was that of Sgr A*. Normally, the algorithm terminates at the r for which $p(r)$ becomes zero.

The more accurate value: $m_{\max} = 4.15 \times 10^6 M_{Sun}$

is achieved when $r_{\max} = 2.8 \times 10^{12} cm.$

The following applies $m_{\max} \approx m_{SgrA*}$

and $r_{\max} = 2.2\, r_{sm}(r_{\max}) \approx 2.2\, r_{sm}(Sgr\ \text{A*}).$

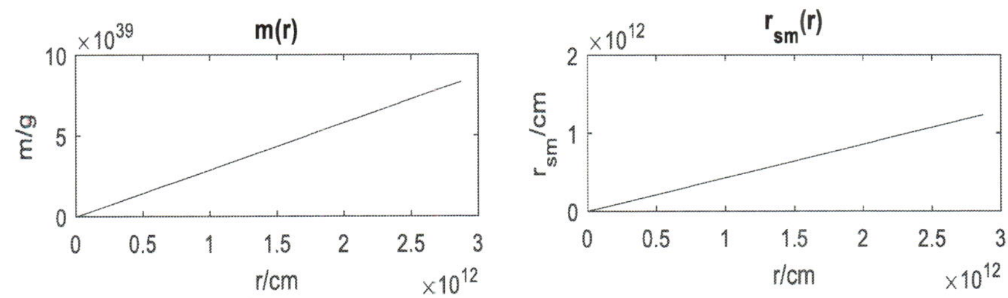

Fig. 21.10: A supermassive object (SMO) with $p = 1/3\,\rho c^2$ and the mass of SGR A*

Supermassive objects (SMO) with core $p = 1/3\,\rho c^2$ and shell $p = const\ \rho^{4/3} c^2$ using the TOV

Next, a more realistic SMO is to be calculated. This MATLAB program starts with the equation of state $p = 1/3\,\rho c^2$ for the core and switches to the equation of state $p = const\,\rho^{4/3}c^2$ for the shell early so that the total mass is approximately that of Sgr A*. In Fig. 21.11 the change occurs for m = 6. 10^39 g. This gives an example of the calculation of SMO's with finite mass of several million M⊙ and finite expansion using the TOV, a purely general relativistic relation. The calculation does not yield an BH but an SMO, because the radius is larger than the associated r_{sm} . The following figures show that m(SGR A*) = 4.3. 10^6 M⊙ is reached when r(SGR A*) = 1.98 10^12 cm. **The radius of this SMO with the mass of SGR A* is thus ~1.56 times larger than the radius of a black hole of the same mass.**

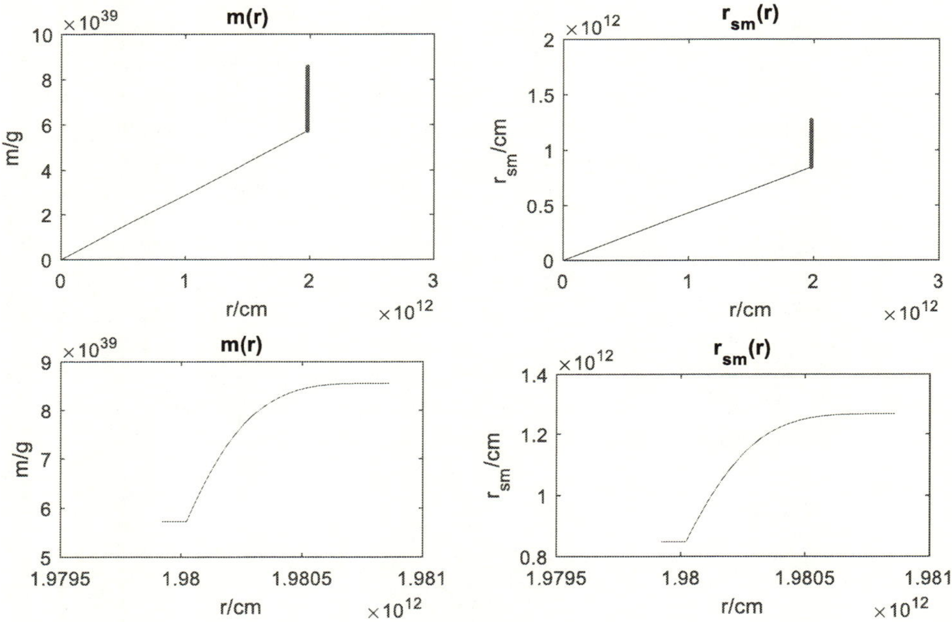

Fig. 21.11: A supermassive object (SMO) with the mass of SGR A*. In the core, the equation of state is $p = 1/3\,\rho c^2$ and in the shell $p = const\,\rho^{4/3}c^2$. The nearly vertical parts of the function in Figs. 1 and 2 show the outer shell with $p = const\,\rho^{4/3}c^2$. They are shown enlarged in Figs. 3 and 4.

With these considerations, one can see the two points of view: Within classical general relativity, collapsing masses of sufficient size reach the radius $r = r_{sm}$ and before they reach the highly relativistic state $p = 1/3\,\rho c^2$ they collapse into a black hole. Within the LI of general relativity, collapsing masses of sufficient size reach the highly relativistic state $p = 1/3\,\rho c^2$ and the application of TOV shows that supermassive objects form, but not black holes. The differences could be explained by observing degenerate objects with a few solar masses (neutron star or quark star), with several million solar masses in the galactic centers using the latest Telescope associations or via gravitational waves.

21.8. Black holes when using the Inner SM and the TOV

After the proof in the previous chapter that for the equation of state $p = 1/3\,\rho c^2$ supermassive objects (SMO) of arbitrary size can exist without forming black holes (BH), in the following we will discuss the possible formation of BH in connection with the inner SM and the Tolman-Oppenheimer-Volkhof equation (TOV) for the equations of state $\rho^{5/3}$ and $\rho^{4/3}$. Possible would be, for example, an infinite gravitational force higher than any other counterforce from the equation of state so that in the event of a gravitational collapse the unphysical state $r_{real} = 0\ \rho = \infty$ would have to arise for LI. It will be shown that this state does not occur for the LI for stellar objects described by the inner SM and the TOV because the counterforces are greater than the gravitational forces - at the latest for degenerate objects. This has implications for the well known relation of (21.56).

For stars of *constant* density, the metric of the inner SM is [357]:

(21.50) $$ds^2 = A_1(r)dr^2 + r^2(d\theta^2 + \sin^2\theta d\phi^2) - B_1(r)c^2dt^2$$

(21.51) with $$A_1(r) = \left(1 - \frac{2M_s r^2}{R^3}\right)^{-1}$$

(21.52) and $$B_1(r) = \frac{1}{4}\left(3\sqrt{1 - \frac{2M_s}{R}} - \sqrt{1 - \frac{2M_s r^2}{R^3}}\right)^2$$

R: Star radius.

$$2M_s = \frac{2GM_{stern}}{c^2}$$

Here, because of the same term $A_1(r)$, $r = R$, the same follows for BH as from SM Assumption (20.4) is now: $R = R_{real} + 2M_s$ and for $R_{real} = 0\ R = 2M_s$ applies, i.e only for $r = R = 2M_s$ and thus $R_{real} = 0$ is there a singularity. The same applies to a static star with non-constant density, since it can be approximated by a static star with constant maximum density.

How is the occurrence of this singularity $R_{real} = 0$ prevented? From the Tolman-Oppenheimer-Volkoff equation (TOV), statements about the gravitational pressure can be derived and compared with the degeneracy pressure. The TOV applies to the interior of the star and reads for constant density [248], [357]. [248]

(21.53) $$p(r) = \rho c^2 \frac{\sqrt{1 - \frac{r_{SM} r^2}{R^3}} - \sqrt{1 - \frac{r_{SM}}{R}}}{3\sqrt{1 - \frac{r_{SM}}{R}} - \sqrt{1 - \frac{r_{SM} r^2}{R^3}}}$$

For the substitution r by $r_{St} + r_{SM}$, which again follows from (20.4), and R by $R_{St} + r_{SM}$, (21.53) changes to

$$p(r_{St}) = \rho c^2 \frac{\sqrt{1 - \frac{r_{SM}(r_{St} + r_{SM})^2}{(R_{St} + r_{SM})^3}} - \sqrt{1 - \frac{r_{SM}}{R_{St} + r_{SM}}}}{3\sqrt{1 - \frac{r_{SM}}{R_{St} + r_{SM}}} - \sqrt{1 - \frac{r_{SM}(r_{St} + r_{SM})^2}{(R_{St} + r_{SM})^3}}} \tag{21.54}$$

Only the area $r_{St} \geq 0$ has physical significance. Index *St* stands for star, *SM* for *outer* Schwarzschild metric.

In this transition to LI, the simple relationship $r = r_{St} + r_{SM}$ was used. One can think of more general forms for the transition between the outer and inner SM, e.g.

$$r \to r_{St} + r_{SM}(r_{St}), \tag{21.55}$$

which only agree with the above assignment for the limiting case $r = R$. For our considerations, however, the above limiting case is decisive and sufficient, because only for $r \approx R \approx r_{SM}$ and thus $r_{St} \approx 0$ must it be proven that a stable star can exist. However, formula (21.55) is important for the discussion of the Robertson-Walker metric (RWM), see Chapter 22.

With these preliminary remarks, we can answer the actual question of whether singularities occur in the TOV and for which stellar radii.

EI: In the usual form of TOV, instability results for $r = 0$, provided $R = (9/8) r_{SM}$, then p becomes infinite, i.e. the wellknown condition applies: A star of radius R is only stable and does not collapse into an BH if

$$R > \frac{9}{8} r_{SM} . \tag{21.56}$$

LI: This condition does not exist for the LI of TOV, as will now be shown. For this, formula (21.54) must be discussed. When does the denominator in (21.54) become zero, or for which R_{St} does the denominator remain >0? R_{St} is the radius of the star in the LI.

With $a = \frac{r_{SM}}{r_{SM} + R_{St}}$ the denominator of (21.24) at the position $r_{St} = 0$ becomes

$$3\sqrt{1 - a} - \sqrt{1 - a^3} \geq 0 \tag{21.57}$$

and after transformations to

$$8 \geq a(9 - a^2) \tag{21.58}$$

Considered as an equation, one has the solutions

$$a_1 = 1$$

and after polynomial division

$$a_{2/3} = -\frac{1}{2} \pm \sqrt{\frac{1}{4} + 8} \tag{21.59}$$

The inequality (21.58) is fulfilled between $a = 1$ and $a = 0$. That is, the denominator of (21.54) is positive for $0 < R_{St} < \infty$ and (21.54) is infinite only if the density is infinite. In other words, the relationship $R > (9/8)r_{SM}$ no longer has any meaning for the LI.

As has been shown, the TOV is singular only for $R_{St} = 0$, i.e. an BH cannot form for finite stellar radii. Now the question is whether a gravitational collapse can progress to $R_{St} = 0$ or whether the counterpressure of matter eventually prevails. For this, it is necessary to estimate the TOV for the case $R_{St} \to 0$ at the location $r_{St} = 0$:

$$\text{(21.60)} \qquad p(r_{St} = 0) = \rho c^2 \frac{\sqrt{1 - \frac{(r_{SM})^3}{(R_{St} + r_{SM})^3}} - \sqrt{1 - \frac{r_{SM}}{R_{St} + r_{SM}}}}{3\sqrt{1 - \frac{r_{SM}}{R_{St} + r_{SM}}} - \sqrt{1 - \frac{r_{SM}^3}{(R_{St} + r_{SM})^3}}}$$

The factor $(R_{St} + r_{SM})^3$ is truncated. Dissolving the powers under the root allows the truncation of R_{St} .

$$\text{(21.61)} \qquad p(r_{St} = 0) = \rho c^2 \frac{\sqrt{R_{St}^2 + 3R_{St}r_{SM} + 3r_{SM}^2} - \sqrt{(R_{St} + r_{SM})^2}}{3\sqrt{(R_{St} + r_{SM})^2} - \sqrt{R_{St}^2 + 3R_{St}r_{SM} + 3r_{SM}^2}}$$

$$\text{(21.62)} \quad \lim_{R_{St} \to 0} p(r_{St} = 0) = \lim_{R_{St} \to 0} \rho c^2 \frac{\sqrt{R_{St}^2 + 3R_{St}r_{SM} + 3r_{SM}^2} - \sqrt{(R_{St} + r_{SM})^2}}{3\sqrt{(R_{St} + r_{SM})^2} - \sqrt{R_{St}^2 + 3R_{St}r_{SM} + 3r_{SM}^2}}$$

$$= \lim_{R_{St} \to 0} \rho c^2 \frac{\sqrt{3r_{SM}^2} - \sqrt{(r_{SM})^2}}{3\sqrt{(r_{SM})^2} - \sqrt{3r_{SM}^2}}$$

$$= \lim_{R_{St} \to 0} \rho c^2 \frac{1}{\sqrt{3}}$$

At the point $r_{St} = 0$, the TOV for the limiting case $R_{St} \to 0$ has a value proportional to the density.

Now we can compare the gravitational pressure with the degenerative pressure. Consider a spherical mass of arbitrary mass and low density. The gravitational pressure causes this object to collapse. The temperature increases and counteracts the gravitational pressure. If this is not enough, the collapse continues and a degenerate state is reached. If there is still no stable state, the supermassive object approaches the limiting case $R_{St} \to 0$. Now the following must apply

$$\lim_{R_{St} \to 0} p(r_{St} = 0) \le \lim_{R_{St} \to 0} p_F .$$

Both tend towards infinity. The gravitational pressure is proportional to the density ρ, the degeneracy or Fermi pressure p_F is in the non-relativistic limiting case proportional to $\rho^{5/3}$ and in the relativistic limiting case to $\rho^{4/3}$ [377] and, independent of the amount of the constant pre-factor of the degeneracy pressure, the Fermi pressure becomes greater than the gravitational pressure - at least for the limiting case $R_{St} \to 0$ a stable state is achieved, as should be shown.

"Cosmologists are often in error, but never in doubt."
SINGH ca. 1960

22. Black holes and conservation of energy of the RWM

22.1. Formulas

First of all, the mathematical relationships between an expanding universe, an exploding dust star and SM will be explained. Expanding universe and exploding dust star have the same metric (RWM), with the difference that dust stars have a center. The connection of dust stars with the SM is illustrated by a simple consideration: Any dust grain of a dust star is in free fall as in a SM, with the center of gravity being formed by all the dust grains that are closer to the center than the dust grain under consideration. The dust grains at a greater distance have no influence, because they form a hollow sphere which, as in Newtonian gravitational theory, does not exert any gravitational forces on the inner particles, more see ch. 22.2.

We take the following formulae from the literature and limit ourselves in the discussion to the case $k = 0$, when simple formulae are important. Except for the designations, there is no difference between STEPHANI [357], LANDAU-LIFSCHITZ [190], LIGHTMAN [198], MISNER, THORNE, WHEELER:[230], D'INVERNO [60]. The RWM reads:

(22.1) $$ds^2 = c^2 d\tau^2 - R(\tau)^2 \left(\frac{d\rho^2}{1-k\rho^2} + \rho^2 \left(d\theta^2 + \sin^2 \theta d\varphi^2 \right) \right)$$

Formula (22.1) can be derived from (19.1), which we will spare ourselves.

τ = proper times of the galaxies or dust grains

$R(\tau)$ = scaling factor ("world radius")

ρ = radial coordinate

$k = +1,\ 0,\ -1$

$r = R(\tau)\rho$: Distance from the center in the ρ, τ - coordinate system

$U = 2\pi R(\tau)\rho$: Measured circumference of a circle of latitude.

$R(\tau)\frac{d\rho}{\sqrt{1-k\rho^2}}$: Measured distance between two near radial points (dust star, galaxy).

For dust stars, the explosion site is the center; for the expanding universe, any galaxy can be considered the center.

At the same time, the SM

(22.2) $$ds^2 = \frac{1}{A} dr^2 + r^2 (d\theta^2 + \sin^2 \theta d\phi^2) - Ac^2 dt^2$$

with $$A = 1 - \frac{2M_S(r)}{r}$$

ds^2 can be negative, positive or zero, as in Minkowski space.

$2M_S(r)$ is the Schwarzschild radius of the mass enclosed by the sphere of radius r.

Thus, the following equations of motion apply, STEPHANI. [357, formula (23.31)]. Their same shape shows the relationship of SM and RWM. The motion of every dust grain or every galaxy in the RWM can be represented as a motion in an SM, see also ch. 22.2 and 22.3.

(22.3) $$\left(\frac{dr}{dc\tau}\right)^2 = -k\rho^2 + \frac{F(\rho)}{r}$$ Gravitational collapse

(22.4) $$\left(\frac{dr}{dc\tau}\right)^2 = \left(E^2 - 1\right) + \frac{2M}{r}$$ Free fall, $E = E_G$

Taking into account $E = E_G$, (22.4) follows directly from (21.5).

The left sides are equal, therefore it follows from the right sides:

(22.5) $$F(\rho) = 2M(\rho)$$

(22.6) $$E(\rho) = \left(1 - k\rho^2\right)^{\frac{1}{2}}$$ due to $E^2 - 1 = -k\rho^2$

One has:

$$E(\rho) = \left(1 - \frac{2M(\rho)}{r_{\max,\rho}}\right)^{\frac{1}{2}}$$ for $k = +1$ because of $\left(\frac{dr}{d\tau}\right)^2 = 0$ for $r = r_{\max,\rho}$

$$E(\rho) = 1$$ for $k = 0$ because of $\left(\frac{dr}{d\tau}\right)^2 = 0$ for $r \to \infty$

$$E(\rho) = \frac{1}{\sqrt{1 - v^2/c^2}}$$ for $k = -1$ because of $\left(\frac{dr}{d\tau}\right)^2 = \frac{v^2}{1 - \frac{v^2}{c^2}}$ for $r \to \infty$

22.2. Lorentz interpretation for exploding dust stars

For the LI, the RWM is an incomplete description of an explosion, this is due to the coordinate τ. τ is the proper time of the exploding particles, what is missing is the time coordinate t for an outside observer. Since the particles move at different speeds, $\tau = f(t, r)$ is a function of t and the location r of the particles. The RWM must be supplemented. Therefore, the coordinate system (r, t) in which the center of the exploding star rests, is added to the co-moving coordinate system (τ, s).

Instead of deriving the corresponding equations together with and from the RWM, it is easier to derive them completely anew, namely from the SM and its equations for free fall. The coordinate system of the outside observer is thus that of the SM.

It shall be shown: The motion of a particle in the RWM, $R = R(\tau)$, is the same as the free fall of the particle in the field of the spherical mass enclosed by $R(\tau)$ according to the SM. This results in the function $\tau = f(t,r)$. It is fixed by the two equations of motion $r = g(\tau)$, formula (21.10), and $r = f(t)$, formula (21.9), for free fall. A special situation shows the connection. The surface of the expanding dust star moves according to the RWM, but also like a free-falling particle in the SM, where the gravitational mass is the total mass of the star. (One can somewhat better imagine the reverse, the collapse of a star with a particle just outside the surface of the star). The corresponding equations are formulae (21.10) and (21.9). Why is this so simple? Why does the SM hold outside the RWM, even though everything is moving and you don't have a center at rest as in the SM? This follows from the Birkhoff theorem. Assuming a centrally symmetrical mass distribution, it applies to the interior of a hollow sphere, see quotation in Chapter 17, even if the particles move radially, and also for radially symmetrical and radially moving mass shells in the massless layers between them. The SM with the mass of the inner spherical region always applies. Inside a hollow sphere one has the SM with $m = 0$, i.e. the Minkowski space. Another quote [299]: "... consider a vacuum zone between concentric spheres S_1 and S_2 with spherically symmetric but possibly radially moving matter both inside S_1 and outside S_2. In between, the Schwarzschild metric applies."

It follows from this: The SM applies not only outside the expanding dust star, in the space surrounding it, but also inside the dust star - with the following restrictions. Think of the dust star as a series of spherical shells of thickness dr. Then the following applies

(22.7) $$\sum dr = R_{Star}$$

On the basis of the Birkhoff theorem none of the outer shells has a gravitational effect on the inner shells. Only the inner shells have a gravitational effect on each shell at a distance r, i.e.

(22.8) $$\sum_{Shell < r} dm = m(r)$$

$m(r)$ is the mass of the residual star enclosed by the shell. The gravitational field of the SM thus acts on the shell r with $m = m(r)$. The shell r is in free fall in the field $m(r)$, as is the surface (shell $r = R_{Star}$) in the field $m_{Star} = m(R_{Star})$, see Fig.22.1.

These ideas must be written down quantitatively and one obtains the above formulae of ch. 21.1, in particular the equation of motion $r = f(t)$, formula (21.9), for free fall, valid for each of the individual shells - of course with different parameter values in each case - and valid for the time t of an outside observer.

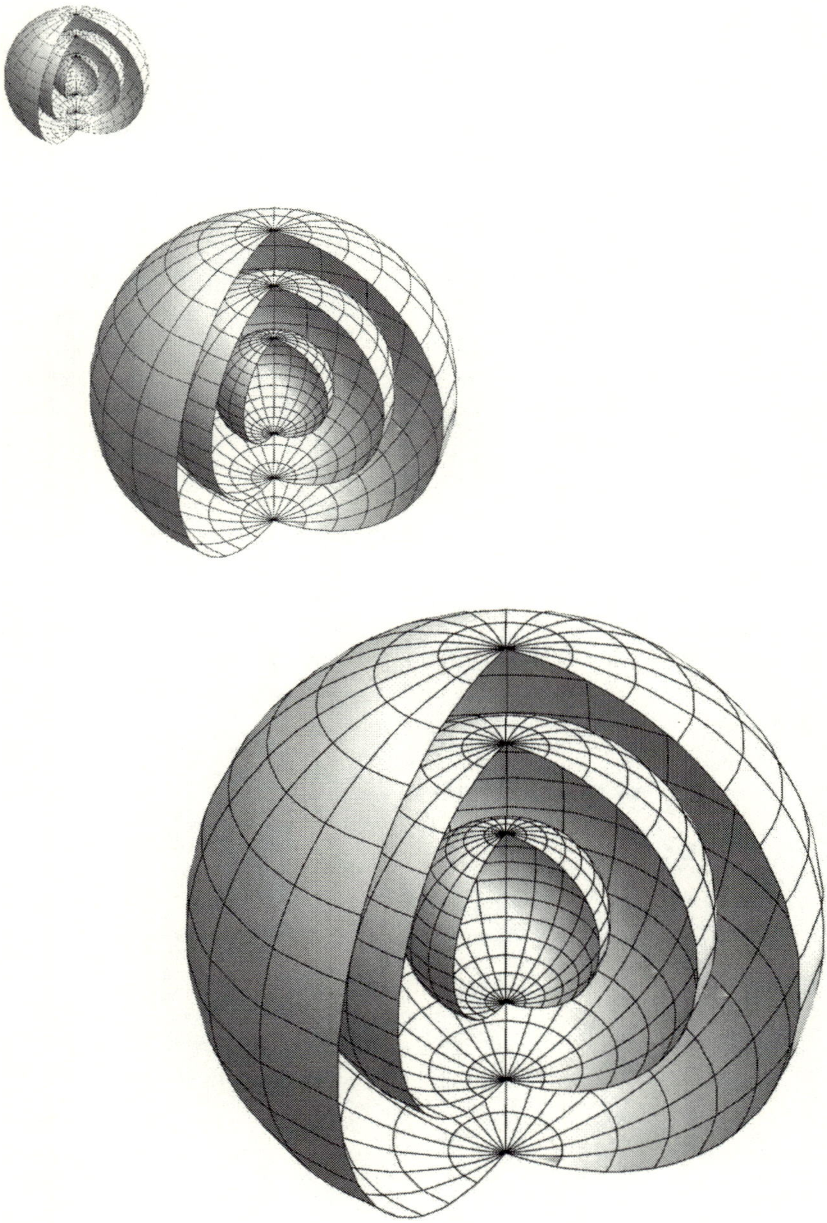

Fig. 22.1 Three time points of an expanding dust star

The shells of an expanding dust stars form hollow spheres. Only the inner shells have a gravitational effect on the outer shells.

22.3. Black holes (BH) and conservation of energy in the RWM

In this chapter it will be proved that the LI of the SM and the LI of the RWM harmonise. For this purpose, the black holes (BH) and the conservation of energy of the RWM are examined more closely and it is confirmed that both are unproblematic for the LI, but it looks different for the EI. Since for freely falling particles, dust stars, the expanding universe the SM always applies as well, the EI or LI of the RWM is easily verifiable. In contrast to dust stars, the expanding universe can be understood differently again according to the EI - namely as expanding space - which must be discussed additionally.

As the following calculation shows, in the EI spaces with $k = 1$ are an BH and should therefore not exist at all. For the proof, the equations for the Friedmann cosmos are used which were needed to calculate the so-called world radius $R = R(\tau)$, see Fig. 19.7. Because of $dR/d\tau = 0$ at maximum expansion $R = R_{\text{max}}$, the two formulae are obtained, see [357, formula (23.38)]:.

(22.9) $$R_{\max} = \frac{8\pi G M^{\sim}}{3c^4}$$

(22.10) $$\mu_{\max} c^2 R_{\max}^3 = M^{\sim}$$

$\mu_{\max}$ is the density at maximum expansion, $M^{\sim}$ is the same constant in both equations. Eliminating them gives

(22.11) $$\frac{2G}{c^2}\frac{4\pi}{3}\mu_{\max} R_{\max}^3 = R_{\max}$$

Since $\mu_{\max}$ is the density at maximum expansion, one obtains

(22.12) $$\frac{2GM_{Uni}}{c^2} = R_{\max}$$ and because of (16.29)

(22.13) $$2M_s = R_{\max}$$

On similar considerations in LANDAU-LIFSCHITZ [190], it says: "At any given time, a certain part of the sphere is already within its event horizon. "

Since $R \leq R_{\max} = 2M_s$, everything is in an BH for $k = 1$. Radial motion does not change this. Their existence in these spaces can be seen as proof that BHs do not exist. As a way out, one could assume that, at least for our universe, only the variant of an expanding space is permissible. The latter does not provide more than a breathing space, one considers inner areas of the universe at earlier times, where the density was higher. BHs should have already formed there, which is not the case.

What is true in EI for finite spaces is correspondingly true for infinite spaces with $k = 0, -1$. Because of the Birkhoff theorem it is true for every cavity that the masses outside do not exert any gravitational forces towards the inside, and the masses inside, despite the radial motion, form an BH, provided that the total mass in the cavity is large enough. This is always achievable if the cavity is chosen large enough. We live in a space with $k = 0$ and $\Lambda \neq 0$ - research status 2009 - and notice nothing of an BH. There are no problems here for the LI. Radial outward motion, as observed, rules out an BH; at most, with the opposite direction, one could - in contradiction to the LI - just experience the phase before the final collapse, $r = 0$, within an BH. Quantitative examination can confirm or refute the LI.

LI, quantitative: Let's take a spherical dust star with any radius r_{real} of dust particles or galaxies. The gravitational field outside is described by an SM with radius $r_{real} + 2M_s$, inside by SM's with decreasing radius, because the associated dust sphere becomes smaller and smaller. The associated Schwarzschild radius for the LI only exists in the description, not in reality. In principle, this can be checked by measurement.

If one measures the radius and circumference of a dust star, the following applies to the

LI: (22.14) $U_{measured} = 2\pi(r_{real} + 2M_s)$

EI: (22.15) $U_{measured} = 2\pi r_{real}$

The measurements are made along a great circle of any inner dust sphere and along the radius. Different measured values for the circumference belong to the same radius r_{real} determined from the radial measurement, as the two formulae (22.14) and (22.15) show. Only one of them can be correct. r_{real} is calculated according to the same formula for both interpretations:

(22.16) $$r_{real} = \sum_i ds_i \sqrt{1 - \frac{2M_i}{r_i}}$$

ds_i: The radially measured thicknesses of the individual shells

$2M_i$: Schwarzschild radii of the associated spheres with radius r_i

Conservation of energy

EI: The total energy for dust stars with $k = +1, 0, -1$ is uniformly predicted in terms of gravitational mass m_{grav} :

(22.17) $$\begin{aligned} m_{grav} &= \frac{4}{3}\pi\mu\rho_0^3 R(c\tau)^3 \\ &= \frac{4}{3}\pi\mu r_0^3 \end{aligned}$$

STEPHANI [357], μ: density. This equation may be regarded as correct, since it is derived from the field equations. One should expect according to the EI:

$$m_{EI} = \int_{\text{alle Schalen}} dm_{gemessen} = \int_{\text{alle Schalen}} \mu_{gem}(c\tau) dV_{gem} \qquad \text{because of } \mu_{gem} = \frac{dm_{gem}}{dV_{gem}}$$

$$= \mu_{gem}(c\tau) \int_{\rho=0}^{\rho=\rho_0} dV_{gem} \qquad \text{because of } \mu(c\tau) = \text{konst}$$

$$= \mu_{gem}(c\tau) \int_{\rho=0}^{\rho=\rho_0} 4\pi\rho^2 R(c\tau)^2 \frac{R(c\tau)d\rho}{\left(1-k\rho^2\right)^{\frac{1}{2}}}$$

(22.18)
$$= 4\pi\mu_{gem}(c\tau)R(c\tau)^3 \int_{\rho=0}^{\rho=\rho_0} \frac{\rho^2 d\rho}{\left(1-k\rho^2\right)^{\frac{1}{2}}}$$

("alle Schalen" same as "all shells", "gem" or "gemessen" same as "measured")

This integral only agrees with m_{grav} for $k = 0$

For the LI applies:

(22.19)
$$m_{LI} = \int_{\text{alle Schalen}} dm_{LI}$$

$$= \int_{\text{alle Schalen}} dN m_0 \left(1-k\rho^2\right)^{\frac{1}{2}} \qquad \text{for all } k$$

dN: Number of particles in a dish,

$m_0(1-k\rho^2)^{\frac{1}{2}}$: real mass of a particle, see (20.6) and (22.6).

On account of

$$\mu_{gem} = \frac{dm_{gem}}{dV_{gem}}$$

$$= \frac{dNm_0}{4\pi\rho^2 R(c\tau)^2 \dfrac{R(c\tau)d\rho}{\left(1-k\rho^2\right)^{\frac{1}{2}}}} = \text{const}$$

for each shell $d\rho$:

(22.21)
$$m_{LI} = \int_{\text{alle Schalen}} \mu_{gem} 4\pi\rho^2 R(c\tau)^2 R(c\tau)d\rho$$

$$= \frac{4}{3}\pi\mu_{gem}\rho_0{}^3 R(c\tau)^3$$

$$= m_{grav}$$

The integral agrees with m_{grav} for all k.

What remains is the thesis of the EI for the expanding universe that the space between the galaxies is expanding. The galaxies therefore do not move, although the distance between them increases. Objections:

(a) The EI must give good reasons why quite different laws should apply here than for SM and dust stars is proven, since the formulae are the same.

(b) Although the galaxies are not moved, but only space expands, the galaxies have gained energy through the expansion. Connect two galaxies by a rope, one end of the rope remains fixed, the other can drive a generator and convert the kinetic energy of the galaxies into electricity. This means energy from nothing, because before the expansion of space began, the galaxies were at rest.

(c) The objections of ch. 19.8.3-5 apply.

However, it is not to be disputed: The idea of an expanding universe means a new kind of mathematical-physical space from which cosmological laws can be read. It illustrates Hubble's law, relocates the origin of microwave radiation (CMB) to a vivid (proper) point in time and has led to the discovery of a positive cosmological constant. Nevertheless, physical reasons must be given to justify this construction; philosophical arguments that it is our spatiotemporal reality are not very convincing. Equal τ, certainly not equal t, could mean an equal physical state in cosmic dimensions (equal particle densities, equal temperature T, equal entropy density) and thus (or in some other way) cosmological simultaneity could be physically justified. This would be the point of view of EINSTEIN and POINCARÉ: "The simultaneity of two events must be defined in such a way that the wording of the laws of nature becomes as simple as possible." PAIS [254]. This brings new physics, as the EI shows, but no new philosophy, as the LI shows.

"Don't be angry with the mirror if you are ugly".
russ. Proverb

23. Comparison of the Einstein and Lorentz interpretations of the General Theory of Relativity

The Einstein (EI) and Lorentz (LI) interpretations of general relativity can be distinguished experimentally, and thus retroactively those of special relativity. Despite the same formulae, the following differences arise.

Einstein Interpretation (EI)

- There are black holes.
- The fall law $r = f(t)$ - formula (21.9) - loses its validity for $r \to 2M$.
- Conservation of energy theorems - formulae (21.3), (22.17) - are correct in terms of formulae, but do not fit all physical processes in the sense of the EI, see ch. 21.3 and 22.3.
- Discrepancy to QM and QF, see ch. 11.7 Quantum ether, and to ch. 24.13 the smallness of the cosmological constant.
- The EI of the SM is incomplete, s. ch. 16.8.
- Particularly important is ch. 21.4, contradictory statements of the EI on the total energy of a particle at rest in the gravitational field.

Lorentz Interpretation (LI)

- Black holes are a theoretical limiting case, what exists are supermassive degenerate stellar objects with physical properties like normal and degenerate stars. (Emission of cosmic rays up to the UHECR, emission of electromagnetic waves, etc.), see ch. 21.3.
- The fall law $r = f(t)$, formula (21.9), remains valid for $r \to 2M$, which is experimentally verifiable, s. ch. 21.3.
- Law of conservation of energy, formula (20.6), remains valid for inner and outer SM as well as for RWM, s. ch. 22.3.
- Because of the relationship quantum ether "the same as" non-empty vacuum of QM and QF "the same as" particle-less ground state of quantum fields, there is a natural relation to QM and QF, s. ch. 11.7. Another example concerning ch. 11.7: The smallness of the cosmological constant, ch 24.13.
- There are first astronomical observations that can be interpreted in favour of the LI, for example, the origin of the UHECR (ultra-high-energy cosmic ray) possibly lies in AGN's (active galactic nucleus). Furthermore: An important experimental confirmation of the LI of the RWM is the observation of different Hubble constants. The LI of the SM is confirmed by the low

spin of Sgr A*. These observations are discussed in Chapter 24.11 "Astronomical Observations".

- Very convincing for the author: The negative gravitational potential of the of the Newtonian theory of gravitation must correspond to a negative mass ($E = mc^2$). Negative masses, however, do not exist. This incomprehensible fact finds a natural explanation in chapter 24.8 with formula (24.9) and in chapter 24.9 with formula (24.18) through the LI.
- The LI resolves the contradictory statements of the EI on the total energy of a particle at rest in the gravitational field, ch. 21.4.

On the author's homepage there are more comparisons of EI and LI. They are mainly:

- From the contradictory statement on the energy of a particle at rest in the gravitational field of the SM, ch. 21.4, a similar situation follows for the RWM. The article "Cosmology and Lorentz-Interpretation (LI) of GRT" deals with this and the solution of the contradiction, [jb.3].
- According to the prevailing view, the wave function of quantum mechanics cannot be real if classical relativity is to hold. But if the Lorentz interpretation of relativity is preferred, it does not hold, then the wave function is real, non-local and the Copenhagen interpretation of quantum mechanics is to be considered correct. This is explained in more detail in the article "Interpretations of quantum mechanics and Lorentz interpretation", [jb.7].
- The two-body problem of GRT is not considered solvable in classical GRT. Wikipedia states the following about the two-body problem: "Since the presence of the two masses changes the space-time structure itself, concepts such as the *center of mass, total energy, angular momentum, etc. are no longer applicable.* Therefore, no reduction of the problem to a one-center problem is possible." In the article "The two-body problem of GRT" it is shown why this need not apply to LI [jb.8].

The author's thoughts on how the two-body problem could be solved, however, are only a suggestion. The mathematically demanding considerations on the detection of gravitational waves go deeper here. If one wants to correctly predict the emitted energy of gravitational waves for mutually orbiting neutron stars or black holes, it must also be possible to calculate the limiting case for large distances, i.e. that no gravitational waves are emitted. But this is precisely the solution to the two-body problem. Obviously, the calculation has succeeded, after all, gravitational waves have been successfully demonstrated. Here is a quotation from KIP S. THORNE [366.2], page 457ff: "Among the application examples for the paradigm of flat space-time [i.e. the LI] is the calculation of the change in mass of black holes and other bodies when gravitational waves are absorbed by them. This includes the calculations of CLIFFORD WILL, THIBAULT DAMOUR and other authors who have shown how neutron stars orbiting each other generate create gravitational waves." "When working in the field of relativity, it is extremely useful to have both paradigms at hand."

From these quotations one may conclude: For the solution of the two-body problem and for the calculation of gravitational waves, one needs the Lorentz interpretation, and furthermore: **The principle followed since the first edition (1994) of treating both interpretations equally has the approval of Nobel Prize winner KIP S. THORNE.**

24. Notes, additions, comments on I. and II.

24.1. Absolute special relativistic effects

To deepen ch. 9, excerpts from the lecture *Interpretations of the Special Theory of Relativity - Born, Einstein, Lorentz,* Dresden 2000, and in the following chapter the contribution to the discussion by W. T. HERING, Munich, [151] and the reply are given.

Content:

1. preliminary remarks

2. overview of the three interpretations

3. rotating reference systems

4. EINSTEIN'S theses on rotating discs

5 Scharnhorst effect

1. preliminary remarks

With regard to certain, simple thought experiments - acceleration of large lengths (and not only of point objects, like elementary particles) and for rotations - there are differences between Born on the one hand, and EINSTEIN and LORENTZ on the other. Whereas, there is agreement between BORN and EINSTEIN in the philosophy of space-time and in the rejection of an ether, which LORENTZ never renounced.

Because of the experimental equivalence of all three interpretations, the following widespread view is contradicted: Every kind of a ether conception leads to an observable contradiction and the four-dimensional space-time continuum is experimentally proven. I refer to the literature [328], [29], [402] as well as chapters 8 - 11.

2. overview of the three interpretations

The Bornian interpretation is the most elegant and generally accepted. It is almost inconceivable that it does not agree with EINSTEIN in all details, yet a distinction can be made between the two experimentally, (but not between EINSTEIN and LORENTZ).

The Bornian (kinematic) interpretation is best illustrated by the following picture:

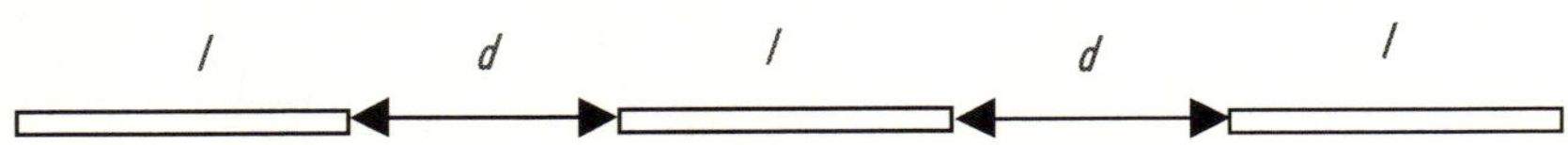

Fig. 24.1 Application kinematic interpretation

On the x-axis are rods of length l at distance d. If one changes the inertial frame, both the rods and their distances are Lorentz-contracted, but no forces can have acted on the mass-free distances in particular. From this follows the Bornian (kinematic) interpretation: All relativistic effects are equivalent to a change of the reference frame and are not associated with forces. The Lorentz interpretation explains the picture differently, since for it the scales transported along changed when the inertial frame was changed. These properties are summarised in the table Fig. 24.2.

	Relativistic effects: physically real or only measuring effects?	Are rel. effects kinematic, i.e. not caused by forces?	Do forces occur in relativistic effects?	Are rel. effects something absolute i.e. independent from inertial system?
Born	**real**	**yes**	**no**	**no**
Einstein	**real**	**yes**	**yes, e.g. for slices set in rotation**	**no**
Lorentz	**real**	**no**	**yes**	**yes, but the principle of relativity restricts the measurability**

Fig. 24.2 Table Born-Einstein-Lorentz interpretation

3. rotating reference systems

Measurable Differences between BORN's and EINSTEIN's interpretations arise for the Ehrenfest paradox and its special cases. They are to be treated quantitatively. Let us first consider the transition from an inertial frame at rest to a rotating reference frame. The corresponding relativistic formula from LANDAU-LIFSCHITZ [190] reads for a rotating coordinate system for circumference U and radius r

(24.1)
$$U = \frac{2\pi r}{\sqrt{1 - \frac{\Omega^2 r^2}{c^2}}}$$

There is a hyperbolic geometry $U > 2\pi r$, as also the *numerical example* shows: U = 20 m, $v = r\Omega = c\sqrt{3}/2 = 0.866$ c, i.e. the root factor is ½,

$$r = 20 \text{ m} / 4\pi = 1.59 \text{ m}.$$

To 20 circumferential elements of the length of 1 m, the radius $r = 1.59$ m.

Now brake the circular ring. The number of circumferential elements does not change and neither does their length - this is the Born-Einstein thesis.. But now

(24.2)
$$U' = 2\pi r'$$

(Euclidean geometry)

The 20 circumferential elements of a length of 1 m now form a circle of radius $r' = 3.18$ m and $r' > r$ applies. This means that if a rotating circular ring is braked, the radius increases, as illustrated in Fig. 24.4.

There are two ways to illustrate this result. Either you draw the rotating circumferential elements shortened, as in fig. 24.4 and the figure remains flat, or the circumferential elements and the radius are not shortened, then you have a curved surface (see fig. 24.3). Braking means leaving the curved surface. Furthermore, if you brake a wheel with gaps, the length of the spokes (radius) does not change and the gaps can close. In fig. 24.5 the 20 circumferential elements of the length 1 m in Euclidean space form a circle with radius 3.18 m, in hyperbolic space this is only possible with gaps.

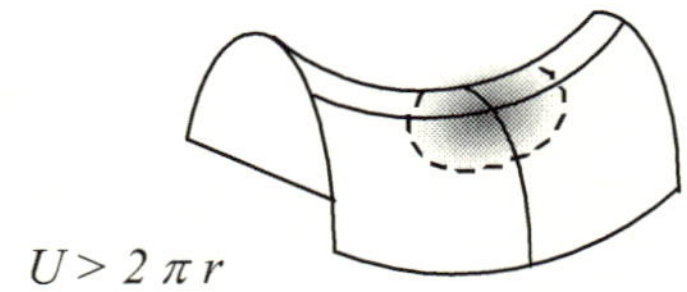

Fig. 24.3 Rotating disc as a hyperbolic surface

For these two examples, *experimental* equivalence of the Lorentz and Einstein interpretations applies. Whether circumferential elements increase or space becomes Euclidean, both ideas are permissible, the measurement results, i.e. the number of length elements, are predicted in agreement. *What is not true, however, is the thesis that relativistic effects are purely kinematic.* For Fig. 24.4, the increase in size is a dynamic effect, i.e. associated with forces, because the circumferential elements are moved away from the center. These forces are perpendicular to the forces of rotational acceleration (and, according to EINSTEIN, should have their cause in curved space-time). *This is the contradiction to Born's interpretation* - that is why the thesis is repeatedly put forward that when rotating discs are braked or accelerated, no forces can arise due to Lorentz contraction, there is no difference to a change of reference systems, where no change of the circumferential elements occurs either. Fig. 24.4 and 24.5 are said to be wrong. (This thesis is held with great conviction by about half of all relativistic physicists - at least that was true at the time of the lecture).

4. EINSTEIN'S theses on rotating discs

EINSTEIN contradicted this [354]:

"It is well to remark that a rigid circular disk at rest must break up if it is set into rotation, on account of the Lorentz contraction of the tangential fibers and the non-contraction of the radial ones. Similarly a rigid disk in rotation (produced by casting) must explode as a consequence of the inverse changes of length, if one puts it at rest."

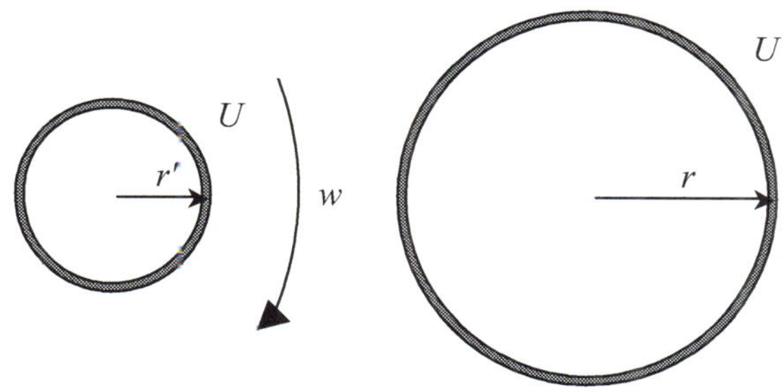

Fig. 24.4 Rotating and stationary circular ring

If a rotating circular ring is decelerated, it increases in size by the factor γ, because the Lorentz contraction is terminated. (Centripetal forces are neglected).

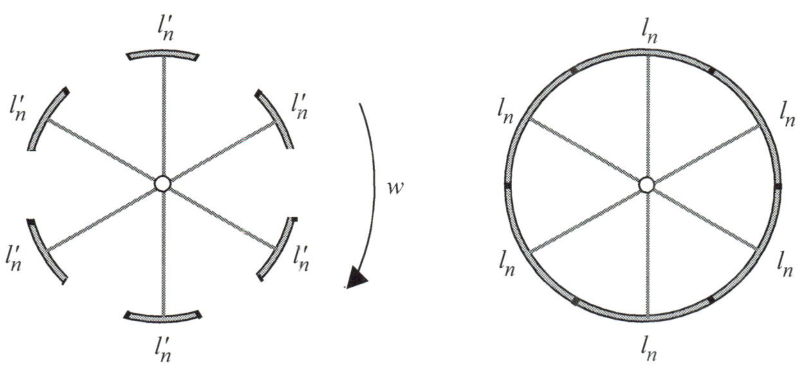

Fig 24.5 Rotating wheel with spokes

If you add spokes to Fig. 24.4, left picture, and increase the radius, you get the left picture of this Fig. and after braking a closed wheel, right picture. ($\gamma = 2$).

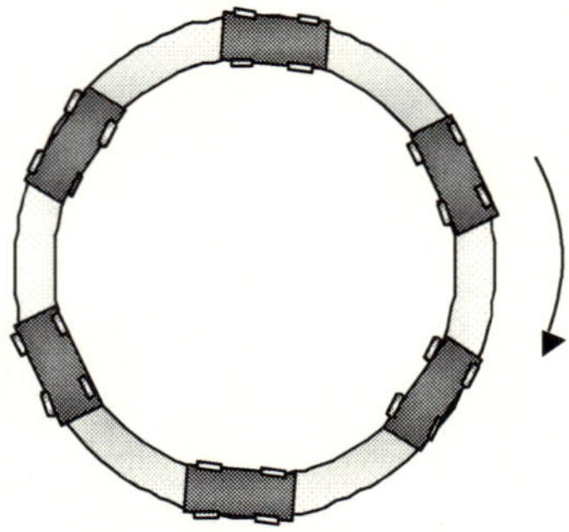

Fig. 24.6 Rail vehicles Lorentz contract, their distances increase.
Since rotating wheels are destroyed at high speeds, imagine a magnetic levitation train.

This letter dates from 1919, another with the same content from 1951. It is not a thoughtless and therefore unpublished utterance. It goes without saying that these serious differences between Born's and Einstein's interpretations could be decided experimentally... or have already been decided (MICHELSON-MORLEY, see below).

5. Scharnhorst effect

The Scharnhorst effect is suitable for postulating true faster-than-light speeds. It is a quantum mechanical effect, but not yet experimentally confirmed, and means the following: As the experimentally confirmed Casimir effect shows, vacuum properties can be changed. In such a modified vacuum (quantum ether), light could also have a higher velocity than c, which is called the Scharnhorst effect in [402]. The thesis of LI is: In a vacuum (quantum ether) with Scharnhorst effect, the Lorentz transformations apply unchanged, only c has a different value in the formulae. The EI would have greater difficulties because the principle of the constancy of the speed of light would be violated.

24.2. Wilhelm T. HERING: Discussion remarks on the lecture by J. BRANDES at the 64th DPG Conference in Dresden

Prof.-Berberich-Str. 34, D-85579 Neubiberg, FRG

21 April 2000 (translated)

The measurements of the occultation times of Jupiter's moons (O. RÖMER) showed that the relative motion of the solar system to the aether had to be < 100 km/s. Today, this limit has been lowered to $v < 0.5$ cm/s via radar measurements on active satellites (which also excludes all theories in which the Earth carries the aether along, since its rotational speed is already 0.46 km/s). Therefore, the earth would have to move through the aether at 300 km/s (orbital speed around the sun). The zero effect of the MM experiment (today reduced to $v < 0.006$ cm/s by laser measurements) is thus in contradiction. The contraction hypothesis of LORENTZ was an attempt to find a dynamic solution to the problem. Today's measurements of the variation in the natural frequency of orbiting resonators, which should increase as a result of L-contraction, show no effect. So the dynamic theory of L-contraction is ruled out. The idea of an aether that tolerates these contradictions, i.e. is unobservable, contradicts the principle of economy (Occam's razor) and is therefore unnecessary. However, if one consults the quantum theory of photons, a neat solution to the aether problem emerges: a light wave is formed from the coherently superimposed wave function amplitudes of the individual photons, which are all in the same state (Bose - Einstein degeneracy). The photons with the inertial mass $m = h/c\lambda$ and $v=c$ carry the light wave with them. This therefore does not need an ether because it is not derived from the classical wave motion. In this respect, the theory of electromagnetic waves no longer belongs to classical physics.

There is no contradiction to the kinematic theory of L-contraction in the tearing apart of a wheel set in rotation: because of the relativity of simultaneity the uniform tangential acceleration of all mass elements dx of the wheel circumference in the laboratory system means that in the local system (in which v, but not the acceleration disappears), each dx is accelerated by the time $dt = (\beta\gamma/c)dx$ later than the preceding and earlier than the following dx (see BERGMANN-SCHÄFER Vol. 1 (11th ed. 1998) p. 935 ff). The circumferential elements dx thus move further and further away from each other. In the local system, however, lie the causes of the dynamic behaviour of a body, the effects of which (shocks, tearing, etc.) remain unchanged with a Lorentz transformation. The wheel is therefore also torn in the laboratory system, whereby the dx still lie on the circumference of the wheel. A paraphrase of this with Euclidean geometry distracts from the true causes.

The forces that tear the wheel depend on the reference system, thus represent relativistic apparent forces, analogous to the inertial force that makes a ball on an accelerated cart run against the direction of motion. A thread with which the ball is attached to the carriage would also break, because the effect of apparent forces cannot be distinguished from dynamically caused forces.

The described physics of the wheel can be well illustrated by the analogous straight-line acceleration of the particle bunches in a linear accelerator (loc. cit.): there, the distance between the bunches in the laboratory system remains the same, while in the local system of the particle beam it grows beyond all limits for $\beta \to 1$. This is noticeable in the fact that the electrical repulsion between charged particles disappears in the direction of motion, since they are arbitrarily far apart in the local system. In the relativistic mechanics of charged particles, this effect occurs as the concentration of the field lines increasing with v in the direction of the plane perpendicular to the direction of motion.

... and a short reply:

This high-level contribution by HERING has helped the author. Nevertheless, he cannot agree with the remarks on OCCAM's razor and on the ether, see also ch. 9 and 10.5.

1.The first section of the discussion concerns Lorentz's contraction hypothesis; as W. T. HERING rightly points out, it is experimentally refuted. On the other hand, not everything is wrong with it, it can be improved and becomes the LI. The LI has learned from EINSTEIN (see earlier chapters and [328]), because of the principle of relativity only relative movements between physical objects are measurable, absolute movements relative to an ether not. The experiments mentioned by W. T. HERING are correctly predicted by the LI. Whether the principle of economy is violated is doubtful, because EI and LI are incomplete theories with regard to quantum gravity. Now, LI fits well with QM and QF and it would be conceivable that it is extendable to quantum gravity and EI is not - because of Occam's razor, however, one would have opted for EI and thus wrongly.

2. The explanations of the second section are very attractive. First: A rotating, rigid disc is destroyed. In Bell's thought experiment applies: "The rope between the vehicles will break" [313]. This means that the Born-Einstein interpretation comes nearer to experimental predictions from the Lorentz interpretation. With it W. T. HERING will not only receive approval from his colleagues. The actual thesis that it is apparent forces that destroy a rotating disc can be agreed with (only) in the result, namely that the disc is destroyed.

3) First of all some doubts against the arguments of W. T. HERING:

(a) Possibly wrongly a circularly accelerated reference frame is taken as a basis, because the formulas of the special relativity theory are valid only for inertial frames. Allowed is every single co-moving inertial frame S_i. One of the n circle elements dx_i rests in it, all others move relative to it. Altogether, the circular ring contracts as for the inertial system S in which the center is at rest. See Fig. 9.4, rolling wheel. (The circular ring is a special case of a disk).

(b) All n circular elements dx_i of the circular ring are considered to be at rest. This implies the simultaneous application of n different inertial systems and contradicts the inertial system requirement, ch. 10.8.6.

(c) In a circular ring, "preceding" and "following" is a contradictory statement. The reference to Bergmann-Schäfer is not entirely correct. There only linear accelerations are treated.

Arguments (a) - (c) summarized mean for the LI: Relative to all inertial frames, especially to S (circle center rests) and S_i (circle element dx_i rests), the rotating circular ring contracts. **They mean for the EI:** The rotating reference frame, in which all dx_i rest, is quite admissible. But it follows the statement contradictory to S that the circumference of the circular ring increases.

In addition, experiment can decide here.

(a) As mentioned in the discussion: The MICHELSON-MORLEY experiment proves not only the Lorentz contraction in inertial frames moving relative to each other. More precisely, the measuring apparatus is located in a rotating reference frame (Earth's surface) and is therefore the direct experimental proof of Lorentz *contraction* on rotating disks.

(b) Further experimental verification is difficult, among other things, because of the additional centrifugal forces. Fig. 24.6 could be promising for large radii. There all massive elements together contract.

24.3. Bell's spaceship paradox - addition

Bell's spaceship paradox was discussed in ch. 9.3. A very nice discussion, but limited to the EI, can be found in WIKIPEDIA [397.1]. There it is stated, just as in ch. 9.3, that relative to the co-moving inertial frame *S'*, in which the rockets are always at rest, forces occur that stretch the rope. This causes the rope to tear and the experiment outcome is correctly explained.

What is not discussed: There is Lorentz contraction relative to the initial inertial frame *S*, the rope tries to contract. Since the distance between the rockets remains the same, it breaks. The difference between *S* and *S'*: In *S*, internal forces occur in the rope because of the Lorentz contraction, in *S'* external forces occur because the distance between the rockets increases. Both cannot be true at the same time, but both inertial systems are equal. Relative to S and S', the measurement results are predicted to be the same and correct, although the physical justifications differ. The LI explains why, the EI does not, see ch. 9.3.

24.4. Historical note on the Lorentz interpretation

"It is noteworthy that EINSTEIN only refers to other authors to a very limited extent in his writings, and only a few critics of the principle of relativity explicitly discussed." TREDER has already pointed this out [327].. A small statement from one of EINSTEIN'S letters to KLEIN may explain this. EINSTEIN wrote: "I do it like everyone else; what I like I accept in silence, what I don't like I mention ..." (24.3.1918)" [327]. The above reproach refers in particular to the general theory of relativity, which is widely regarded as EINSTEIN'S sole work because of the few references to it in the literature. "Accordingly, it must be stated that EINSTEIN was considerably influenced by the Göttingen discussion. EINSTEIN also admitted this in letters to third parties." [327]. "For the further development of the Göttingen discussions, WIECHERT's effort to accept the fundamentals of Einstein's results on the one hand, but on the other hand to hold on to the existence of an ether just as LORENTZ did." [327]

E. WIECHERT is equal to H. A. LORENTZ in the clarity of his theses, as can be seen very clearly from the work on the history of science by W. SCHRÖDER [327]. At this point, SOMMER [346] should be mentioned. Since the Göttingen mathematician HILBERT was the first to discover the so-called Einstein field equations, the change of name to Hilbert-Einstein equations is historically justified.

Further literature on this topic: Wikipedia, the articles by F. WINTERBERG and DANIELLA WÜNSCH. There it is explained that Hilbert was to be deprived of his priorities in the Hilbert-Einstein equations with the intention of forgery.

"Superstition, idolatry and hypocrisy have great reward,
but truth goes begging."
Martin LUTHER

24.5. Critics of the Lorentz interpretation

In the following, objections to the LI, mostly from gravitational physicists, are reproduced. Not because the objections have convinced the author, but to help the reader form his or her own opinion. As a rule, the criticism exhausts itself in pointing out the deviation from the classical theory of relativity. What deviates is said to be wrong. For example: "Black holes swallow, they do not spit", therefore they cannot be the origin of UHECR. Or: A freely falling particle retains its rest mass, as classical general relativity teaches, therefore the approach, formula (20.6), that the rest mass changes is wrong. For this, a better counter-argument (this time a student): One must not look at the law of falling 'naively'. The total energy of a freely falling body becomes almost infinite (for local inertial systems in the vicinity of an BH), therefore the one-body problem, as treated by the SM, becomes a two-body problem and thus the collapse to the BH does occur after a finite coordinate time t. But: With this, one disputes the validity of the classical general theory of relativity, because one asserts a different formula for free fall, while the LI retains the formula.

Another gravitational physicist fundamentally disputes LI: "Physically important is a distance measure that also carries a direct physical meaning. But there is only one of these, namely the one given by the generally non-Euclidean spacetime." The LI needs two metrics, the non-Euclidean one, which describes what is measurable, and in addition the Euclidean one of the distinguished inertial frame, which is needed, for example, for the law of conservation of energy, formula (21.15). This is one metric for the inner and one for the outer curvature of a surface. So something like this should be forbidden. For this critic, everything about the LI is wrong from the start; he then did not have to comment on the verifiable consequences, e.g. on the law of conservation of energy, and has not done so. A criticism in full contradiction to bimetric theories [8-6].

No clear agreement has been found with the author's lectures with philosophical objections, without scruples our existence is denied, as if one could say: "I think, therefore I am not after all". On these issues, non-physicists have no difficulty in agreeing. Also enjoyable, the lecture on the cosmological constant. ch. 24.13, has been classified by an elementary particle physicist as technically convincing and worthy of discussion, theoretical astrophysicists nevertheless reacted negatively, some arguments see [vor23].

Another critique of the LI can be found in chapter 24.2, Ehrenfest Paradox. There it is inappropriately argued with OCCAM's razor - the LI is an extension and thus naturally more complex than the initial theory.

The discussion with a very well-known gravitational physicist about the Weylian argument came to an abrupt end when the author wanted to explain his counter-arguments, s. ch. 16.8.

No argument, but quite common, is the accusation that one has not understood the theory of relativity. This can happen to anyone. For example, even the famous J. S. BELL was accused of not having understood the theory of relativity because of his proximity to the LI, see [18]. Another variant is to silence counter-arguments, so the arguments of the world-famous gravitational physicist SEXL ch. 7.4, remain completely unheeded. For further information see. Homepage of the author, [vor29], [vor30].

24.6. Fermat's Principle in General Relativity

The EI explanation for the deflection of light in gravitational fields is based on the idea that light has mass and is therefore deflected like a particle: $m = E/c^2$ and $E = \hbar\omega$. The considerations of LANDAU-LIFSCHITZ[190] and STEPHANI[357] show the completely different variant necessary for LI and for light because of (20.6). The equations of motion for light in gravitational fields follow from a generalisation of Fermat's principle

$$\int dt = \frac{1}{c}\int \sqrt{\ldots}\,dt = \frac{1}{c}\int \frac{dl}{\sqrt{A}} = \text{Extr.} \tag{24.3}$$

In the derivation of (24.3) ω is constant, LANDAU-LIFSCHITZ [190]. This is self-evident, because in optically dense media light does not change its frequency.

STEPHANI [357, p. 108] writes: "The variation principle (24.3) can also be read in such a way that the three-dimensional space has a refractive index $n = \sqrt{A}$, which is caused by gravity (and also contributes to the deflection of light), and that the speed of light v is correspondingly $c = nv$ reduced." [spelling of formulae adapted] and continues quite in the spirit of EI: "v is the speed of light with respect to the coordinate time t and therefore, like t itself, has no immediate physical significance."

LI: This is precisely why the EI cannot do anything with the generalised Fermat principle. Only in the LI do t and v have a real meaning. What is the point of an extreme value for t if t does not exist? For the speed of light in the radial and tangential directions, it follows from (24.3) with $d\theta = 0,\ d\phi = 0$ and $dr = 0$:

$$c_{rad} = cA \text{ resp. } c_{\tan} = c\sqrt{A} \tag{24.4}$$

which corresponds to the earlier formulae (16.31) and (16.32). For LI, formula (24.3) reads as follows: The 'really' elapsed time, the quotient of 'really' travelled distance and 'really' velocity, has an extremum.

24.7. Weight force, gravitational force

Is the weight force, more generally the gravitational force, there first and must be counteracted, or does it only arise when the geodesic movement in curved space-time, i.e. the uniform motion, is changed? The gravitational force exists when there is space and time (LI), it arises when there is curved space-time and therefore curved space-time is not only a description of gravitational fields (EI).

Applied to free fall in the SM, the EI says: As long as you have the stone in your hand, there is the gravitational force, because the stone is prevented from moving in 4-dimensional space-time on a geodesic. If you let go of the stone, the gravitational force disappears, because now the stone is moving on a 4-dimensional geodesic, i.e. 'uniformly rectilinear'.

LI: The gravitational force is always there, when you carry the stone you feel it, when you let go of it, it causes the 3-dimensional acceleration.

Without further consideration, one must regard both as equivalent, although only one of the two can be correct.

24.8. The negative Newtonian gravitational potential

The gravitational potential ϕ of the SM results from general considerations of RINDLER [299] to:

(24.5) $\varphi = \frac{c^2}{2}\ln\left(1 - \frac{r_{SM}}{r}\right)$. Series expansion shows agreement with the negative Newtonian gravitational potential

$$\varphi = -\frac{c^2}{2}\frac{2M_S}{r} + \ldots$$
$$= -\frac{GM_{Star}}{r} + \ldots$$

Similar to this: WOODHOUSE [408.1].

Formula (20.6) can be confirmed with a thought experiment. Move a mass $m(r)$ with negligible speed from $r = r_A$ to $r = \infty$. The work expended E is

$$E = \int_{r=\infty}^{r=r_A} \frac{d\phi}{dr} m(r) dr \qquad (24.6)$$

because: Energy = force times distance, force = field strength times mass, field strength = gradient of the potential - quite analogous to classical mechanics and special relativity.
Because of (24.5) one obtains

$$\frac{d\varphi}{dr} = \frac{GM_{Star}}{r^2}\frac{1}{1 - \frac{2M_S}{r}} \qquad (24.7)$$

(24.7) together with (20.6) inserted into (24.6) and integrated, yields (including the correct factor c^2)

(24.8) $$E = m_0 c^2 (1 - (1 - \frac{2M_s}{r_A})^{\frac{1}{2}})$$, suitable for

(24.9) $$E_{total} = E + m(r_A)c^2 = m_0 c^2$$

All the work expended served to increase the rest mass $m(r_A)$ to m_0.
As you can see: The negative Newtonian gravitational potential times m_0 is nothing other than the amount by which the rest mass in the gravitational field decreases.

For the author, the *explanation* of the negative gravitational potential, which is consistent with special relativity, is particularly impressive. What does the EI have to counter this?

In (24.7), the relativistic law of gravity corresponding to Newton's law of gravity can be recognised for the special case $M_{Star} >> m_0$:

(24.10) $$K = \frac{d\varphi}{dr} m(r) = \frac{GM_{Star}\, m_0}{r^2} \frac{1}{\left(1 - \frac{2M_S}{r}\right)^{1/2}}$$

(EI s. [299], [408,1], for other metrics and similar restriction: [190, §88]). In the Newtonian limit case (large radii), both laws agree.

Neither for the general case nor for the virial theorem is the relativistic generalisation known to the author. (For Coulomb forces there is the relativistic virial theorem, s. [190, §35]). On the other hand, one cannot doubt that two masses at a distance r exert a measurable force on each other and that such laws exist, but the formulae are missing because the orbits curved by gravitational forces in normal space are described by force-free, geodesic (i.e straight) orbits in curved space and thus gravitational forces are eliminated from general relativity. An approach to make gravitational forces descriptive again can be found in FLIEßBACH [113], formulae (11.9), (19.17), (20,27), as well as RINDLER [299], formula (9.29). The mathematics is the same for both cases.

An argument that gravitational forces also emanate from particles with zero rest mass can be found in T. A. MOORE [234.1], Box 20.1. However, no matter how fast a particle falls, the total energy remains constant, see formulae (21.6) - (21.8).

24.9. EINSTEIN's law of conservation of energy

The Einstein law of conservation of energy for gravitational fields is [367]

(24.11) $$U = \iiint \left(\tilde{T}_4^4 + \tilde{t}_4^4\right) dxdydz$$

Here U is the total energy, $\tilde{T}_4^4$ the energy density of the particles and $\tilde{t}_4^4$ the density of the potential energy of the particles in the gravitational field. If one applies this formula to an isolated system, one obtains "EINSTEIN's very satisfactory result" TOLMAN [367] after extensive, complicated calculation.

(24.12) $$\frac{U}{c^2} = m$$

m = total mass of the isolated system. Applying (24.11) to a spherical ideal fluid, we obtain [367, formula (97.7)]:

(24.13) $$U = \int \rho_{00} dV_0 + \int \rho_{00} \Psi dV_0 + 3\int p_0 dV_0 + 3\int p_0 \Psi dV_0$$

ρ_{00} = density of the liquid, Ψ = potential of the gravitational field.

From this follows the special case of a particle at rest in the gravitational field ($p_0 = 0$) :

(24.14) $$\frac{U}{c^2} = M_{Star} + m_0 + m_0 \Psi(r)$$

(Particle with rest mass m_0 at position r in the gravitational field of a star of mass M_{Star}). For the SM applies [60]:

(24.15)

i.e. Ψ is the Newtonian gravitational potential. Inserted in (24.14)

(24.16) $$\frac{U}{c^2} = M_{Stern} + m_0 \left(1 - \frac{GM_{Stern}}{c^2 r}\right)$$

("Stern" the same as "Star", "Feld" the same as "Field")

The total energy is the sum of stellar mass, particle mass and (negative) potential energy. The same result follows for the LI: From hypothesis (20.6) follows for the reduced rest mass of the particle

(24.17) $$\begin{aligned} m_{0,Feld} &= m_0 \left(1 - \frac{2M}{r}\right)^{\frac{1}{2}} \\ &\cong m_0 \left(1 - \frac{GM_{Stern}}{c^2 r}\right) \end{aligned}$$

and thus for the total energy as the sum of both masses again:

(24.18) $$\frac{U}{c^2} = M_{Stern} + m_0 \left(1 - \frac{GM_{Stern}}{c^2 r}\right)$$

Thus, hypothesis (20.6) for the SM is consistent with EINSTEIN's law of conservation of energy and justifies it. It is not compatible with the principles of special relativity. The *total energy* $E = mc^2$ is equal to the *total mass* except for the factor c^2, and it is not permissible to subtract the potential energy because no mass corresponds to it. In LI, on the other hand, the particle in the field has a lower mass and, in accordance with special relativity, it remains with the formula $E = mc^2$. *The negative Newtonian gravitational potential times m_0 is nothing other than the amount by which the rest mass decreases in the gravitational field.*

As in ch. 24.8: For the author, the explanation of the negative gravitational potential, which is consistent with the special theory of relativity, is particularly impressive. What does the EI have to counter this?

24.10. Are we living in a black hole?

For $k = 1$ see chapter 22.3. According to the current state of research, our universe is infinitely extended ($k = 0$) and since radial motion (escape motion of galaxies) does not prevent the formation of a black hole, the only question is where is the lower limit for $2M$. Because of

$$r_{SM,Uni} = 2M = \frac{2GM_{Uni}}{c^2} \quad \text{and} \tag{24.20}$$

$$M_{Uni} \cong \frac{4}{3}\pi \rho_0 (2M)^3 \quad \text{applies:} \tag{24.21}$$

$$2M \cong \sqrt{\frac{3c^2}{8\pi G \rho_0}} \tag{24.22}$$

M_{Uni}= mass of the universe within the radius $2M$

ρ_0= today's average density of the universe

$$\Omega_M = \frac{\rho_0}{\rho^0_{crit}} \tag{24.23}$$

$$\Omega_M = 0.25$$

$$\rho^0_{crit} = 1.9\,10^{-29} h^2 \left[\frac{g}{cm^3}\right]$$

$$h = 0.6 - 0.8$$

All numerical values from [22]. This calculates

$$\rho_0 = 0.25\,1.9\,10^{-29} 0.7 kg^2 \left[\frac{g}{cm^3}\right]$$

$$= 2.3\,10^{-27} \left[\frac{kg}{m^3}\right]$$

$$2M = \sqrt{\frac{3(3\,10^8)^2}{8\pi 6.67\,10^{-11}\;2.3\,10^{-27}}}$$

$$= 2.6\,10^{26}\,\text{m} \tag{24.24}$$

$$= 8.5\,10^3\,\text{Mpc}$$

From $r = 8.5\;10^3$Mpc the associated Schwarzschild radius is larger than r, one remains in a black hole. Objects up to $z \cong 6$, $\sim 10^4$Mpc, are observed, already the visible part of the universe would have to be a black hole if the EI is right. However, this is not the case, since within an BH there is a point singularity and therefore there can be no outward radial motion, see ch. 21.6. For a circumference measurement at $z = 6$ one would have for the EI $U_{SL} = 2\pi\ 2M$ and for the LI $U_r = 2\pi(r + 2M)$, see formula (22.14) and (22,15), the circumference measurement at $z = 6$ could additionally differentiate between EI and LI.

A positive cosmological constant Λ of the magnitude observed slightly reduces the Schwarzschild radius. This is shown by TOLMAN's formula (82.10). For infinite *r,* it gives rise to further problems for the EI.

24.11. Astronomical observations

24.11.1 Observation options

There are a number of astronomical observations, which could develop to a confirmation of the LI. There are flares in the galactic core of the Milky Way, UHECR and energy problems with supernovae, with gamma bursts as well as with jets from AGNs. To the dark matter the LI can contribute to my knowledge nothing clarifying. But if it should be: The particle-less ground state, n = 1/2, could exert attractive forces, different for each type of particle. To the dark energy (smallness of the cosmological constant) s. ch. 24.13.

Flares in the galactic core of the Milky Way: Flares with a period of 27 minutes are observed [127] in the galactic core of the Milky Way. For the EI they are co-rotating phenomena at the inner accretion disk around the BH of the galactic core. For the LI, the flares are on the surface of a supermassive degenerate star-like object that has a rotation period of 27 minutes that can change (differential rotation). Flares in galactic nuclei and their formation therefore do not differ in principle for the LI from the corresponding processes on our Sun. They have their possible cause in magnetic reconnection [mag], [vor13], [vor15]. Quite analogously, the same laws apply to jets from AGN's as in small-scale star formation, [108], [112]. Precondition for their formation are accretion and angular momentum. For BH's jets have to be explained in a modified way, which can lead to verifiably different predictions. Today's (2010) observations in the center of the Milky Way fit the idea of a supermassive, degenerate star-like object. This also includes the winds from the center, which are faster than 1000 km/s [85]. If one lets the LI hold, there is no clear observation that can only be caused by an BH.

Origin of the UHECR: It was already moved to the centers of the galactic nuclei by the LI before the widely known results of the Pierre Auger Observatory [266], [178] [29]. It remains to be seen how well the assignment to the AGN's succeeds and whether one can therefore exclude BH's there, as the LI predicts.

The cosmological constant is probably proved (and the earlier attempt of an alternative explanation, 3rd ed., appendix 15.4, is superfluous). Contrary to expectations, the cosmological constant is an argument in favor of the LI, s. ch. 24.13.

Other ways of testing the LI experimentally are simulations. The simulation of supernova, gamma bursts and jets from AGN's has not been successful to date (2009). [250] although it is carried out with high professional effort. With the LI, changed approaches result. Since no BH can form during collapse, stars with arbitrarily high central density are allowed to form in the simulations and so-called soft equation of state which allow a deeper oscillation and later reversal of the compression wave with less heat loss. This could mean that the reaction $p + e^- \rightarrow n + \nu$ is less energetic overall and there is more kinetic energy in the returning compression wave. Likewise, some of the conditions for the formation of the heavy elements could change. In addition, pair instability supernovae, [397.2], are possible for the LI because of the higher core temperatures, while for the EI black holes would have to form, [411]. For the LI, it is not a fundamental problem if a star of 40 solar masses collapses into a magnetar, as observed, [411]. but the EI expects a black hole and therefore new assumptions about stellar evolution become necessary for them RITCHIE[296.1], [411]. Applying astrophysics to both interpretations should result in a number of observable differences. [18.1], [407.1].

Another, quite different approach, [vor21], [vor22], but one that has not met with any favor, arises from the following considerations. Supernovae and gamma bursts mean the collapse of a massive star into a neutron star, see Fig. 24.7. During the free fall, the rest mass and kinetic energy of each particle is partially converted into wave energy, as discussed in Chapter 21.3 and shown in Figure 21.4. Now, one can postulate a novel effect of quantum gravity and assume that this wave energy is released as radiant energy when the particles impact and form a neutron star. The radiant energy is probably light; gravitational waves cannot be ruled out.

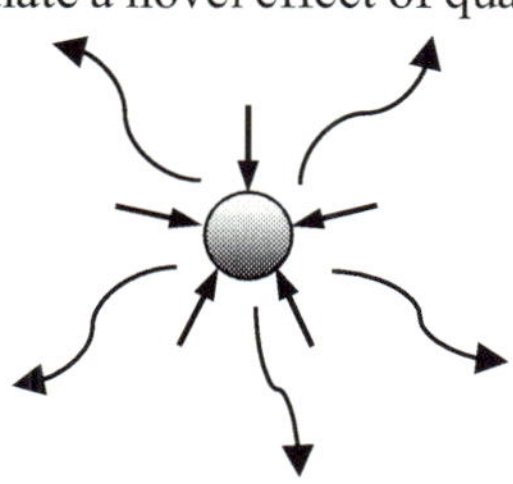

Fig. 24.7 Gamma burst. Collapse of a massive star into a neutron star.

Approximately, the free fall of the particles takes place in a spherically symmetric field as described by the Schwarzschild metric. This allows a quantitative estimation, see Fig. 21.4.

Numerical example:

If a proton falls on a neutron star of mass $M = 1\ \mathrm{M_{Sun}}$, R =15 km = 5 x $2M_s$ then

the kinetic energy: 6.4%

the wave energy: 4.2%,

the gravitational mass: 89.4% of the total energy.

The total energy is equal to the rest energy of the proton (100%). With this hypothesis, the wave energy of 4.2% of the total energy is released directly as radiant energy (light) upon impact with the star.

In the formation of a neutron star $M = 1\ \mathrm{M_{Sun}}$, R =15 km = 5 x $2M_s$, approximately the same percentages apply. The gamma burst GRB000301C has a radiation output, if isotropic, of ~2 M_{Sun}. This roughly fits the following values: $M = 100\ M_{Sun}$, final radius $R = 1500$ km = 5 x $2M_s$. Then you have the radiant energy of ~4,2 M_{Sun} and correspondingly less if the explosion is not isotropic.

These hypothetical assumptions do not contradict the equivalence principle, a novel effect of quantum gravity at gravitational collapse is always conceivable. With the help of formula (21.15), it can be calculated quantitatively. Such an assumption could be helpful for the above simulations and perhaps otherwise.

The above considerations have been taken over from the fourth edition. More attractive seem at present the considerations about the turbulent boundary layer from chapter 21.6 and the observation of different Hubble parameters. More details on the author's homepage [jb.15] and in the following.

24.11.2 Contradictory measurements of the Hubble parameter

contradictory measurements of the Hubble parameter (Hubble constant) lead to two values, namely the value 74.03 (km/s)/Mpc and the 9% lower value of 67.4 (km/s)/Mpc. In [147.1] one reads: "The new estimate of the Hubble parameter is 74.03 kilometers per second per megaparsec. The number indicates that the Universe is expanding at a rate about 9 percent faster than that implied by Planck's observations of the early Universe, which give a value for the Hubble parameter of 67.4 kilometers per second per megaparsec."

This contradicts the idea of an expanding universe of the EI, because different Hubble parameters because of

$$H(t) = a'(t)/a(t)$$

mean that one has different scale factors *a(t)* and/or different derivatives *a'(t)* at one and the same time with the consequence that the expansion speed of the universe has two different values at one and the same time, which is impossible. Similarly, it is impossible for the observed redshift z of one and the same galaxy to have two different values at one and the same time. Hubble's law (19.5) does not apply simultaneously to two different Hubble parameters.

For the LI, the expansion of the universe is nothing other than the expansion of a dust star, or a metagalaxy, ch. 19. In this case, different Hubble parameters mean direction-dependent and/or object-dependent propagation velocities of an only *approximately* isotropically exploded metagalaxy. If these observations remain, the BH problem of ch.24.10 arises again for the EI. At the beginning of the expansion of the metagalaxy, our universe was extremely dense, so that a black hole should have formed from the dust star. None of this is observed, so black holes do not exist, as the LI predicts.

Further observations have the same consequences. In [jb.15, Fig. 5 and 6] the directionality of the Hubble parameter is illustrated. With the EI, on the other hand, one expects a constant value in all directions. Furthermore, the observations of SHAMIR justify it [339.1], [jb.15], to speak of a rotating universe: "Since the spin patterns of a galaxy as visible from Earth is also an indication of the actual spin direction of the galaxy, the large-scale patterns in the distribution of the spin directions can be an indication of a rotating universe". Again, with the consequence: instead of the expansion of the universe, in reality an expanding and rotating metagalaxy and no black hole, although the density was high enough at the Big Bang and afterwards.

24.11.3 Questionable predictions of Sgr A* characteristics

This chapter reproduces the talk of the author *Questionable predictions of Sgr A* characteristics by the EHT group* [jb 20]. The main observational fact: The EHT-group predicts a high angular momentum or spin of Sgr A* but observation confirms a low one.

1.) Introduction and summary

On the one hand, there is the famous EHT picture of Sgr A*, Fig. 24.9. On the other hand, there are at least three questionable predictions by the EHT group on the properties of Sgr A* that contradict observations: a*=0.9375 vs. a*=0.15; spin direction "face-on" vs. "edge-on"; variability of radiation produced by the accretion disk vs. variability of radiation produced by stellar winds during accretion. Moreover, unconvincing predictions by BRODERICK et al. exist of a "UV bump" that is supposed to be absent only in black holes. In fact, it must also be absent in degenerate supermassive objects that are not black holes. We start with the arguments of BRODERICK et al.

2.) The arguments of Broderick et al. to prove an event horizon

The main idea to prove an event horizon: If there is no black hole but some supermassive object then accreting matter hits the surface of the supermassive object and the kinetic energy becomes thermalized and radiates away. This radiation results in an "UV bump" (which is not observed). But if there is a black hole then the accreted matter passes the event horizon without losing kinetic energy by radiation and without creating an UV bump (which agrees with observation).
The non convincing argument: If there is a *degenerate* object e. g. similar to white dwarfs, neutron or quark stars then the accreting matter is taken to enlarge the degenerate object and this consumes kinetic energy because the Fermi edge has to be overcome and no UV bump will arise, too. [jb 18] and chapter 21.7 give an example of a supermassive, degenerate object having no event horizon.
In more detail: At first some citations of BRODERICK et al.
“That Sgr A* is indeed a black hole, i.e., contains a horizon, is implied by its spectral energy distribution (SED), which lacks the thermal bump associated with accretion onto a photosphere (Broderick & Narayan 2006; Broderick et al. 2009).” **This argument became part of the EHT papers of M87* and SGR A*. [412]**
“Accretion onto compact objects with a surface, e.g., white dwarfs, neutrons stars, results in the formation of a boundary layer in which any remaining kinetic energy contained within the accretion flow is thermalized and radiated. In contrast, gas

accreting onto a black hole is free to advect any excess energy across the horizon without further observational consequence. If the mass accretion rate can be independently estimated, this difference provides an observational means to distinguish between the presence of a surface, or more accurately a "photosphere," and a horizon" [34.4]
"Therefore, Sgr A* must have an event horizon behind which the kinetic energy of the infalling accreting gas is hidden." [34.4]
"We consider a model in which Sgr A*, the 3.5x10^6 M_{sun} supermassive black hole candidate at the Galactic Center, is a compact object with a surface. Given the very low quiescent luminosity of Sgr A* in the near infrared, the existence of a hard surface, even in the limit in which the radius approaches the horizon, places severe constraints upon the steady mass accretion rate in the source, requiring $dM/dt <$ 10^-12 M_{sun} /yr. This limit is well below the minimum accretion rate needed to power the observed submillimeter luminosity of Sgr A*. We thus argue that Sgr A* does not have a surface, i.e., it must have an event horizon. The argument could be made more restrictive by an order of magnitude with microarcsecond resolution imaging, e.g., with submillimeter VLBI." [34.2, 34.3]
These citations prove that the existence and creation for degenerate objects is not taken into consideration. But they also prove that **the theory of BRODERICK et al. has a very convincing kernel.** It proves that an alternative to black holes cannot be some ordinary stellar objects since the measured temperatures are not high enough. It proves that alternatives to black holes if there are ones have to be degenerate objects. This is the case for the alternative of the Lorentz interpretation [jb 18] and chapter 21.7.
When degenerate objects (neutron stars, white dwarfs, or possibly quark stars) grow you will need an energy supply. In order to occupy higher energy levels, accretion energy of the infalling matter is required. If the volume of the degenerate object does not change, energy levels become occupied, all of which are greater than the Fermi momentum p_F.. Only if the volume of the degenerate object increases then there are new, unoccupied energy levels less than the Fermi momentum p_F or the Fermi Energy E_F..
The calculation of supermassive, degenerate objects by the author [jb 18] and chapter 31.7 using the TOV (Tolman-Oppenheimer-Volkoff equation) shows both effects. As the total mass increases, the radius in Schwarzschild units remains nearly constant (1.56 r_{sm}), but this means an increase in volume in usual units. On the other hand, the central density n_e and thus p_F also increases as the total mass grows, see formula (1). As density n_e increases, p_F rises. If the density of a degenerate object becomes arbitrarily large and this assumption is obvious, if one assumes that a black hole does not form, then p_F becomes arbitrarily large and there is no energy left for the thermalisation.
In both cases, accretion energy is consumed and not thermalised. This effect has not been investigated in [412], [34.2] - [34.4]. But **the formation of degenerate objects has some properties of black holes**. The latter absorbs kinetic energy from particles that pass the event horizon, but degenerate objects transform kinetic energy into excited states of the degenerate object. In both cases, the thermalisation of the falling matter is completely or partially avoided.

(1) $$n_e = \frac{8\pi}{3 \cdot h^3} p_F^3$$

The formular (1) shows: As fermion density n_e increases, the Fermi momentum p_F increases as well. If the density of a degenerate object becomes arbitrarily large and this assumption is obvious, if one assumes that a black hole does not form, then p_F becomes arbitrarily large and there might be no energy left for the thermalisation.

3.) Derivation of formula (1)

The formula (1) for p_F follows from general considerations. The author follows HANSLMEYER, Chapter 9.1.5 [143.2]:
Fermions include particles with half-number spin, e. g. electrons or other elementary particles such as quarks and core components. For them, the *Pauli principle* applies: Each quantum cell of a 6-dimensional phase space

(2) $(x,y.z,p_x,p_y,p_z)$

shall not contain more than two fermions, in our case electrons.
The volume of such a quantum cell (or phase space cell) is:

(3) $h^3 = dp_x\, dp_y\, dp_z\, dV$

So if we look at a shell $[p,p+dp]$ in the momentum space, then there are $4\pi p^2 dV/h^3$ quantum numbers that do not contain more than $8\pi\, p^2 dV/h^3$ electrons. So, from quantum mechanics, the condition follows:

(4) $f(p)\, dp\, dV <= 8\pi\, p^2 dV/h^3$

$f(p)$ gives the number of particles in the range $dp\, dV$. This is derivable by quantum mechanics using the formula of the Fermi-Dirac distribution.
The state in which all electrons have the lowest energy without violating the *Pauli principle* is that in which all phase space cells are equipped with two electrons up to the momentum p_F. All other phase space cells, however, are empty:

(5) $f(p) = 8\pi\, p^2/h^3 \qquad p <= p_F$

(6) $f(p) = 0 \qquad p > p_F$

From this, it is to be inferred:

(7) $$n_e\, dV = \int_0^{pF} \frac{3\pi P^2\, d\rho}{h^3} = \frac{8\pi}{3h^3} p_F^3\, dV$$

That's the formula (1) above.

4.) At least three Sgr A* characteristics predicted by the EHT group contradict with astronomical observations

As explained in [jb 17], there are a number of astronomical observations that contradict with the properties of black holes in classical general theory of relativity. Three of them relate to characteristics of Sgr A* as predicted by the EHT group. These are

1.) a*=0.9375 against a*=0.15;
2.) spin direction “face-on” against “edge-on”;
3.) accretion radiation variability arising with accretion disks against variability arising with accretion wind.

These details are presented in chapter 13 of [jb 17] together with a series of illustrations and should not be repeated here. If, for the sake of simplicity, the independent astronomical observations are assumed as correct, which could change by more detailed examination, then the too high spin a*=0.9375 has special significance. A high spin value fits with the observed radius of the light ring seen in the EHT image of Sgr A* (and also of M87*, see fig. 24.9). The light rings in fig. 24.9 are explained to be the "innermost stable circular orbit" with radius r_{isco} of a "thin accretion disk". Figure 3 of [jb 19] explains the quantitative relationship between r_{isco} and spin. If the spin is nearly zero, as the observation shows, $r_{isco} \sim 6\ r_{sm}$ is too large to be the radius of the light ring of the EHT image. However, since it is the innermost area of the accretion disc, which glows brightly due to special, frictional effects and thus explains the brightness of the black holes in the galactic centers, this explanation for the bright glow of the galactic black holes is ruled out. The mapping of r_{isco} and brightness of Sgr A* is standard, see [413].

The three contradictory observations combined exacerbate this problem. They contradict the standard assumption that the accretion of matter to black holes is carried out by means of a "thin accretion disk" and is therefore radiation intensive. Observation 1.) excludes that the light ring in the Sgr A* image is the innermost area of an accretion disc, because r_{isco} is too large. Observation 2.) gives an accretion disc the wrong direction, it should be "edge on". In the same way, observation 3.) contradicts a heating and radiation due to friction like processes of the accretion disc because 3.) means accretion in larger chunks. The black hole does not feed with "porridge" but devours "fresh flesh". [277.1].

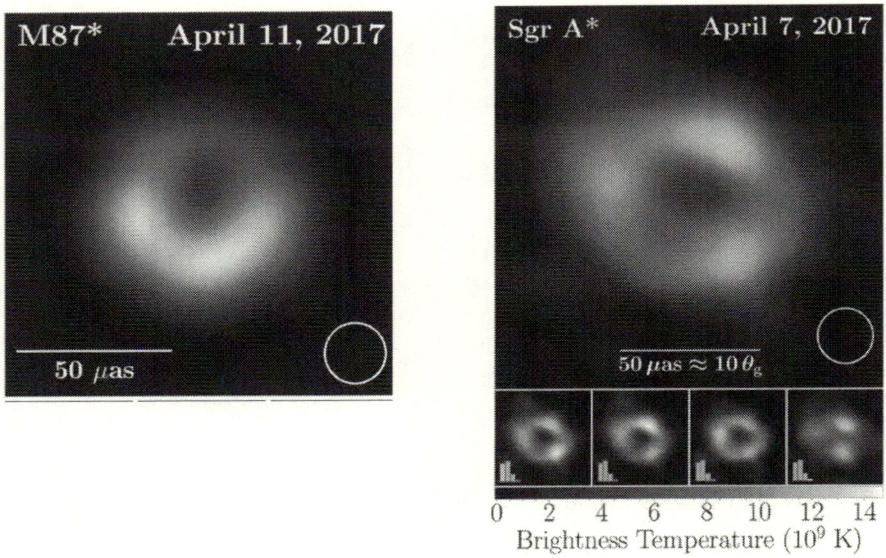

Fig. 24.9 Famous EHT images of M87* and Sgr A*.

The EHT group explains both images as follows: the bright ring is the radiating inner ring of the accretion disk, the central, circular dark area is the black hole plus scattered radiation. **The alternative explanation:** the accreted matter hitting the supermassive object produces a hot, bright ring, much brighter than the unhit inner region of the supermassive object.

24.12. Cosmic Microwave Background (CMB) and the so-called WMAP cold spot

EI: The cosmic microwave background (CMB) is the present-day observation of recombination radiation during the formation of H atoms from protons and electrons at the time $\tau \cong 300000$ years after the Big Bang.

LI: Recombination could still be taking place today in distant regions that are still moving away from the point of origin of the Big Bang at speeds sufficiently close to the speed of light (time dilation, gravitational redshift, differences between t and τ , see ch. 19.8.4, cosmological simultaneity).

With both models, the observed isotropy of the CMB can be explained.

EI: The Sachs-Wolfe effect, which e.g. is included in the important observations of the WMAP cold spot [312], has its cause in frequency changes of light when it has to fight against a gravitational potential.

LI: The Sachs-Wolfe effect cannot be explained by different gravitational potentials as is the case for the EI, since light changes its wavelength and speed but not its frequency, see ch. 21.1. The effect arises because there is a potential difference between the point of origin and the point of reception right from the start and the atoms emitting the light oscillate slower or faster. With the models indicated above, there should be no difference in the prediction of effects from the CMB spectrum: On the one hand, expanding space and cosmological simultaneity of the EI, and on the other hand, an exploding 'dust star' and our location in space (approximately) as the center of the Big Bang for the LI. If one assumes purely radial expansion processes, a present-day 'cold spot' also already existed when the CMB radiation was emitted and the Sachs-Wolfe effect is in a way inevitable - and what has not yet been taken into account: Since the time of flight of the light has changed, it is compared with light that was emitted at a different time. Both models make the same predictions, they are mathematically equivalent. If one considers external influences or transversal movements and time-varying density fluctuations, e.g. through the formation and development of galaxies and through the expansion of the 'dust star' as a whole, differences in predictions of EI and LI should not be ruled out.

24.13. The smallness of the Cosmological Constant

The Robertson-Walker metric (RWM) describes both the explosion of a dust star and the expansion of the universe, see ch. 18.5. This led to the question: Why is the Big Bang the emergence of space and time and not, as would seem obvious, the explosion of a supermassive object, i.e. the emergence of a metagalaxy in space and time? As it turns out, the smallness of the cosmological constant is closely correlated with this question. In the first case, the non-empty vacuum must be co-created; in the second case, it already exists and would only have to be modified. "To explain the smallness of the cosmological constant Λ is one of the most outstanding challenges in modern theoretical physics" because

(24.25) $\Lambda(\text{observed}) = 10^{-122}\,\Lambda(\text{theoretical})$

and "thus wrong by 122 orders of magnitude" See BERGSTRÖM and GOOBAR [22].

A possible solution is given by the second variant of the RWM to describe the expansion of a dust star. In this case, the non-empty vacuum together with infinite space already exists before the big bang and the following applies:

Λ(observed) = Λ(theoretical, changed by big bang) - Λ(theoretical, before big bang)

Such a difference can be arbitrarily small for a dust star and is even zero if Λ(theoretical) is not changed by the big bang. However, the usual reading of the RWM means, i.e. the EI requires, that continuously during the expansion empty space together with non-empty vacuum must be created out of nothing and this in turn means that one would have to observe Λ(theoretical) and not Λ(observed) as the above difference. The problem of the smallness of Λ - the contradiction between Λ(theoretical) and Λ(observed) - remains in this case.

For the LI, Λ(theoretical) holds and the real problem at present is merely to give a mechanism that slightly changes Λ(theoretical) at the big bang or, assuming that Λ(theoretical) does not change, to find out something quite different as the cause of Λ(observed). The author's paper [vor24] describes - using BERGSTRÖM and GOOBAR [22] - the convincing quantum mechanical reasons for the enormous size of Λ(theoretical) and explains why for the EI this must indeed be called a towering problem, while for the LI the smallness of Λ does not mean any fundamental difficulty, but initially only an open question like many others. *On the contrary, if the LI had such a problem with the quantum field theory and the EI did not, the author would regard the LI as refuted.*

25. LUDWIG NEIDHART - Review

Review of 'Special and general theory of relativity for physicists and philosophers' by J. Brandes und J. Czerniawski, Karlsbad 4th edition 2010

Dr. habil. Ludwig Neidhart

28.1.2016

Written largely by Jürgen Brandes, with a one-chapter contribution from Jan Czerniawski, this book is a pro-found and a technically in-depth representation of the „Lorentzian Interpretation of the theory of relativity" (LI), which has thus far been recognised by only a few, and is set against the Einsteinian standard interpretation. It is the latest stance in a „work in progress", which since 1994 has now appeared in its 4th extended edition. The book captivates with its mathematically exact, and yet for the most part, generally comprehensible representation. It also impresses the reader with physically as well as philosophically transparent arguments and a wealth of ideas. It includes revisions of earlier approaches, new ideas and critical views also with regard to arguments within the Lorentzian camp (thus, for example, the claim of some Lorentz supporters on the difference between the one-way speed of light and average speed of light on closed paths is rejected on p. 75 and p. 107).
Overall, the reader gains insights into a very lively and innovative research area and school of thought, beyond the mainstream. In addition, one also learns something through suitably selected and generally lesser-known quotes from Einstein and other researchers to underpin what has been stated (especially statements relating to LI in ch. 7). In light of these many positive features one can look beyond the fact that perhaps the one or the other formulation appears unclear, and several marginal errors in the text as well as in some formulas are noted. However, in none of these cases does it affect the correctness of the results and conclusions.

The remarkable and special feature of this book in comparison to other textbooks on the theory of relativity is that the topic of „The theory of relativity" is dealt with more comprehensibly and more refined than usual. In this way, physical and philosophical aspects are simultaneously illuminated (ch. 11). It clearly differentiates between facts (cf. ch. 5 on experimental proof) and interpretations, and as the readers are introduced step by step into the theory of relativity they also learn about the two most important competing interpretations (LI and EI) of the theory of relativity.

Although the authors clearly plead for the LI (cf. in summary ch. 12), they avoid one-sidedness and also present the EI fairly so that readers can form their own judgment.

Finally, alongside the special theory of relativity, to which the standard representations of the LI are usually limited, the general theory of relativity and cosmology are also included (ch. 14 – 22). The extension of the LI to the general theory of relativity and cosmology is partly a new development of the LI, where the authors themselves play a significant role. This extension is based on the idea of Poincaré, that the abstract space-time curvature of the EI can be explained as the effect of a real benchmark distortion.

For the EI, absolute space and time lengths do not exist, and there is neither rest nor movement nor simultaneity in the absolute sense. For the LI however, all of these do exist because here the existence of a preferred absolute stationary reference system is postulated (even though this system cannot be detected through measurement). This postulate, a hallmark of the LI, is simply added to Einstein's postulates, which do not have to be negated (compare chapter 6.6, 13 and 20), whereas the special feature of the EI is that it negates precisely this postulate.

Both interpretations are in accordance with observations to date and can thus be counted as experimentally equal as already Sexl and Mansouri had ascertained in their "test theory of special relativity" (ch. 7.4). Experimental decisions for the one or the other theory are therefore not possible, or at least so it seems.

In the peripheral zone of the theory however, especially in its extension to the general theory of relativity and cosmology there are still experimentally testable differences between the predictions of both theories. And so it is shown that in applications of the general theory of relativity and cosmology, there are indeed principal experimental testable differences (ch. 21.2, 21.5, 24.8 and 24.11). Furthermore, in this relation, it is important that according to the LI (but not to the EI) forces arise in the Lorentz contraction, and that in the LI-version of the general theory of relativity (differently to the EI-version thereof) it is easy to explain energy conservation and the negative Newtonian gravitational potential. Moreover, there are differences in the solution of the paradoxes, which in the LI often appear simple and thus convincing, whereas based on the EI, there are sometimes different (and mutually exclusive) solutions provided.

Generally known and also lesser known paradoxes and their solutions are also illustrated here in detail: the paradoxes of Bell (9.3 and 24.3), Wood (10.3), Ehrenfest (10.5), Sagnac (10.7) and generally the problem of circular accelerating bodies (ch. 9.4 and 24.1.3) as well as the Garage and Lid paradox (10.2 and 10.4). The best known of these is the clock or twin paradox (ch. 10.8) presented with different solutions in particular detail.

Of special interest for the debate between philosophy and the natural sciences are the "philosophical contributions" of EI and LI in chapter 11, where various aspects of the phenomenology of times are discussed and various interpretations are developed with regards to the Minkowski space, the Ether concept and the interpretation of physical quantities in EI and LI. Lastly, philosophical reflections are also relevant here because the two interpretations had been mainly philosophically motivated by Einstein and Lorentz.

Also enlightening is chapter 8, written by Czerniawski, where initially a general kinematic transformation formula is deduced from four plausible assumptions. Remarkably, the linearity of the formula does not belong to the presupposed assumptions, but is the result thereof. Depending on the specification of certain parameters, the general transformation formula goes across into the non-relativistic Galilean transformation or the relativistic Lorentzian transformation. The addition of two further assumptions, namely Einstein's principle of relativity and the principle of the constancy of the speed of light then ensures the accuracy of the Lorentz transformation.

Czerniawski defends the absolute reality of relativist effects and distances himself from more radical critics of Einstein, who, for example, deny Einstein's postulates. He believes that the actual core of the LI is the postulate of non-invariant, absolute sizes and a thereby related privileged reference system. In favour of this postulate, Czerniawski points to the experience of time flow and the correspondence principle (whereby a newer physical theory is meant to include the old theory as a special or extreme case). Czerniawki believes this principle is violated if one sees the Minkowski timespace as real, as it appears in the EI; because in this timespace there is no element that corresponds with the absolute time of the non-relativist physics.

Czerniawski finally differentiates a material ether (whose existence is hypothetical) from the "protophysicalist" geometric ether, required by LI, which is identical with the special frame of reference. Should a material ether exist, one can view it as the special frame of reference in this ether resting frame of reference. Otherwise, with regard to clocks, one can determine absolute simultaneity through suitable adjustments free from the influence of gravity and movement, and define a special reference system through the Weltlinien, which run orthogonally to the hyper levels of absolute simultaneity."

All in all this book can be recommended to philosophers of science and expert physicists, but also to all those laypersons interested in a modern physical view of the world. Also those who are critical of the stance of the authors will be able to profit from reading it because the facts and arguments laid down here can certainly also provide food for thought in many aspects, and is definitely suitable for broadening the horizons of the reader.

26. Directories

26.1. Useful and hypothetical figures

Speed of light *c:* $2.998\ 10^{8}$ m/s

Gravitational constant *G:* $6.67\ 10^{-11}$ m^3 /(kg s $)^2$

One light year: $0.946\ 10^{16}$ m

One parsec, pc: 3.o86 10^{16} m = 3.26 light years

One year: 3,156 10^{7} s

Mass of the sun: 1.99 10^{30} kg

Radius of the sun: 6.96 10^{8} m

Hubble constant H_0: 70.5±1.3 (km/s)/Mpc and 67.4 (km/s)/Mpc. (The EI of the RWM allows only one value, s. ch. 24.11.).

$1/H_0$: 1.4 10^{10} years

Delay parameter $q_0 = (4\pi G\rho_0 - \Lambda)/3H_0^2$: -0.59

(negative q_0 means: the expansion speed of the universe is increasing)

Accelerated expansion has existed since $z \leq 1$, which is about 50% of the world age,

Delayed expansion until $z \geq 1$

The universe persists to this daytime t_0: 1.35 10^{10} years

Density of the universe today, ρ_0

Baryon density (normal matter): 4.3 10^{-28} kg/m ^{3}

incl. dark matter: 2.5 10^{-27} kg/m ^{3}

cosmological constant $\Lambda = \Omega_\Lambda 3H_0^2$: 1.14 10^{-35} s^{-2}

($\Lambda/3$ times the distance is the acceleration by Λ , [22]. The theoretical value ofΛ is 10^{122} -times greater).

Numerical values from: [28], [357], [230], [20], [189.1].

Hypothetical, calculated for a finite universe with the previous values for $H_0 = 1/(1.8\ 10^{10}$ years), $q_0 = 1$ and $k = 1$:

Maximum expansion of the universe R_m : 3.6 10^{10} Light years

It is reached at time t_m : 5.7 10^{10} years

(The universe will last twice as long)

Time passed until today t_0 10^{10} years

Present-day extent R_0	1.8 10^{10} Light years
Today's density:	1.2 10^{-26} kg/m^3

(40 times greater than the value from galaxy observations).

Today's volume $V_o = 2\,\pi^2 R_0^{\,3}$:	9.7 10^{79} m^3
Total mass of the universe:	11.6 10^{53} kg
which corresponds to:	6 10^{23} solar masses

(Of the hypothetical values, t_0 agrees well with the value calculated from the age of rocks and stars, but the density was determined too high. This led to the search for "missing matter" [23]. Those who preferred a Euclidean universe ($k = 0$) also had to search for missing matter. The proof of a positive cosmological constant (i.e. of dark energy) has solved this problem. The universe is Euclidean ($k = 0$), consists of visible and dark matter as well as dark energy and expands exponentially. A finite universe that is only *almost* Euclidean cannot be ruled out).

26.2. Glossary: Special and General Theory of Relativity

Antinomy: Unsolvable internal contradiction of a theory. Often eliminated by minor refinements of the theory, see also paradox.

Big Bang: The formation of the universe about 15 billion years ago through a violent explosion, as called for in the Friedmann cosmos and confirmed by the Hubble law (Lorentz interpretation). The Einstein interpretation sees this as the beginning of the expansion of space and the beginning of time and not an explosion. The physical difference between the two theses lies in the definition of simultaneity, ch. 19.8.4.

Cosmology: the study of the origin and development of the universe

cosmological principle: see Fig. 19.1. The universe has the same structure in all directions and at all distances. A simplified assumption for the development of space models.

cosmological constant: Introduced by EINSTEIN, then pronounced dead by him, is experiencing a renaissance. A positive cosmological constant - it is confirmed by observations of galaxy movements - means a far-reaching force in the universe that opposes gravity. The smallness of the cosmological constant is a major problem for the EI, see ch. 24.13.**dark energy:** The form of energy that is associated with the cosmological constant and causes an acceleration of the expansion of the universe. Not the form of energy that corresponds to normal matter and means mass attraction.

dark matter: in the past, the "missing matter" that would have made the universe finite. Today, the invisible matter that only makes itself felt through gravity, e.g. in the rotation curves of galaxies. Its mass is several times that of visible matter.

Dimension: see Fig. 11.1. property of spaces; indicates the number of independent coordinates with which the position of a point in the associated space is fixed.

Distinguished or excellent inertial system: An inertial system that is supposed to differ from others in certain properties. Thus, one can theoretically assume that the one-way velocity of light only has the value c in one inertial frame and not in all of them and check how this is compatible with the principle of relativity.

Einstein convention for the definition of simultaneity of clocks in inertial systems: Spatially distant clocks run synchronously according to this convention if the velocities of light measured with them yield the value c.

Einstein (EI) and Lorentz (LI) interpretation of special relativity: The physical equations of special relativity can be interpreted differently with regard to space and time, but the two interpretations do not differ in the prediction of physically measurable phenomena. For the Einstein interpretation there is no absolute simultaneity, for the Lorentz interpretation there is, but it is not measurable.

Einstein (EI) and Lorentz (LI) interpretation of general relativity: The various metrics in general relativity can be interpreted differently with regard to space and time; in the prediction of physically measurable phenomena, the two interpretations do not differ in principle, but in verifiable details. For the Lorentz interpretation, curved spaces describe changes in scale and not space and time, for the Einstein interpretation, scales do not change in gravitational fields, but space and time are curved there, see also Poincaré's thesis.

Einstein field equations ... remain and the Schwarzschild metric (SM) is the basis for both interpretations, see also Hilbert-Einstein equations.

Equation of state: The state of a gas is its pressure p, its volume V and its temperature T. Their dependence on each other describes the equation of state. For ideal gases, this is the ideal gas equation $pV/T = const$. For degenerate states, e.g. in neutron stars, the pressure in matter (neutrons) is only a function of its density.

Euclidean geometry: The congruence theorems for plane triangles apply, as learned in school. This is fulfilled in three-dimensional Euclidean space for every plane. Three-dimensional Euclidean space is defined by its metric, formula (18.6).

Field: Spatial area in which forces or other physical quantities occur. Thus one speaks of gravity fields or gravitational fields as the areas in which gravitational forces are present and correspondingly of electric or magnetic fields.

Four-dimensional space-time continuum: Not space by itself and time by itself are real, but only the union of both into a four-dimensional continuum or four-dimensional space ("continuum" is another term for "space in its more general meaning"). Because of length contraction and time dilation, there are no lengths and time intervals independent of inertial systems. There is only a four-dimensional spatial-temporal concept of distance that has a meaning independent of the inertial system and justifies the concept of a four-dimensional space.

Friedmann cosmos: The - finite or infinite - cosmos as it can be derived from the cosmological principle with the simplest physical assumptions. It arose in a big bang and continues to expand, either to infinity or only to a maximum size, after which it shrinks and finally collapses again.

Galilean transformation: see Lorentz transformation

Gaussian coordinate system: Curvilinear coordinate system as required for curved spaces, see Fig. 16.8. Longitude and latitude circles of the earth's surface also form a Gaussian coordinate system.

Geometry(s): Instead of curved spaces, we can speak of non-Euclidean geometries. Thus, spherical geometry is the geometry in the curved space of a spherical surface.

General theory of relativity: Newtonian theory of gravitation further developed by Einstein. An extended principle of relativity that applies to reference systems at rest in gravitational fields and to accelerated reference systems, and not only to inertial systems, allows improved predictions, e.g. about the motion of planets and the movement of clocks in gravitational fields.

Gravitational field: see field

Gravitational collapse: Collapse of a cooled star under the influence of gravity. Theoretically, a neutron star or black hole is formed.

Hilbert-Einstein equations: HILBERT was the first to discover the so-called Einstein field equations, which should be expressed in the name, see SOMMER [346].

Inertial system: (inertia, lat. the inertia) A reference system that moves in a straight line and uniformly like inertial masses that are not subject to forces. Reference frame is the abstraction for the measuring place one is working on and an important property is its velocity relative to other measuring places.

Interpretation: see Einstein and Lorentz interpretation

Length contraction: Bodies moving relative to an inertial system are (or appear) shortened in the direction of motion. The proof is provided by the famous MICHELSON-MORLEY EXPERIMENT. Length contraction is the starting point of various paradoxes.

Lorentz transformation: Physical formulae for converting position and time coordinates within inertial systems. They are the basis of special relativity and replace the classical Galilean transformation, for which there is a uniform time $t = t'$ for all inertial frames.

Metric: Measure, important term in physical and mathematical curved spaces. The metric determines the distance between two (closely spaced) points. Depending on the metric, points on a spherical surface, for example, have the length of their chords or their circular arcs as their distance.

Minkowski space: s. four-dimensional space-time continuum

Newton's law of gravitation: NEWTON'S assumption about the magnitude and direction of gravitational forces. It allows the calculation of planetary orbits and is of great importance as a limiting case for the development of general relativity.

Newtonian gravitational potential: It is negative because energy must be expended to remove a body from the gravitational field. Because of $E = mc^2$, this means negative mass or mass decrease, see ch. 2.3.

One-way speed of light: Measurement of the speed of light in one direction (and not back and forth) using two clocks. It has great theoretical significance, but is not feasible because the clocks must first be synchronised, which in turn requires light signals or an equivalent method.

Paradox, paradox: Seemingly contradictory conclusion from a theory, not an antinomy. In special relativity, they arise from time dilation (twin paradox) and length contraction (e.g. garage paradox).

Phenomenology, phenomenological: branch of philosophy, or philosophical direction, which makes statements about the essence and mode of being of objects. Thus physical and mental processes are real, they exist, thought and imagination contents are thought or imagined, mathematical objects are ideal, and real if they can be part of reality, such as spherical surfaces in real space. The philosophical problem of special relativity is: Is real space a four-dimensional space-time continuum?

Poincaréan model: see Fig. 18.12. Representation of hyperbolic geometry in the normal (Euclidean) plane.

Poincaréan thesis: For it, a curved physical space can always be interpreted as normal (Euclidean) space, in which scales and objects are subject to change instead.

Principle of equivalence: The starting point of the equivalence principle is the observation that all bodies fall at the same speed, i.e. the 'equality of heavy and inert mass'. One does not feel any gravitational forces during free fall and one does not observe any on the falling bodies or light signals. From this follows the principle of equivalence: "In local inertial frames, i.e. in free fall, the laws of special relativity apply." The variant for the LI: see ch. 21.6.

Principle of constancy of the speed of light: The speed of light has the value *c in* all inertial systems, or it is measured this way in all inertial systems because of the principle of relativity.

Principle of relativity No inertial system is distinguished from others by physical measurements. All laws of nature have the same form in inertial systems. (The Lorentz interpretation adds: As far as the laws refer to measurable and not to so-called hidden parameters, s. ch. 6.5).

Proper time, proper time interval: The time indicated by a clock connected to a moving or stationary body. An elementary physical measurement that must be correctly predicted by any theory.

relativistic and Lorentzian interpretation: designation for Einstein (EI) and Lorentz interpretation (LI) in the earlier editions.

Rest mass, rest length: Amount of the mass, the length of a body, measured in the inertial frame in which it is at rest.

Robertson-Walker Metric: The Metric for Spaces of the Friedmann Cosmos

Schwarzschild metric: The metric for spaces in centrally symmetric gravitational fields, e.g. field of the sun.

Space: Multiple meaning: a) The normal three-dimensional space surrounding us, our reality, the real space. b) Mathematical spaces. Generic term for everything that can be imagined as consisting of points and for whose points a distance can be defined. In this usage, the spherical surface is a space (with two dimensions). The distance between two points is the length of their circular arc. c) Mathematical-physical spaces: mathematical spaces with physical meaning, e.g. the surface of the earth as a spherical surface. They are also called real spaces in contrast to b.)

Spacetime: Generic term for spaces that have coordinates of place and time. Thus, the four-dimensional space-time continuum is a space-time like the curved spaces of the general theory of relativity including time.

Special theory of relativity: EINSTEIN'S consistent application of the principle of relativity with revolutionary predictions about moving lengths and clocks, as well as the relationship between mass and energy. It is fundamental knowledge of physics.

Time dilation, time stretching: Moving clocks run slower, a space traveller remains younger than his twin on Earth. It is the starting point of the clock or twin paradox.

26.3. Bibliography and references

Attached page references refer to the source.

[1] Achuthan, P.; Shankara, T. S.; Venkatesan, K.: Ether – as advocated by Einstein and others. Speculations Sci. and Technol. (Schweiz) 2, 277 (1979)

[2] Alley, C.: Maryland-Experiment 1975, in: [337], Page 23ff, (experiment of [143] in higher accuracy).

[3] American Association of Physics Teachers: Special Relativity Theory. Selected Reprints. New York: American Institute of Physics 1962 (References clock paradox)

[4] Anderson, R. S. J.; Stedman, G. E.: Distance and the conventionality of simultaneity in special relativity. Found. of Phys. Lett. 5, 199 (1992)

[5] Anderson, R. S. J.; Stedman, G. E.: Spatial measures in special relativity do not empirically determine simultaneity relations. A reply to Coleman and Korte. Found. of Phys. Lett. 7, 273 (1994)

[6] Anderson, R., Vetharaniam, I., Stedman, G. E., Conventionality of synchronisation, gauge dependence and test theories of relativity, Physics Reports 295, 93–180 (1998) Literature references for both interpretations.

[7] Arminjon, M.: Accelerated Expansion as Predicted by an Ether Theory of Gravitation, gr-qc/9911057v2 6 Dec 1999.

[8] Arminjon, M.: *Ether theory of gravitation: why and how?* arXiv:gr-qc/0401021 v1 7 jan 2004

[9] Arp, H. Evolution of quasars into galaxies and its implications for birth and evolution of matter in: F. Selleri, ed., Open questions in relativistic physics, (1998) Published by Apeiron 4405, rue Dominique, Montreal, Quebec H2W 2B2 Canada P. 265–273

[10] Aspden, H.; Eagles, D. M.: Aether Theory and the Fine Structure Constant. Physics Letters 41A, 423 (1972)

[11] Atkinson, R. d'E.: Acceleration and the Lorentz transformation. Am. J. Phys. 48, 581 (1980)

[12] Audretsch, J.: Gravitationstheorie in flacher Raum-Zeit. in: Audretsch, J.; Mainzer, K. (Hrsg.): *Philosophie und Physik der Raum-Zeit.* Grundlagen der Exakten Naturwissenschaften, Bd. 7. Mannheim, Wien, Zürich: BI Wissenschaftsverlag 1988 (Seite 76)

[13] Bader, F.; Dorn, F (Hrsg.): *Physik* – Hannover: Schroedel-Verlag 1986 (As an example for Baden-Württemberg. The special theory of relativity belongs to the fundamentals of physics. Its Einstein interpretation is/was already well taught in school). (page 415)

[14] Bailey, J. et al.: Nature 268, 301 (1977) (very accurate measurement of the lifetime of mesons in the storage ring).

[15] Baving, B.: *Ergebnisse und Probleme der Naturwissenschaften – Eine Einführung in die heutige Naturphilosophie.* 10. Aufl. S. Hirzel Verlag Zürich 1954 S. 134 (Preferential position of the present)

[16] Beauregard, L. A.: The sui generis conventionality of simultaneity. Philos. Sci. 43, 469 (1976)

[17] Bell J.S.: *How to teach Special relativity.* Progress in Scientific Culture 1, No 2 (1976); repr. in: Bell J.S., Speakable and Unspeakable in Quantum Mechanics, Cambridge Univ. Press, Cambridge 1987.

[18] Bell, J. S., How to teach special relativity, repr. in: Bell, S. J., *Speakable and Unspeakable in Quantum Mechanics,* Cambridge Univ. Press, Cambridge 1987

[18.1] Bennet, J., Donahue, M., Schneider, N., Voit, M.: *Astronomie – Die kosmische Perspektive*, Pearson Studium 2010

[19] Berenda, C. W.: The Problem of the Rotating Disk. Phys. Rev. 62, 280 (1942)

[19.1] Berends, F.: *Hendrik Antoon Lorentz: his role in physics and society,* J. Phys.: Condens. Matter 21 (2009) 164223 (8pp)

[20] Bergmann; Schaefer, *Lehrbuch der Experimentalphysik,* Berlin; New York: de Gruyter, Bd. 1 – Bd. 8. Daraus: Raith, W. (Hrsg.), Blome, H. J., Feitzinger, J., Hoell, J., Priester, W., Scheffler, H., Speer, F., *Sterne und Weltraum,* Bd 8, 1997 S. 314

[21] Bergmann; Schaefer, *Lehrbuch der Experimentalphysik,* Berlin; New York: de Gruyter, Bd. 1 – Bd. 8. Daraus: T. Dorfmüller, W. T. Hering, K. Stierstadt Bd. 1. *Mechanik, Relativität, Wärme* 11. Aufl. 1998 (Treatment of Bell's rocket example.)

[21.3] Bergmann; Schaefer *Lehrbuch der Experimentalphysik,* Berlin; New York: de Gruyter, Bd. 1 – Bd. 8. Daraus: Bd.3, S.1362

[21.4] Bergmann; Schaefer, *Lehrbuch der Experimentalphysik,* Berlin; New York: de Gruyter, Bd. 1 – Bd. 8. Daraus: Bd.4, S. 922

[22] Bergström, L.: Goobar, A.: *Cosmology and Particle Astrophysics.* 2. Ed., Springer Berlin, Heidelberg, New York 2004, S. 116

[23] Berry, M.: *Kosmologie und Gravitation.* Stuttgart: B. G.. Teubner 1990, (Seite 98,107,140,142ff, 149)

[24] Berzi, V.; Gorini,V.: Reciprocity Principle and the Lorentz Transformations. J. Math. Phys. 10, 1518 (1969) (Derivation from reciprocity principle)

[25] Birk, G. T., Lesch, H., Schopper, R., Wiechen, H.: Particle acceleration in AGN magnetic reconnection regions. Astroparticle physics 11, 63, (1999)

[26] Biskamp, D.: *Magnetic reconnection in plasmas*. Cambridge University Press 2000

[27] Born, Hedwig, Born, M., *Der Luxus des Gewissens – Erlebnisse und Einsichten im Atomzeitalter,* München: Nymphenburger Verlagsbuchhandlung 1969 Seiten 131,133

[28] Born, M.: *Die Relativitätstheorie Einsteins.* 4. Aufl. Berlin, Heidelberg, New York: Springer Verlag 1964 5. Aufl. 1969 (p 219, an impressively beautiful account of rel. int.).

[29] Brandes, J., *Die Einstein´sche und lorentzianische Interpretation der Relativitätstheorie am Beispiel der Ehrenfestparadoxie*, Tagungsband und CD Didaktik der Physik 1999 Ludwigsburg, S. 411

[30] Brandes, J.: *A new evaluation and restrictive interpretation of the Lorentz transformation.* Internationaler Kongress für Relativität und Gravitation, München 1988

[31] Brandes, J.: *The Test of the Second Principle in Special Relativity and the Proof of the Lorentz Theory.* Bericht – Physikalisches Staatsinstitut der Universität Hamburg 1968 (unpublished)

[32] Bridgman, P. W., *A Sophisticate's Primer of Relativity*, Introduction by A. I. Miller, Wesleyan University Press, Middletown, Connecticut, USA 1983
Vom gleichen Autor: *Reflections of a Physicist*, Philosophical Library, New York, Jahr: ?.

[33] Brinkmann, K.: *Grundfehler der Relativitätstheorie.* Tübingen, Zürich, Paris: Hohenrain-Verlag 1988 (Exemplary and to be fair, a - brilliantly written - rejection of relativity theory, see also Theimer).

[34] Broad, C. D.: *Scientific Thought* – London: Routledge and Kegan, Paul 1923 und New York: Humanities Press 1969 (Broad recognized and discussed the equivalence of both interpretations, but preferred the Einstein interpretation).

[34.2] Avery E. Broderick and Ramesh Narayan: On The Nature of the Compact Dark Mass at the Galactic Center, 31.3.2006, arXiv:astro-ph/0512211v2

[34.3] Avery E. Broderick, Vincent L. Fish, Michael D. Johnson, Katherine Rosen-feld, Carlos Wang, Sheperd S. Doeleman, Kazunori Akiyama, Tim Jo-hannsen, and Alan L. Roy: MODELING SEVEN YEARS OF EVENT HORI-ZON TELESCOPE OBSERVATIONS WITH RADIATIVELY INEFFICIENT ACCRETION FLOW MODELS, The Astrophysical Journal, 820:137 (16pp), 2016 April 1

[34.4] Avery E. Broderick (1,2), Ramesh Narayan (3), John Kormendy (4,5,6), Eric S. Perlman (7), Marcia J. Rieke (8), Sheperd S. Doeleman The Event Horizon of M87 arXiv:1503.03873v2 [astro-ph.HE]

[35] Bronstein, I. N.; Semendjajew, K. A.: *Taschenbuch der Mathematik* Thun, Frankfurt am Main: Verlag Harri Deutsch 1987 (Seite 32)

[36] Brown, G. B.: Reply by the author. Am. J. Phys. 46, 1071 (1978) (reply to Erlichson)

[37] Brown, Harvey R.: *Physical Relativity – Space-time Structure from a Dynamical Perspective.* Clarandon Press Oxford 2005, p. 121, Swann and others.

[38] Brown, M. R., Cothran, C. D., Landreman, M., Schlossberg, D., Matthaeus, W. H.: Experimental observation of energetic ions accelerated by three-dimensional magnetic reconnection in a laboratory plasma. Ap. J. 577, L63 (2002)

[39] Browne, P. F.: Relativity of rotation. J. Phys. A: Math. Gen. 10, 727 (1977)

[40] Büchel, W.: Die Struktur wissenschaftlicher Revolutionen und das Uhren-"Paradoxon". Z. allg. Wiss. theor. 5, 218 (1974)

[41] Budo, A., *Theoretische Mechanik*, Berlin: Deutscher Verlag der Wissenschaften 1956 (S.106)

[42] Builder, G.: Ether and Relativity. Australian J. Phys. 11, 279 (1958), sowie in [3]

[43] Buras, R, Rampp, M., Janka, H.-Th., Kifonidis, K.: *Improved models of stellar core collapse and still no explosions: What is missing?* arXiv:astro-ph/0303171 v1 7. März 2003

[44] Buth, M.: Zum Verhältnis von Protophysik und spezieller Relativitätstheorie. PN 20, 213 (1983)

[45] Cantoni, V.: Comments on: "Do Metric Standards Contract?" Found. of Phys. 10, 809 (1980) (reply to Phipps)

[46] Capek, M.: Relativity and the Status of Becoming. Found. of Phys. 5, 607 (1975)

[47] Carmeli, M.: The Dynamics of Rapidly Rotating Bodies. Found. of Phys. 15, 889 (1985)

[47.1] Carroll, S., M.: *Lecture Notes on General Relativity*, arXiv:gr-qc/9712019

[48] Cavalleri, G., Tonni, E. Nuovo Cimento B 112, 897 (1997) und G. Cavalleri, E. Tonni, Hadronic J. 20, 173 (1997)

[49] Cavalleri, G.: On Some Recent Papers Regarding the Ehrenfest's Paradox. Lettere Nuovo Cimento 3, 608 (1973)

[50] Cavalleri, G.; Bernasconi, C.: Invariance of light speed and nonconservation of simultaneity of separate events in prerelativistic physics and vice versa in special relativity. Nuovo Cimento B 1O4B, 545 (1989)

[51] Chang, T.: A Space-Time Theory with a Privileged Frame. Found. of Phys. 13, 1013 (1983)

[52] Chang, T.: A suggestion to detect the anisotropic effect of the one-way velocity of light. J. Phys. A 13, 207 (1980)

[53] Chang, T.; Torr, D. G.: A Modified Lorentz Theory as a Test Theory of Special Relativity. Found. of Phys. Lett. 1, 353 (1988)

[54] Cialdea, R.: A New Test on the Second Postulate of Special Relativity Sensitive to First-Order Effects. Lett. Nuovo Cimento 4, 821 (1972)

[55] Claybourne, J. P.: Why an ether is positively necessary and a candidate for the job. Galileian Electrodyn. (USA) 4, 38 (1993)

[56] Clifton, R. K.: Some recent controversy over the possibility of experimentally determining isotropy in the speed of light. Philos. Sei. 56, 688 (1989)

[57] Cole, G. H. A., Dunning-Davies, J. D.: Describing states of super-dense matter, in: Physical interpretations of relativity theory VI, British Society for the Philosophy of Science, London 11.–14. 9. 1998 M. C. Duffy, ed., University of Sunderland GB p. 84

[58] Craig, J., Litvinenko, Y. E.: Particle acceleration scalings based on exact analytic models for magnetic reconnection. Ap. J. 570, 387 (2002)

[59] Cremer, T.; Interpretationsprobleme der speziellen Relativitätstheorie. Thun, Frankfurt am Main: Verlag Harri Deutsch 1988

[60] d´Inverno, R. (Hrsg. G. Schäfer, Jena), *Einführung in die Relativitätstheorie*, Weinheim, New York, Basel, Cambridge, Tokio: VCH 1995 (Mach's Principle, pp. 249, 290)

[60.1] d´Inverno, R. (Hrsg. G. Schäfer, Jena), *Einführung in die Relativitätstheorie.2. Aufl.*, Weinheim, 2. Aufl. New York, Basel, Cambridge, Tokio: VCH 1995, Formel (16.12)

[61] Davies, P. A.; Ashworth, D. G.: Angular velocity in rotating systems. J. Phys. A: Math. Gen. 10, L147 (1977) (reply to Browne)

[62] Davies, P. A.; Jennison, R. C.: Experiments involving mirror transponders in rotating frames. J. Phys. A: Math. Gen. 8, 1390 (1975)

[62.1] Davis, T. M. et al., Scrutinizing exotic cosmological models using ESSENCE supernova data combined with other cosmological probes, arXiv:astro-ph/0701510v1, (2007)

[63] de Abreu, Rodrigo, Guerra, Vasco: *Relativity – Einstein's lost frame*. Veröffentlicht von]extra muros[2005 www.luzboa.com (A recommendable book on the Lorentz interpretation of special relativity).

[64] de Gouveia dal Pino, Elisabete M., Lazarian, A.:Ultra-high-energy cosmic-ray acceleration by magnetic reconnection in newborn accretion-induced collapse pulsars. Ap. J. 535, L31 (2000)

[65] De Ritis, R.; Guccione, S.: Can Einstein's definition of simultaneity be considered a convention? Gen. Rel. and Grav. 17, 595 (1985) (Criticism of Reichenbach - Grünbaum)

[66] Demoulin, P., Klein, K.-L.: Structuring of the solar plasma by the magnetic field, in: J.-P. Rozelot, L. Klein, J. C. Vial (Eds), Transport and energy conversion in heliosphere, Konferenz 1998, S. 99, Springer Verlag 2000

[66.1] Department of Astrophysics: *Lifting the veil on the black hole at the heart of our Galaxy,* 21 January 2019, https://www.ru.nl/astrophysics/news-agenda/news/news-ru/lifting-veil-black-hole-heart-our-galaxy/

[67] Deser, S.: *Self ineraction and gauge invariance.* Gen. Rel. & Grav. I, 9-18, 1970 (from [230], p. 425 and 436)

[68] Dessauer, F., *Naturwissenschaftliches Erkennen – Beiträge zur Naturphilosophie*. 2. Aufl. Frankfurt: Verlag Joseph Knecht 1960 S. 363

[69] Dewan, E.; Beran, M.: Note on Stress Effects due to Relativistic Contraction. Am. J. Phys. 28, 517 (1960)

[70] Dieks, D.: Time in special relativity and its philosophical significance. Eur. J. Physics (UK) 12, 253 (1991)

[71] Dingle, H.: Don't bring back the ether. Nature 14, 113 (1967) (Article factual, by the way.)

[72] Dingle, H.: The twin paradox of the theory of relativity. Antenna (Italien) 53, 17 (1981) (H. Dingle's own summary shortly before his death).

[73] Dirac, P. A. M.: Is there an Äther? Nature 169, 906 (1951)

[74] Dittus, H., Lämmerzahl, C.: *Die Pionier-Anomalie – das größte von Menschen durchgeführte Experiment und sein Rätsel.* Physik J. 5 (2006) 25

[75] Dittus, H.: *Die Pioneer-Anomalie und andere ungeklärte Beobachtungen in unserem Sonnensystem.* Vortrag Physikalisches Kolloquium Karlsruhe 07.07.2006

[76] Dörling, J.: Did Einstein need General Relativity to solve the Problem of Absolute Space? Or had the Problem already been solved by Special Relativity? Brit. J. Philos. Sci. 29, 311 (1978)

[77] Dudley, H. C.: Is there an ether? Multidisciplinary Res. (USA) 4, 6 (1976)

[78] Dudley, H. C.: Phenomenological Model of the Fitzgerald-Lorentz Contraction Bases on Interactions with the Neutrino Sea. Lettere Nuovo Cimento 5, 641 (1972)

[79] Duffy, M. C., Levy, J., Editors: Ether space-time and cosmology Band 1-3 PS Publications (January 6, 2008), C. Roy Keys Inc (January 15, 2009) Apeiron (November 29, 2009) (Many contributions with alternative interpretation of relativity)

[80] Duffy, M. C.: Ether formulations of relativity. Indian J. of Theor. Phys. 28, 295 (1980) (Literaturhinweise) (Seite 310)

[81] Duffy, M. C.: Misconceptions about the ether. Indian J. of Theor. Phys. 28, 141 (1980) (Literaturhinweise)

[82] Duffy, M. C.: *The ether, disclosing models and objective reality.* 2005, michael.duffy9@btopenworld.com

[83] e.g. *Proceedings of Physical interpretations of relativity theory V,* British Society for the Philosophy of Science, London Sept. 1996 M. C. Duffy, ed., University of Sunderland GB

[84] Earman, J.: Space-Time, or how to solve philosophical problems and dissolve philosophical muddles without really thinking. J. Philos. 67, 259 (1970)

[85] Eckart, A.*: Super-Massive Black Holes at the Centers of Galaxies: The Case of Sagittarius A* at the Center of the Milky Way.* Hauptvortrag SYBH 1.4 DPG-Verhandlungen Bonn 2010

[86] Efimov, N. W.: *Höhere Geometrie.* Berlin: Deutscher Verlag der Wissenschaften 1960, (Seite 160, 516)

[87] Ehrenfest, P.: Gleichförmige Rotation starrer Körper und Relativitätstheorie. Phys. Z. 10, 918 (1909)

[88] Ehrenfest, P.: Gleichförmige Rotation starrer Körper und Relativitätstheorie. Phys. Z. 10, 918 (1909)

Einstein did not know the solution, see [92], later he accepted the theses of Herglotz et al [150].

[89] Einstein, A.: *Grundzüge der Relativitätstheorie.* 6. Aufl. Braunschweig, Wiesbaden: Friedr. Vieweg u. Sohn 1990

[90] Einstein, A.: Äther und Relativitätstheorie. Rede Mai 1920, Leiden in: *v.* Meyenn, K.: *Albert Einsteins Relativitätstheorie – Die grundlegenden Arbeiten.* Braunschweig: Friedr. Vieweg u. Sohn 1990 (Seite 116f, 119)

[91] Einstein, A.: Die Grundlage der allgemeinen Relativitätstheorie. Ann. d. Phys. 49, (1916) in: [207]. (Seite 122)

[92] Einstein, A.: Zum Ehrenfestschen Paradoxon. Phys. Z. 12, 509 (1911) (Einstein comments on the reality of the Lorentz contraction in a comment to the paper [Varicak])

[93] Einstein, A.: *Dialog über Einwände gegen die Relativitätstheorie.* Die Naturwissenschaften 6 (1918) S. 102 (Attempt to explain the twin paradox with the general Relativitätstheorie).

[94] Einstein, A.: Kosmologische Betrachtungen zur allgemeinen Relativitätstheorie. Sitzungsberichte d. Preussischen Akademie d. Wissenschaften 1919. In: [207]

[95] Einstein, A.: *Über die spezielle und die allgemeine Relativitätstheorie.* 23. Aufl. Braunschweig, Wiesbaden: Friedr. Vieweg u. Sohn 1992, (Seite 90, 107)

[96] Einstein, A.: Zur Elektrodynamik bewegter Körper. Ann. d. Phys. 17 (1905) in: [207] (Einstein's famous justification of the special theory of relativity)

[97] Einstein, A: Über den Äther. Verh. Schweiz. Naturforsch. Ges. 105, Teil II, 85 (1924) aus [302]

[98] Eisele, A. M.: On the behaviour of an accelerated clock. Helv. Phys. Acta 60, 1024 (1987) (Decay times of elementary particles are independent of acceleration).

[99] Eisenstaedt, J. (Hrsg.): *Studies in the history of general relativity* – (Einstein studies, Bd. 3). Basel, Berlin, Stuttgart: Birkhäuser 1992, (Bd.3 Seite 286, Bd. 1 Seite 218)

[100] Ellis, R.; Bowmann, P.: Conventionality in distant simultaneity. Philos. Sci. 34, 116 (1967)

[101] Erlichson, H.: A comment on Ruderfer's letter. Am- J. Phys. 43, 279 (1975)

[102] Erlichson, H.: Comment on "A Unidirectional Test of Special Relativity?". Am. J. Phys. 41, 1298 (1973)

[103] Erlichson, H.: Einstein synchronization. Am. J. Phys. 46, 1071 (1978)

[104] Erlichson, R.: The Lorentz-Fitzgerald Contration Hypothesis and the Combined Rod Contraction - Clock Retardation Hypothesis. Phil. Sci. 38, 605 (1971)

[105] Erlichson, R.: The Rod Contraction-Clock Retardation Ether Theory and the Special Theory of Relativity. Am. J. Phys. 41, 1068 (1973)

[105.1] Event Horizon Telecope Results
First M87 Event Horizon Telescope Results. I. The Shadow of the Supermassive Black Hole The Event Horizon Telescope Collaboration *et al.* 2019 *ApJL* **875** L1
First M87 Event Horizon Telescope Results. V. Physical Origin of the Asymmetric Ring The Event Horizon Telescope Collaboration *et al.* 2019 *ApJL* **875** L5

First M87 Event Horizon Telescope Results. IV. Imaging the Central Supermassive Black Hole The Event Horizon Telescope Collaboration *et al.* 2019 *ApJL* **875** L4
First M87 Event Horizon Telescope Results. III. Data Processing and Calibration The Event Horizon Telescope Collaboration *et al.* 2019 *ApJL* **875** L3
First M87 Event Horizon Telescope Results. II. Array and Instrumentation The Event Horizon Telescope Collaboration *et al.* 2019 *ApJL* **875** L2

[106] Evett, A. A.; Wangsness, R. K.: Note on the Separation of Relativistically Moving Rockets. Am. J. Phys. 28, 566 (1960) (Kommentar zu (82))

[107] s. Selleri, F

[108] Fahr, H., J., Willerding, E., A.: *Die Entstehung von Sonnensystemen – Eine Einführung in das Problem der Planetenentstehung.* Spektrum Akademischer Verlag Heidelberg Berlin 1998

[109] Falk G.; Ruppel, W.: *Mechanik Relativität Gravitation.* Berlin, Heidelberg, New York: Springer-Verlag 1983 (Seite 319)

[110] Feenberg, E.: Conventionality in Distant Simultaneity. Found. Of Phys. 4, 121 (1974)

[111] Feinberg, E. L.: Can the relativistic change in the scales of length and time be considered the result of the action of certain forces? Sov. Phys. Usp. 18, 624 (1976)

[112] Fendt, C., Camenzind, M.: *On collimated stellar jet magnetospheres*. Astronomy & astrophysics 313, 591-604, 1996

[113] Fließbach, T.: *Allgemeine Relativitätstheorie*. Mannheim, Wien, Zürich: BI Wissenschaftsverlag 1990, (Seite 153, 191)

[114] Fock, V.: *Theorie von Raum, Zeit und Gravitation.* Berlin: Akademie-Verlag 1960. (Seite 264)

[115] French, A. P.: *Die spezielle Relativitätstheorie*. Braunschweig, Wiesbaden: Friedr. Vieweg u. Sohn 1986, (Seite 225)

[116] Frisch, D. H.; Smith, J. H.: Measurement of the Relativistic Time Dilation Using Mesons. Am. J. Phys. 31, 342 (1963)

[117] Fritsch, A.: Zur Theorie der nicht einsteinisch-poincaréschen Relativitätstheorie. In.. Kritik und Fortbildung der Relativitätstheorie, Bd. 11, Hrsg.: K. Sapper, Akademische Druck- und Verlagsanstalt Graz, Österreich 1962

[118] Gardner, M.: Mathematical Garnes. Sc. Am. 232(4), 126 (1975) Längenparadoxie

[119] Genz, H.: *Die Entdeckung des Nichts. Leere und Fülle im Universum*. München, Wien: Carl Hanser Verlag 1994 (Deals with quantum mechanical effects in a generally understandable way, which are not the subject of our book). (page 256)

[120] Genz, H.: Das Zwillingsparadoxon. Physik in unserer Zeit 33, 226 (2002) sowie Das Zwillingsparadoxon (Leserbriefe) Physik in unserer Zeit 34, 48 (2003)

[121] Genz, H.: *Wie die Zeit in die Welt kam – Die Entstehung einer Illusion aus Ordnung und Chaos*. Wien: Carl Hanser Verlag 1996

[122] Genzel, R. et al.: Near-infrared flares from accreting gas around the supermassive black hole at the Galactic Center. Nature **425**, 934 (2003). (*Periodizität von 17 Minuten)* http://dx.doi.org/doi:10.1038/nature02065

[122.1] Genzel, R, Spiegelgespräch mit Nobelpreisträger Genzel vom 10. 10. 2020 in: Zeitschrift „Der Spiegel"

[123] Gerthsen, Ch; Kneser, H. 0.; Vogel, H: *Physik.* 15. Aufl. – Berlin, Heidelberg, New York: Springer Verlag 1986, 1999

[124] Ghosal, S. K.; Chakraborty, P.; Mukhopadvhyay, D.: Conventionality of distant simultaneity and light speed invariance. Europhys. Lett. (Schweiz) 15, 369 (1991) (critical of Reichenbach)

[125] Ghosal, S. K.; Nandi, K. K.; Chakraborty, P.: Passage from Einsteinian to Galilean Relativity and Clock Synchrony. Z. Naturforsch. 46a, 256 (1990)

[126] Giannoni, C.: Clock Retardation, Absolute Space, and Special Relativity. Found. of Phys. 9, 427 (1979)

[127] Gillessen, St.: *Eine Nacht im Zentrum der Milchstraße.* Sterne und Weltraum 2 (2009) p. 52 (Flares with a period of 27 minutes.)

[128] Golden, S., Non-Kinematicity of the Dilation-of-time Relation of Einstein for Time-intervals, Z. Naturforsch. 55 a, 563 (2000)

[129] Greenberg, M. J.: *Euclidean and Non-Euclidean Geometries.* 2. ed. San Francisco: W. H. Freeman u. Co. 1980, (Seite 327)

[130] Grieder, A.: Relativity, Causality and the 'Substratum'. Brit. J. Philos. Sci. 28, 33 (1977)

[131] Grøn, Ø.: Relativistic description of a rotating disk. Am. J. Phys. 43, 869 (1975)

[132] Grøn, Ø.: Relativistic Description of a Rotating Disk with Angular Acceleration. Found. of Phys. 9, 353 (1979) (second solution)

[133] Grøn, Ø.: Special-Relativistic Resolution of Ehrenfest's Paradox: Comments on Some Recent Statements by T. E. Phipps, Jr.. Found. of Phys.

[134] Grøn, Ø.: The Rotating Disk: Reply to Grünbaum and Janis. Found. of Phys. 10, 499 (1980)

[135] Grøn, Ø.; Nicola, M.: The Consistency of the Postulates of Special Relativity. Found. of Phys. 6, 677 (1976)

[136] Grove, D. L; Fox, J. G.: Phys. Rev. 90, 378 (1953) (Messung von *m(v)* für Protonen mit großer Genauigkeit)

[137] Grünbaum, A.: Why I am afraid of absolute space, Australian J. Philos. 49, 96 (1971)

[138] Grünbaum, A.; Janis, A. I.: The Rotating Disk: Reply to Grøn. Found. of Phys. 10, 494 (1980)

[139] Guilini, D., Straumann, N.: Das Rätsel der kosmischen Vakuumenergiedichte und die beschleunigte Expansion des Universums, Phys. Bl. 56, Heft 11, 41 (2000)

[140] Günther, H.: *Grenzgeschwindigkeiten und ihre Paradoxa – Gitter, Äther, Relativität,* Stuttgart, Leipzig: B. G. Teubner Verlagsgesellschaft 1996

[141] s. Weyl, H.

[142] Häfele, J. C.; Keating, R. E.: Around-the-World Atomic Clocks: Predicted Relativistic Time Gains. Science 177, 166 (1972)

[143] Häfele, J. C; Keating, R. E.: Around-the-World Atomic Clocks: Observed Relativistic Time Gains. Science 177, 168 (1972)

[143.2] Arnold Hanslmeyer, Einführung in Astronomie und Astrophysik, 2. Aufl. Spektrum Verlag 2011, Chapter 9.1.5

[144] Havas, P.: Simultaneity, Conventialism, General Covariance, and the Special Theory of Relativity. Gen. Rel. and Grav. 19, 435 (1987)

[145] Hawking, S. W. *Eine kurze Geschichte der Zeit – Die Suche nach der Urkraft des Universums,* Rowolth, Reinbek bei Hamburg (1991)

[146] Hayden, H. C.: Yes, moving clocks run slowly, but is time dilated? Galileian Electrodyn. (USA) 2, 63 (1991)

[147] Heckmann, O. *Theorien der Kosmologie,* berichtigter Nachdruck, Heidelberg, Berlin: Springer Verlag 1968 The book also presented Einstein's theory of relativity during the National Socialism.

[147.1] heic1908 - Science Release, *Latest Hubble Measurements Suggest Disparity in Hubble Constant Calculations is not a Fluke* https://www.spacetelescope.org/news/heic1908/?lang

[148] Held, A. Hrsg.: *General Relativity and Gravitation. One Hundred Years After the Birth of Albert Einstein.* Bd. 1, 2. New York, London: Plenum Press 1980

[149] Hentschel, K.: *Interpretationen und Fehlinterpretationen der speziellen und allgemeinen Relativitätstheorie durch Zeitgenossen Albert Einsteins.* Science Networks Bd. 6. Historical studies. Basel, Boston, Berlin: Birkhäuser Verlag 1990 (extensive literature data mainly on the Einstein interpretation)

[150] Herglotz, G.: Über den vom Standpunkt des Relativitätsprinzips aus als «starr" zu bezeichnenden Körper. Ann. Physik 31, 393 (1910) (Vertiefung von [88])

[151] Hering, in: Brandes, J.: *Interpretationen der speziellen Relativitätstheorie – Born, Einstein, Lorentz.* Frühjahrstagung Didaktik der Physik 2000 (Dresden), Tagungs-CD, Redaktion: V. Nordmeier, Münster

[152] Herrmann, F., Job, G. : *Altlasten der Physik.* Aulis Verlag Deubner, 2002, Seiten 152, 219ff

[153] Hermann, G., May, E., Vogel, Th.: *Die Bedeutung der modernen Physik für die Theorie der Erkenntnis – Drei mit dem Avenarius-Preis ausgezeichnete Arbeiten.* Hirzel-Verlag Leipzig 1937

[154] Herneck, F., *Einstein und sein Weltbild – Aufsätze und Vorträge,* Berlin: Buchverlag Der Morgen 1976 Seiten: 313, 197, 304, 308

[155] Herrmann, R. A.: Special relativity and a nonstandard substratum. Special Sci. Technol. (UK) 17, 2 (1994)

[155.1] HESS Collaboration: *Localising the VHE -ray source at the Galactic Center,* (2009) http://de.arxiv.org/abs/0911.1912v2

[156] Hirayama, T.: Magnetic morphologies of solar flares in: Y. Uchida, K. Tanaka, Flare physics in solar activity maximum 22, Tokyo 1990, Springer Verlag 1991

[157] Hoffmann, B. Negative Mass and the Quasars, 1966 (M. Arminjon, this conference, private communication)

[158] Hogan, C. J., Kirshner, R. P., Suntzeff, N. B. Die Vermessung der Raumzeit mit Supernovae, Bild der Wissenschaft, p. 40- 45, März 1999

[159] Holst H.: Voortfysike Verdensbilkede oy Einsteins Relativitetsheori. Kopenhagen 1920 s. [368] (Seite 221), Z. f. Physik 1 108 (1920)

[160] Hoyer, U.: Theorie der Lorentztransformationen. Z. Allg. Wiss. Theo. 19, 28 (1988)

[161] Hrvoje Nikolic, Relativistic contraction of an accelerated rod, Am. J. Phys. 67, 1007 (1999)

[162] Hunt, B. J.: The origins of the FitzGerald contraction. B. J. Hist. Sci. (UK) 21, 67 (1980) (historisch)

[163] Huntington, E. V., A New Approach to the Theory of Relativity, Phil. Mag. 23, No. 136, 494 (1912) (Probably the first systematic derivation of the Lorentz interpretation).

Linear acceleration is discussed in [69], circular acceleration in [187].

[163.1] S. Issaoun, M. D. Johnson, L. Blackburn, C. D. Brinkerink, M. Mo´scibrodzka, A. Chael, C. Goddi, I. Mart´ı-Vidal, J. Wagner, S. S. Doeleman, H. Falcke, T. P. Krichbaum, K. Akiyama, U. Bach, K. L. Bouman, G. C. Bower, A. Broderick, I. Cho, G. Crew, J. Dexter, V. Fish, R. Gold, J. L. G´omez, K. Hada, A. Hern´andez-G´omez, M. Janßen, M. Kino, M. Kramer, L. Loinard, R.-S. Lu, S. Markoff, D. P. Marrone, L. D. Matthews, J. M. Moran, C. M¨uller, F. Roelofs, E. Ros, H. Rottmann, S. Sanchez, R. P. J. Tilanus, P. de Vicente, M. Wielgus, J. A. Zensus, and G.-Y. Zhao: *The Size, Shape, and Scattering of Sagittarius A* at 86 GHz: First VLBI with ALMA,* arXiv:1901.06226v1 [astro-ph.HE] 18 Jan 2019

[164] Ivert, P.-A.: Poincare's principle determines the behaviour of moving particles and clocks. Acta Phys. Acad. Sci. Hung. 48, 439 (1980)

[165] Ives, H. E.: Derivation of the Lorentz Transformations. Philosoph. Mag. 36, 392 (1945)

[166] Ives, H. E.: Impact of a Wave-Packet and a Reflecting Particle. J. Opt. Soc. Am33, 163 (1943) (Derivation from the law of conservation of energy)

[167] Ives, H. E.: Theory of the Double Fizeau Toothed Wheel. J. Opt. Soc. Am. 29, 472 (1939) (presumably first solution of the Ehrenfest paradox according to Lorentz)

[168] Ives, H. E.: The Clock Paradox in Relativity Theory. Nature 168, 246 (1951) (Ergänzung zu ([225])

[169] Ives, H. E.; Stilwell, G. R.: An Experimental Study of the Rate of a Moving Atomic Clock. J. Opt. Soc. Am. 28, 215 (1938); (Experimental improvement in: J. Opt. Soc. Am. 31. 369 (1941))

[170] s. Brandes, J

[171] s. Lévy, J.

[172] Jammer, M.: *Concepts of force*. Cambridge: Harvard University Press 1957. Zitiert in: [96].

[173] Jammer, M.: *Das Problem des Raumes*. Darmstadt: Wissenschaftliche Buchgesellschaft 1960. (Seite 182f, 203)

[174] Janich, P.: *Grenzen der Naturwissenschaft: Erkennen als Handeln.* München: C. H. Beck'sche Verlagsbuchhandlung 1992

[175] Janossy, L.: Theory of relativity based on physical reality. Budapest 1971

[176] Jefimenko, O. D., On the Experimental Proofs of Relativistic Length Contraction and Time Dilation, Z. Naturforsch. 53 a, 977 (1998)

[177] Jefimenko, O. D., Comment on the Comment by G. Schäfer on the Paper "Non-kinematicity of the Dilation-of-time Relation of Einstein for Time-intervals" by S. Golden, Z. Naturforsch. 55a, 846 (2000)

[178] João R. T.: Recent Results of the Observatory Pierre Auger de Mello Neto Instituto de Física Universidade Federal do Rio de Janeiro S.Tomée Príncipe, S.Tomée Príncipe, September2009

[179] Joos, G.: *Lehrbuch der theoretischen Physik*. 15. Aufl. – Wiesbaden: Aula-Verlag 1989

[180] Joseph, G.: Geometrie and special relativity. Philos. Sci. 46, 425 (1979)

[181] Kaufmann, W.: Über die Konstitution des Elektrons. Ann. d. Phys. 4. Folge, Bd. 19, 487 (1906) (measurement of m(v) for electrons, probably first measurement of m(v) ever)

[182] Kelly, E. M.: Interferometer ether-drift experiment nullified by resynchronisation of observers clock. Opt. Lett. (USA) 11, 697 (1986) (critical of Marinov)

[183] Kelly, E. M.: Resolution of the Ehrenfest Paradox by New Contraction (Expansion) Criteria. Z. Naturforsch. 43a, 865 (1988)

[184] Kennedy, R. J.; Thorndike, E. M.: Experimental Establishment of the Relativity of Time. Phys. Rev. 42, 400 (1932)

[185] Kliem, B.: *Current sheets in solar corona.* in: A. O. Benz, A. Krüger (Eds.) Coronal magnetic energy releases, Konferenz 1994, S. 93, Springer Verlag 1995

[186] Kornack, T. W., Sollins, P. K., Brown, M. R.: Experimental observation of correlated magnetic reconnection and Alfvénic ion jets. Phys. rev. E 58, R36 (1998)

[186.1] Kopczyński, W. & Trautman, A., Spacetime and Gravitation, Wiley & Sons, Chichester 1992.

[187] Kraus, K.: Note on Lorentz Contractions and the Space Geometry of the Rotating Disc. Z. f. Naturforschung 25a, 792 (1970)

[188] Kuijpers, J.: Particle acceleration., in C. Chuideri, G. Einaudi (Eds) Plasma Astrophysics, Lectures 1994, Springer Verlag 1995

[189] Kundt, W. ed., *Jets from stars and galactic nuclei,* (1995) Springer Verlag Berlin

[189.1] Lahav, Liddle, *cosmological parameters, (2010)* arXiv:1002.3488v1

[190] Landau, L. D.; LIFSCHITZ, E. M.: *Lehrbuch der theoretischen Physik,* Bd. 2. Berlin: Akademie-Verlag 1967, 1990 (Seite 23, 190, 302, 381, §90 d. 4. Aufl.,)

[191] Langevin, P.: L'evolution de l'espace et du temps. Scientia 10, 31 (1911) (presumably discoverer of the twin paradox)

[192] Laugwitz, D.: *Differentialgeometrie.* 3. Aufl. Stuttgart: B. G. Teubner 1977, (Seite 116)

[192.1] Leibundgut, B., supernovae and cosmology, arXiv:0802.4154v1 (2008)

[193] Leonard, P. J. T., Bonnell, J. T. gamma-ray bursts of doom, Sky & Telescope, Feb. 1998, P. 28–34

[194] Lesch, H., Birk, G. T.: Particle acceleration ..., A&A 324, 461 (1997)

[195] Lesch, H.: Electron heating in accretion disks of active galactic nuclei. A&A 245, 48 (1991)

[195.1] Lévy, J. *From Galileo to Lorentz ... and beyond.* Montreal: Apeiron 2003 p. 63

[196] Liebscher, D.-E.: *Kosmologie. Einführung für Studierende der Astronomie, Physik und Mathematik.* Leipzig: Heidelberg: Johann Ambrosius Barth 1994 (Seite 158, 215)

[197] Liebscher, D.-E.: *Relativitätstheorie mit Zirkel und Lineal.* Braunschweig, Wiesbaden: Friedr. Vieweg u. Sohn 1977, (Seite 28)

[198] Lightman, A., P., Press W., H., Price, R., H., Teukolsky, S., A.: Problem book in relativity and gravitation. Princeton University Press Princeton, New Yersey 1975

[199] Lipschutz, M.: *Differentialgeometrie.* 2. Aufl. Düsseldorf. McGrawHill 1980, (Seite 242)

[200] Lodge, Sir O., The Geometrisation of Physics, and its Supposed Basis on the MICHELSON-MORLEY Experiment, Nature 106 No. 2677, 795 (1921)

[201] Logunov, A. A.; Chugreev, Y. u. V.: Special theory of relativity and centrifuge experiments. Fiz. Astron. (USSR) 43, 3 (1988) (exp. Vorschläge für rotierende Scheiben)

[202] Lorentz, H. A., The MICHELSON-MORLEY Experiment and the Dimensions of Moving Bodies, Nature 106, No 2677, 793 (1921)

[203] Lorentz, H. A.: Alte und neue Fragen der Physik, Physikalische Zeitschrift XI, 1234 (1910)

[204] Lorentz, H. A.: Der Interferenzversuch Michelsons. Aus: *Versuch einer Theorie der elektrischen und optischen Erscheinungen in bewegten Körpern* (Leiden 1895) in: [207]

[205] Lorentz, H. A.: Elektromagnetische Erscheinungen in einem System, das sich mit beliebiger, das Licht nicht erreichter Geschwindigkeit bewegt. Proc. Acad. Sc. Amsterdam 6, 809 (1904) in: [207]

[206] Lorentz, H. A.: *The Theory of Electrons and its Application to the Phenomena of Light and Radiant Heat*. New York: 1909, 1915

[207] Lorentz, H. A.; Einstein, A.; Minkowski, H.: *Das Relativitätsprinzip. Darmstadt:* Wissenschaftliche Buchgesellschaft 1958

[208] Lorenzen, P.: Relativistische Mechanik mit klassischer Geometrie und Kinematik. Math. Z. 155, 1 (1977)

[209] Lorenzen, P.: Zur Revision der Einsteinschen Revision. PN 16, 382 (1976/77)

[210] Lowry, E. S.: The clock paradox. Am. J. Phys. 31, (1963), 59

[211] MacGregor, M. H.: Do Dewan-Beran Relativistic Stresses Actually Exist? Lett. Nuovo Cimento 30, 417 (1981)

[212] Maciel, A. K. A.; Tiomno, J.: Experimental Analysis of Absolute Space-Time Lorentz Theories. Found. of Phys. 19. 521 (1989)

[213] Maciel, A. K. A.; Tiomno, J.: Experiments to detect possible weak violations of special relativity. Phys. Rev. Lett. 55, 143 (1985)

[214] Maciel, A. K. A.; Tiomno, J.: Reply to "Nonequivalence of ether theories and special relativity. Comments on recent interpretation of Lorentz' ether theory". Found. of Phys. Lett. 2. 601 (1989)

[215] Mansouri, R.; Sexl, R. U.: A Test Theory of Special Relativity: I. Simultaneity and Clock Synchronisation. Gen. Rel. and Grav. 8, 497 (1977) (The definitions for the measurement of the one-way speed of light are correct, but unusual, therefore criticism of [273]) (page 503, 512)

[216] Mansouri, R.; Sexl, R. U.: A Test Theory of Special Relativity: II. First-Order Tests. Gen. Rel. and Grav. 8, 515 (1977)

[217] Mansouri, R.; Sexl, R. U.: A Test Theory of Special Relativity: III. Second-Order Tests. Gen. Rel. and Grav. 8, 809 (1977)

[218] Maran, S. P. ed., *The astronomy and astrophysics encyclopedia,* (1991) Cambridge university press, Cambridge, Melbourne, Sidney, J. Krolik, H. p. 8

[219] Marder, L.: *Reisen durch die Raum-Zeit.* Braunschweig, Wiesbaden: Friedr. Vieweg u. Sohn 1979 (References Clock Paradox)

[220] Martin, A.; Roy-Keys, C.: The ether revisited. In: Frontiers of Fundamental Physics, Ed. by M. Barone F. Selleri, New York: Plenum Press 1994 (S. 209)

[221] Martin, J. L.: *General Relativity*. Chichester Ellis Horwood 1988

[222] Maund, B.: The conventionality of temporal relations in relativity theory. Philos. Sci. 41, 394 (1974)

[223] May, E.: *Am Abgrund des Relativismus*, Berlin. Dr. Georg Lüttke Verlag 1941 Preisgekrönt von der Preußischen Akademie der Wissenschaften, ***(only the title is out of date)*** (pp. 74ff, 86, p. 87, p. 210)

[224] McCrea, W. H.: Rotating Relativistic Ring. Nature 234 399 (1971) ch., 7.5

[225] McCrea, W. H.: The Clock Paradox in Relativity Theory. Nature 167, 680 (1951)

[226] Meschkowski, H.; Laugwitz, D. (Hrsg.): *Meyers Handbuch über die Mathematik:* Mannheim, Wien, Zürich: BI Allgemeiner Verlag 1967

[227] Meschowski, H.: *Nichteuklidische Geometrie.* 2. Aufl. Braunschweig, Wiesbaden: Friedr. Vieweg u. Sohn 1961

[228] Michelson, A. A.; Morley, E. H.: On the Relative Motion of the Earth and the Luminiferous Ether. Am. J. Sci. 33 (1887) (Preliminary tests took place in Berlin in 1881).

[229] Minkowski, H.: Raum und Zeit. Vortrag 1908. In: [207]

[230] Misner, Ch. W.; Thorne, K. S.; Wheeler, J. A.: *Gravitation.* San Francisco: W. H. Freeman u. Co. 1973. (Seite 603, 665, 721, 738)

[231] Mitsopoulos, T. D.: The isotropy of velocity of light and the clock paradox. Phys. Essays (Canada) 6, 233 (1993)

[232] Mittelstaedt, P.: Conventionalism in Special Relativity. Found. of Phys. 7, 573 (1977)

[233] Mittelstaedt, P.: *Der Zeitbegriff in der Physik.* Mannheim, Wien, Zürich: BI Wissenschaftsverlag 1989

[234] Mohorovicic, S.: Raum und Zeit in: *Kritik und Fortbildung der Relativitätstheorie*, Bd. 1, Hrsg.: K. Sapper, Akademische Druck- und Verlagsanstalt Graz, Österreich 1958 (gute Darstellung des Ätherbegriffs)

[234.1] Moore, Thomas A.: *A general relativity workbook.* Mill Valley, Carlifornia: University Science Books 2013, Formel (10.4)

[235] Møller, C.: *The general theory of relativity.* Oxford: Clarendon Press 1972. (Seite 294)

[236] Montanus, H.: Special relativity in an absolute Euclidean space-time. Phys. Essays (Canada) 4, 350 (1991)

[237] Morris, T.: The limiting nature of light-velocity as the causal factor underlying relativity. In: Frontiers of Fundamental Physics, Ed. by M. Barone, F. Selleri, New York. Plenum Press 1994 (S. 203)

[238] Mova, S.: Post-relativistic theory of the ether and its impact with some experimental verifications of the special theory of relativity and with recent cosmological observations. G. Fis. (Italy) 38, 163 (1987)

[239] Mueller, G. O.: Forschungsprojekt „95 Jahre Kritik an der speziellen Relativitätstheorie (1908-2003)“ www.ekkehard-friebe.de

[240] Mueller, G. O.: Über die absolute Größe der speziellen Relativitätstheorie. CD 2002 (Very extensive bibliography on critical studies of relativity theory).

[241] Müller, A.: *Die philosophischen Probleme der Einsteinschen Relativitätstheorie.* Braunschweig: Friedr. Vieweg u. Sohn 1922. (Seiten 127, 132)

[242] Mundy, B.: The Physical Content of Minkowski Geometry. Brit. Philos. Sci. 37, 25 (1986)

[243] Niklowitz, K.-H., Die Lorentz-Transformation als Folge der physikalischen Grenzgeschwindigkeit, Mathematisch-Physikalische Korrespondenz Nr. 177, 9 (1994) Hrsg. G. Unger, P. Gschwind, Benedikt Hugiweg 18, CH 4143 Dornach (The Lorentz interpretation finds gratifying approval among anthroposophical natural scientists).

[244] Niklowitz, K.-H., Die Lorentz-Transformation und die Ganzheitsidee, w. o. Nr. 199 (1999)

[245] Nolting, W.: Grundkurs Theoretische Physik 5/1 Quantenmechanik – Grundlagen. 6. Aufl. Springer-Verlag Berlin Heidelberg, New York 2004, S.199

[246] Oberhummer, H.: *Kerne und Sterne.* Leipzig, Berlin, Heidelberg: Johann Ambrosius Barth 1991, (Seite 85)

[247] Ohrstrom, P.: Conventionality in distant simultaneity. Found. of Phys. 10, 333 (1980)

[248] Oppenheimer, J.R., Volkoff, G. M.: *On massive neutron cores.* Phys.Rev.55(1939)374

[249] O'Rahilly, A., *Electromagnetic Theory – a critical examination of fundamentals*, Volume II, Chapt. XIII The Aether, Dover Publications, New York 1940(?)

[250] Ott, C. D., Schnetter, E., Burrows, A., Livne, E., O'Connor, E., Loeffler, F.: *Computational Models of Stellar Collapse and Core-Collapse Supernovae.* (2009) arXiv:0907.4043

[251] P. O. Mazur, E. Mottola, Gravitational condensate stars: An alternative to black holes. arXiv: gr-qc/0109035 v5 27 Feb 2002

[252] Pachner, J.; Miketinac, M. J.: Exact equations for the stationary equilibrium of relativistic rotating objects. Phys. Rev. D 6, 1474 (1972)

[253] Pais, A., *Ich vertraue auf Intuition – Der andere Albert Einstein,* Heidelberg, Berlin, Oxford: Spektrum Akademischer Verlag 1998

[254] Pais, A.: *Raffiniert ist der Herrgott ... Albert Einstein. Eine wissenschaftliche Biographie.* Spektrum Akademischer Verlag Heidelberg Berlin 2000 S. 124, 243

[255] Palacios, P.: Relatividad. Madrid: 1960 (Palacios developed a theory of relativity without time dilation because of the clock paradox).

[256] Parnell, Claire E.: Magnetic reconnection: Classical aspects. in: J.-P. Rozelot, L. Klein, J. C. Vial (Eds), Transport and energy conversion in heliosphere, Konferenz 1998, S. 61, Springer Verlag 2000

[257] Pauli, W in: [314], analog: Herglotz, G. [150], Einstein in [353]

[258] Peat, H.: A revolution in Physics. Multidisciplinary Res. (USA) 4, 37 (1976)

[259] Peat, H.: Can some "anomalous" structural interaction be explained by an excitable ether"? Multidisciplinary Res. (USA) 4, 44 (1976)

[260] Pecht, A.: Remarks on steady rotation about a fixed axis in relativity. Acta Mech. (Österreich) 29, 283 (1978)

[260.1] Perlmutter, S., Schmidt, B. P.: Measuring cosmology with Supernovae. arXiv:astro-ph/0303428v1 (2003)

[261] Petkov, V.: Relativity and the nature of spacetime. Springer-Verlag Berlin Heidelberg, New York 2005 S. 219

[262] Pfarr, J.: Zur Interpretation des Michelson Versuchs. In: Nitsch, L; Pfarr, J.: Stachow, E. W. (Hrsg.): *Grundlagenprobleme der modernen Physik* Mannheim, Wien, Zürich: BI Wissenschaftsverlag 1981

[263] Phipps, T. E. Jr.: Do Metric Standards Contract? – A Reply to Cantoni. Found. of Phys. 10, 811 (1980)

[264] Phipps, T. E. Jr.: Light on Light: A Response to Grøn. Found. of Phys. 11, 633 (1981)

[265] Phipps, T. E.: Kinematics of a "Rigid" Rotor. Lett. Nuovo Cimento 9, 467 (1974)

[266] Pierre Auger Collaboration: *Correlation of the Highest-Energy Cosmic Rays with Nearby Extragalactic Objects.* Science 318 (2007) S. 938

[267] Pierseaux, Y., *La "structure fine" de la théorie de la Relativité Restreinte,* L'Harmattan, Paris, France ISBN 2-7384-8057-8, Dissertation,

[268] Pierseaux, Y., The "fine structure" of special relativity: with or without hidden variable? Vortrag: Physical Interpretations of Relativity Theory VII British Society for the Philosophy of Science London, 15-18 September 2000

[269] Pocci, G.; Sjödin, T.: Outline of a relativistic theory for non-Lorentzian edler. NC B OB, 601 (1981)

[270] Podlaha, M.: The axiomatic foundations of the theory of special relativity – reply to Stiegler. Int. J. Theor. Phys. 11, 69 (1974)

[271] Podlaha, M. F.: Critique of Some Statements in the Theory of Relativity. Nuovo Cimento 66B 9 (1981)

[272] Podlaha, M. F.: Length contraction and time dilatation in the special theory of relativity – real or apparent phenomena? Indian J. Theor. Phys. 23, 69(1975)

[273] Podlaha, M. F.: Light signal synchronisation and clock transport synchronisation in the theory of relativity. Brit. J. Philos. 30, 376 (1979)

[274] Podlaha, M. F.: Notes about relativity and about liberty in science. British Society for Philosophy of Science: Physical Interpretations of Relativity Theory 11. Proceedings, London 3.-8. Sept. 1990 (Sunderland, UK: Sunderland Polytechnic 1990) 222-63 vol. 2

[275] Podlaha, M. F.: On the impossibility to measure the one-way velocity of light (complementary remarks to Sjödin's paper). Lett. Nouvo Cimento 28, 216(1980)

[276] Podlaha, M. F.: Some new aspects of relativity: remarks on Zahar's Paper. Brit. J. Philos. Sci. 27, 261 (1976)

[277] Podlaha, M. F.; Sjödin, T.: An Explanation of the Clock Problem and of the First-Order Experiments from Nonrelativistic Premisses. Lett. Nuovo Cimento 20, 593 (1977)

[277.1] Podbregar, Nadja, 27. Juni 2022, https://www.scinexx.de/news/kosmos/milchstrasse-schwarzes-loch-speist-rohkost/

[278] Pohl, R. W *Elektrizitätslehre*. 21. Aufl. Berlin, Göttingen, Heidelberg: Springer Verlag 1975, 1992

[278.1] Port, K.: Philosophische Schriften, Gesamtausgabe Bd. I - IV, Bd.I Wertphilosophie und Ethik, Esslingen: Port-Verlag 1993 - 1996

[279] Pound, R. V., Rebka, G. A.: *Apparent weight of Photons*. Phys. Rev. Lett. 4 (1960). 539

[280] Priest, E. R., Forbes, T. G.: The magnetic nature of solar flares. A&A Rev 10, 313 (2002)

[281] Priest, E. R.: *Solar Flares* in: Flares and Flashes: proceedings of IAU Colloquium 151, Springer Verlag 1995

[282] Priest, E.: Forbes, T.: *Magnetic reconnection – MHD theory and applications.* Cambridge University Press 2000

[283] Podlaha, M. F.: De Broglie waves, length contraction and time dilatation. Indian J. Theor. Phys. 25, 37 (1977)

[284] Prokhovnik, S. J.: Did Einsteins Programme supersede Lorentz's? Brit. J. Philos. Sci. 25, 336 (1974)

[285] Prokhovnik, S. J.: An introduction to the non-Lorentzian relativity of Builder. Speculations Sci. and Technol. (Schweiz) 2, 225 (1979)

[286] Prokhovnik, S. J.: The Operation of the Relativistic Doppler Effect. Found. of Phys. 10, 197 (1980)

[287] Prokhovnik, S. J.: The Physical Interpretation of Special Relativity – a Vindication of Hendrik Lorentz. Z. Naturforsch. 48a, 925 (1993)

[288] Prokhovnik, S. J.: The Twin Paradoxes of Special Relativity: Their Resolution and Implications. Found. of Phys. 19, 541 (1989)

[289] R. Schödel et al.: A star in a 15.2-year orbit around the supermassive black hole at the center of the Milky Way. Nature **419**, 694 (2002)

[290] Raab, I.: Untersuchungen über das Uhrenparadoxon, Dissertation. Innsbruck 1963

[291] Raab, I.: Der Einfluss der Metrik auf die Definition der trägen Masse in der Relativitätstheorie. Acta Physica Austriaca 28 216 (1968)

[292] Ragot, B. R., Schlickeiser, R.: The acceleration of energetic particles by transit-time damping, Astroparticle physics 9, 79, (1998)

[293] Raschewski, P. K.: *Riemannsche Geometrie und Tensoranalysis.* Berlin: Deutscher Verlag d. Wissenschaften 1959, (Seiten 87, 160)

[293.1] Rebhan, E.: *Theoretische Physik: Relativitätstheorie und Kosmologie*. Berlin: Spektrum Akademischer Verlag 2012, Seite 245

[294] Rembielinski, J.: The relativistic ether hypothesis. Phys. L. 78A, 33 (1980)

[295] Renninger, M: Die Lorentz-Transformationen als Ausdruck der Superposition gegenläufiger Raum-Zeit-Beziehungen, Preisgabe des Relativitätsprinzips. Ann. Physik 44, 378 (1987)

[295.1] Riess, Adam G. et al.: A Redetermination of the Hubble Constant with the Hubble Space Telescope from a Differential Distance Ladder, arXiv:0905.0695 (2009)

[296] Rietdijk, C. W.: Thought experiments relevant to special and general relativity – and some theses. British Society for Philosophy of Science: Physical Interpretations of Relativity Theory II. Proceedings, London 3.-8. Sept. 1990 (Sunderland, UK: Sunderland Polytechnic 1990) 246 vol. 2

[296.1] Ritchie, B. W. et al.: A VLT/FLAMES survey for massive binaries in Westerlund 1: II. Dynamical constaints on magnetar progenitor masses from the eclipsing binary W13, A&A August 11, 2010

[297] Rindler, W.: Einstein's Priority in Recognizing Time Dilation Physically. Am. J. Phys. 38, 1111 (1970)

[298] Rindler, W.: *Introduction to special relativity,* 2. Aufl., Oxford: Clarendon Press 1990

[299] Rindler, W.:Relativity – special, general, and cosmological. Oxford University Press 2001 S. 24, S. 231

[300] Robertson, H. P.: Postulate versus Observation in the Special Theory of Relativity. Rev. Mod. Phys. 21, 378 (1949)

[301] Robertson, H. P.: Geometrie als Zweig der Physik. S. 221 In Schilpp, A.

[302] Rodrigues, W. A. jr.: The Standard of Length in the Theory of Relativity and Ehrenfest Paradox. Nouvo Cimento 74B, 199 (1983)

[303] Rodrigues, W. A. jr.; Tiomno, J.: On Experiments to Detect Possible Failures of Relativity Theory. Found. of Phys. 15, 945 (1985)

[304] Roscoe, D. F.: Geometrized mass as the relativistic aether. British Society for Philosophy of Science: Physical Interpretations of Relativity Theory II. Proceedings, London 3.-8. Sept. 1990 (Sunderland, UK: Sunderland Polytechnic 1990) 256 vol. 2

[305] Rossi, B.; Hall, D. B.: Variation of the Rate of Decay of Mesotrons with Momentum. Phys. Rev. 59, 223 (1941) (The temporally first experiment with particle decays)

[306] Roulet, B., Viennot, L., Rotating rigid rods or incompressible fluids: How to determine the stress? Am. J. Phys. 66, 75 (1998)

[307] Rowlands, P.: The abstract ether theory of Oliver Lodge. British Society for Philosophy of Science: Physical Interpretations of Relativity Theory II. Proceedings, London 3.-8. Sept. 1990 (Sunderland, UK: Sunderland Polytechnic 1990) 259-63 vol. 2

[308] Ruark, A. E.: The Physical Rationale for Special Relativity. Found. of Phys. 5, 21 (1975)

[309] Ruder, H.; Ruder, M.: *Die spezielle Relativitätstheorie.* Braunschweig, Wiesbaden: Friedr. Vieweg u. Sohn 1993

[310] Ruderfer, M.: Remarks on Erlichson's discussion on one-way light tests. Am. J. Phys. 43, 279 (1975)

[311] Ruderfer, M.: Comments on "A New Experimental Test of Special Relativity". Lett. Nuovo Cimento 3, 658 (1970)

[312] Rudnick, L., Brown, S., Williams, L. R.: *Extragalactic Radio Sources and the WMAP Cold Spot.* arXiv: 0704.0908v2 astro-ph (2007)

[313] s. [Bergmann Bd. 1], S. 937

[313.1] s. [113], S. 32, [334]

[314] Sama, N.: On the Ehrenfest Paradox. Am. J. Phys. 40, 415 (1972)

[315] Schäfer, G., Comment on the Paper "Non-kinematicity of the Dilation-of-time Relation of Einstein for Time-intervals" by S. Golden, Z. Naturforsch. 55a, 845 (2000)

[316] Schaffner, K. F.: The Lorentz Electron Theory of Relativity. Am. J. Phys. 37, 408, (1969) (historisch)

[317] Schaifers, K.; Traving, G *Meyers Handbuch über das Weltall.* 3. Aufl. Mannheim, Wien, Zürich: BI Meyers Lexikonverlag 1973

[318] Schilpp, A. Hrsg.: *Albert Einstein als Philosoph und Naturforscher.* Stuttgart: W. Kohlhammer Verlag 1951.

[319] Schleichert, H.: Lösungsversuche für das Uhrenparadoxon, erkenntnislogisch betrachtet. PN 9, 326 (1966) (using formal logic)

[320] Schock, R. The inconsistency of the theory of relativity. Z. allg. Wissenschaftstheorie 12, 285 (1981)

[321] Schöne, N.: *Differentialgeometrie.* Thun, Frankfurt am Main: Verlag Harri Deutsch 1978, (Seite 51)

[322] Schonefeld, W.: *Protophysik und spezielle Relativitätstheorie,* Würzburg: Könighausen und Neumann, 1999

[323] Schönfelder, V. Gammastrahlung aus dem Kosmos, Phys. Bl. 54, 325 (1998)

[324] Schopper, R., Lesch, H., Birk, G. T.: Magnetic reconnection and particle acceleration in active galactic nuclei, A&A 335, 26 (1998)

[325] Schreiber, P.: *Grundlagen der konstruktiven Geometrie.* Berlin: Deutscher Verlag der Wissenschaften 1984, (Seite 134)

[326] Schröder, U. E.: *Spezielle Relativitätstheorie* - Thun, Frankfurt am Main: Verlag Harri Deutsch 1987, 1994

[327] Schröder, W., *Emil Wiechert, Physiker, Geophysiker, Wissenschaftsorganisator,* Mitteilungen des Arbeitskreises Geschichte der Geophysik 19, Heft 1-2, (2000), ISSN

0179-5658 darin: W. Schröder, Ein Beitrag zur frühen Diskussion um den Äther und die Einsteinsche Relativitätstheorie, auch: Annalen der Physik 7. Folge, Band 47, 475 (1990) Seiten 107, 112, 129

[327.1] Schutz, B.:*Gravity from the ground up*. Cambridge University Press 2003, Formel (18.11)

[328] F. Selleri, J. Brandes, J. Czerniawski, U. Hoyer, K. Wohlrabe, *Die Einstein'sche und lorentzianische Interpretation der speziellen und allgemeinen Relativitätstheorie.* Karlsbad: VRI 1998

[329] Selleri, F. (Editor), *Open Questions in Relativistic Physics,* Proceedings, Apeiron, Montreal, Canada,ISBN 0-9683689-1-3, Artikel v. Selleri, F., Brandes, J. Goy, F., Wesley, J. P., u. a.

[329.2] Selleri, F.: Space-time Transformations in Ether Theories. Z. Naturforsch. 46a, 419 (1990/91)

[330] Selleri, F.: Special relativity as a limit of ether theories. British Society for Philosophy of Science: Physical Interpretations of Relativity Theory II. Proceedings, London 3.-8. Sept. 1990 (Sunderland, UK: Sunderland Polytechnic 1990) 508 vol. 3

[331] Selleri, F.: Weak relativity – the physics of space and time without paradoxes. Preprint 2009

[332] Sen, A.-. How Galileo could have derived the special theory of relativity. Am. J. Phys. 62, 157 (1994)

[333] Sexl, R. U.: Die Hohlwelttheorie. Der mathematische und naturwissenschaftliche Unterricht (MNU) 36, 453 (1983) (about a nonsensical theory, philosophically significant) (page 459)

[334] Sexl, R. U.; Urbantke, H. K.: *Gravitation und Kosmologie.* Mannheim, Wien, Zürich: BI Wissenschaftsverlag 1990 (Seite 328)

[335] Sexl, R. U.; Urbantke, H. K.: *Relativität, Gruppen, Teilchen. Spezielle Relativitätstheorie als Grundlage der Feld- und Teilchenphysik.* 3. Aufl. Wien, New York: Springer Verlag 1992 (page 45f, descriptive description of the interpretation by Ehrenfest, 1913) (page 35, 45)

[336] Sexl, R.; Schmidt, H. K.: *Raum Zeit Relativität – Relativistische Phänomene in Theorie und Beispiel,* 4. Aufl. – Braunschweig, Wiesbaden: Fr. Vieweg & Sohn 2000

[337] Sexl, R.; Schmidt, H. K.: *Relativitätstheorie – Grundkurs Physik in der Sekundarstufe II,* Stuttgart: J. B. Metzlersche Verlagsbuchhandlung 1988

[338] Sexl, R.; Sexl, H.: *Weiße Zwerge – Schwarze Löcher.* 2. Aufl. Braunschweig, Wiesbaden: Friedr. Vieweg u. Sohn 1979, 1999. (Seite 28ff)

[339] Shamir, J., Fox, R.: A New Experimental Test of Special Relativity. Nuovo Cimento 62 B. 258, (1969) (reply s. [311)

[339.1] Shamir, L. (2020) *Multipole alignment in the large-scale distribution of spin direction of spiral galaxies* https://arxiv.org/pdf/2004.02963.pdf

[340] Sherwin, Ch. W.: New experimental test of Lorentz's theory of relativity. Physical Rev. A 35, 3650 (1987) (not a test of a current theory)

[341] Shimmin, W. L.: A conjecture regarding changes in dimensions of bodies moving through the ether. Galileian Electrodyn. (USA) 5, 55 (1994)

[342] Shupe, M. A.: The Lorentz-invariant vacuum medium. Am. J. Phys. 53, 122(1985)

[343] s. z. B. Fließbach, Seite 32; Sexl, Urbandtke

[344] Simhony, M.: Structure of matter and space in view of Michelson-Morley, Rutherford and Anderson experiments. Specul. Sci. Technol. (UK) 13, 155 (1990)

[345] Sjödin, T.: A Note on Poincarés Principle and the Behaviour of Moving Bodies and Clocks. Z. Naturforsch. 35a, 997 (1979) (historisch) (Seite 997)

[346] Sommer, K., P.: *Prioritätsstreit zwischen Hilbert und Einstein – Wer entdeckte die allgemeine Relativitätstheorie?* Phys. Unserer Zeit 36 (2005) 231 (Sommer recommends the language Hilbert-Einstein equations instead of Einstein field equations, in order to do justice to HILBERT's priority).

[347] Somov, B. V., Kosugi, T.: Collisionless reconnection and high-energy particle acceleration in solar flares. Ap. J. 485, 859 (1997)

[348] Spavieri, G.: Comments on Chang and Torr's "Dual properties of spacetime under an alternative Lorentz transformation". Comparison of ether theory predictions with the result of a recent optical experiment. Found. of Phys. Lett. 2, 61 (1989)

[349] Spavieri, G.: Nonequivalence of ether theories and special relativity. Physical Rev. A 34, 1708 (1986)

[350] Spavieri, G.: The Arago experiment as a test for modern ether theories and special relativity. Nuovo Cimento 91B, 143 (1986)

[351] Spavieri, G.; Bergamaschi, J.: Observation of starlight with coupled telescopes: a first-order test of the isotropy of space. Nuovo Cimento B 104B, 497(1989)

[352] Sperner, E.: *Einführung in die analytische Geometrie und Algebra.* Teil 1. Göttingen: Vandenhoeck u. Ruprecht 1969

[353] Stachel, J.: Einstein and the Rigidly Rotating Disk. In [148] Bd. 1, S. 1-15, (Seite 7)

[354] Stachel, J.: *Einstein and the rotating disk.* in A. Held, Hrsg.: General Relativity and Gravitation, New York, London: Plenum Press 1980, Bd. 1, S. 1-15

[355] Stedman, G. E.: A Unidirectional Test of Special Relativity. Am. J. Phys. 40, 782 (1972) (enthält weitere Literaturhinweise)

[356] Stedman, G. E.: Reply to Erlichson: Is the Apparent Speed of Light Independent of the Sense in Which It Traverses s Closed Polygonal Path? Am. J. Phys. 41, 1300 (1973)

[357] Stephani, H.: *Allgemeine Relativitätstheorie.* 4. Aufl. Berlin: Deutscher Verlag der Wissenschaften 1991, (Seite 105, 108, 109, 117, 198, 218, 222, 240f, 248) 10,27 (23.21), (23,38)

[358] Straumann, N.: *General Relativity and Relativistic Astrophysics.* Berlin, Heidelberg, New York: Springer-Verlag 1984

[359] Strauss, M.: Rotating frames in special relativity. Int. J. Theor. Phys. 11, 107 (1974)

[360] Suresh, A.: The elastic rotating cylinder. Int. J. Eng. Sci. (GB) 18, 885 (1980)

[361] Süßmann, G.: Begründung der Lorentz-Gruppe. Z. Naturforsch. 24a, 495 (1969)

[362] Syrovatskii, S. I.: On the Problem of the "retardation" of relativistic contraction of moving bodies. Sov. Phys. Usb. 19, 273 (1976)

[363] Taylor, E. F., Wheeler, J. A.: *Physik der Raumzeit.* Spektrum Verlag 1994

[364] Terletzkii, Y. P.: *Paradoxes in the Theory of Relativity.* New York: Plenum Press 1968 (Seite 41)

[365] Tetens, H.: *Was "revolutioniert" die Relativitätstheorie? Wissenschaftstheoretische Überlegungen zur speziellen Relativitätstheorie.* Aus: Relativitätstheorie und Philosophie, Akademie-Vorträge, Bd. 19, Katholische Akademie Schwerte 1985 (Ein Schüler v. H. Dingler, P. Lorenzen)

[366] Theimer, W.: *Die Relativitätstheorie. Lehre – Wirkung – Kritik.* Bern, München: Francke Verlag 1977 (Quote page 51: "Einstein kills two birds with one stone ... The third fly he kills is the theory of relativity itself ..." but: many, useful literature references.)

[366.1] Kip S. Thorne (1994): *Black holes and time warps. Einsteins outrageous legacy*, New York, London: Norton, page 402, 403

[366.2] Kip S. Thorne (1994): *Gekrümmter Raum und verbogene Zeit. Einsteins Vermächtnis.* München Droemer Knaur 4. Auflage 1994, Seite 462, 463

[366.3] M. Thürauf, Vortrag Astrophysik, SS 2011, https://www.lsw.uni-heidelberg.de/users/mcamenzi/SL_Thuerauf.pdf

[367] Tolman, R.: C.: Relativity, Thermodynamics and Cosmology. Dover Publications, INC. New York 1987, reprint 1934, Formel (88.4), S. 322

[368] Treder, H.-J., *Philosophische Probleme des physikalischen Raumes – Gravitation, Geometrie, Kosmologie und Relativitat,* Akademie-Verlag Berlin 1974, ch. 26, S.7, in Englisch: Treder, H.-J., Global and Local Principles of Relativity, Foundations of Physics 1, 77 (1970)

[369] Treder, H.-J.: Aktive und passive Verallgemeinerungen der Lorentz-Poincaré-Transformationen und das Licht- und das Relativitätsprinzip von Einstein. I.+II. Teil. Experimentelle Technik der Physik 23, 113+211 (1975) (Seite 211, 217)

[370] Trudeau, R. J.: *The Non-Euclidean Revolution.* Basel, Berlin, Stuttgart: Birkhäuser 1987, (Seite 237)

[371] Tyapkin, A. A.: On the Impossibility of the First-Order Relativity Test. Lett. Nuovo Cimento 7, 760 (1973) (Tyapkin clearly recognized the hopelessness of such experiments because of the Lorentz interpretation.)

[372] Ungar, A. A.: Formalism to deal with Reichenbach's special theory of relativity. Found. of Phys. 21, 691 (1991)

[373] Ungar, A. A.: The Lorentz transformation group of the special theory of relativity without Einstein's isotropy convention. Philos. Sci. 53, 395 (1986)

[374] Ungar, A. A.: The abstract Lorentz transformation group. Am. J. Phys. 60 (1992)

[375] Ungar, A. A.: Thomas precession and its associated grouplike structure. Am. J. Phys. 59, 824 (1991)

[376] Ungar, A., Ether and the one-way speed of light, Am. J. Phys. 56, 814 (1988)

[377] Unsöld, A., Baschek, B.: *Der neue Kosmos, Einführung in die Astronomie und Astrophysik.* 7. Aufl. Paris, Tokyo: Springer 2002

[378] Unsöld, A., Baschek, B.: *Der neue Kosmos.* 4. Aufl. Paris, Tokyo: Springer 1988, S. 238

[379] Unsöld, A.; Baschek, B.: *Der neue Kosmos.* 5. Aufl. Berlin, Heidelberg, New York: Springer Verlag 1984, (Seite 382)

[380] v. Meyenn, K.: *Die großen Physiker.* Bd. II, München: Beck´sche Verlagsbuchhandlung 1997

[381] Vargas, J. G.: Relativistic experiments with signals on a closed path (reply to Podlaha). Lett. Nuovo cimento 28, 289 (1980)

[382] Vargas, J. G.: Revised Robertson's Test Theory of Special Relativity. Found. of Phys. 14, 625 (1984)

[383] Vargas, J. G.; Torr, D. G.: Testing of the line element of special relativity with rotating systems. Physical Rev. A 39, 2878 (1989)

[384] Varicak, V.: Zum Ehrenfestschen Paradoxon. Phys. Z. 12, 169 (1911)

[384.1] v. Baltz, R.: Letter to President of DPG, 1.7.2013, www.grt-li.de/vBaltz.pdf

[385] Weber, T. A., A note on rotating coordinates in relativity, Am. J. Phys. 65, 486 (1997)

[386] Weigert, A., Wendker, H. J. *Astronomie und Astrophysik,* 3. Aufl. (1996) VCH Weinheim

[387] Weinberg, H.; Mosel, M.: Theory for a Unidirectional Interferometric Test of Special Relativity. Am. J. Phys. 39, 606 (1971)

[388] Weinberg, S.: *Gravitation and Cosmology.* New York: John Wiley u. Sons 1972

[389] Weinstein, D. H.: Ehrenfest's paradox. Nature 232, 548 (1971)

[390] Welling, H., Zawischa, I., Der Gravitationswellendektor GEO 600, Phys. Bl. 55, Nr. 7/8 S. 69 (1999)

[391] Wesley, J. P.: Comments on Prokhovnik's critique of Marinov's experiment. Found. of Phys. 10, 803 (1980) (With a rejoinder from Prokhovnik).

[392] Westphal, W. H *Physik.* 26. Aufl. Berlin, Heidelberg, New York: Springer Verlag 1970

[393] Wetterich, Ch., Quintessenz – die fünfte Kraft. Physik Journal 3 Nr. 12, 43 (2005)

[393.1] Weyl, H., *Raum, Zeit, Materie.* Berlin: Springer-Verlag 1920, (pp. 254f, 257) (H. Weyl left Germany before the war in protest against the injustice against his Jewish colleagues).

[394] Whealton, J. H.: Illustrations of a Dynamical Theory of the Ether. Foundations of Phys. 5, 543 (1975)

[395] Whittaker, E.: A History of the Theories of Aether and Electricity, Bd. I + B. The History of Modern Physics 1800–1950 Vol. 7. Reprint: Tomash Publishers; American Institute of Physics, USA 1987

[396] Wiechert, E., Relativitätsprinzip und Äther, Phys. Z. 12, 698 (1911)

[397] Wikipedia, Internet-Enzyklopädie – Artikel über bellsche Raumschiff-Paradoxie

[397.1] Wikipedia, Internet-Enzyklopädie – Artikel über Zwillingsparadoxie und die Diskussion v. 17.12.2007.

[397.2] Wikipedia, Internet-Enzyklopädie – Artikel über Supernova

[398] Wilhelm, E.: Physical foundations and implications of Lorentz transformations in comparison with experiments and absolute space-time physics. Phys. Essays (Canada) 6, 420 (1993)

[399] Wilhelm, H. E.: Lorentz Transformation as a Galilei Transformation with Physical Length and Time Contractions. Z. Naturforsch. 43a, 859 (1987)

[400] Winnie, J. A.: The Twin-Rod Thought Experiment. Am. J. Phys. 40, 1091 (1972)

[401] Winterberg, F. Statistical Mechanical Interpretation of Black Hole Entropy, Z. Naturforsch. 49a, 1023–1030 (1994)

[402] Winterberg, F. *The Planck Aether Hypothesis – An Attempt for a Finistic Theory of Elementary Particles,* Herausgeber und Anhang: J. Brandes, Karlsbad: VRI 2000, S. 38
Die Idee positiver und negativer Massen und die Idee einer nichtlokalen Zwangskraft (eine Anregung v. H. J. Treder) haben F. Winterberg zu einer vielleicht bedeutenden Äthertheorie geführt. Kritisch: G. Cavalleri [48].

[403] Winterberg, F., Resolution of the Ehrenfest Paradox in the Dynamic Interpretation of Lorentz Invariance, Z. Naturforsch. 53a, (1998)

[404] Winterberg, F.: A Crucial Test for Einstein's Special Theory of Relativity Against the Lorentz-Poincaré Ether Theory of Relativity. Z. Naturforsch. 41a, 1261 (1986)

[405] Winterberg, F.: Nonlinear relativity at high energies. Atomkemenerg. Kerntech. 44, 88 (1984)

[406] Winterberg, F.: Lorentz Invariance as a Dynamic Symmetry. Z. Naturforsch. 42a, 1428 (1987)

[407] Winterberg, F.: Possible Evidence for Weak Violation of Special Relativity. Z. Naturforsch. 42a, 1374 (1987)

[407.1] Wischnewski, E.: *Astronomie in Theorie und Praxis*, 24568 Kaltenkirchen: 2018

[408] Wood, R. W.: Physical Optics – MacMillan, 2nd ed. 1911 (Seite 690) Längenparadoxie

[408.1] Woodhouse, N., M., J.: *General Relativity*. London: Springer-Verlag 2007, ch. 7.4

[409] Woosley, St., Janka, Th.: *The physics of core-collapse supernova.* arXiv:astro-ph/0601261

[410] Yaglom, I. M.: *A Simple Non-Euclidean Geometry and Its Physical Basis.* Berlin, Heidelberg, New York: Springer-Verlag 1979

[411] http://www.astrotreff.de/topic.asp?TOPIC_ID=105729 - Artikel: Neues aus der Astronomie – Wieviel Masse braucht man für ein Schwarzes Loch? - Autor Caro – Diskussion xlorentz.

[412] First Sagittarius A* Event Horizon Telescope Results. VI. Testing the Black Hole Metric, chapter 4.
First M87 Event Horizon Telescope Results. VI. The Shadow and Mass of the Central Black Hole, chapter 9.4

The above sources divided into subject areas (until the 3rd edition):

[all] **General relativity, introductions and textbooks**: Audretsch, Berry, Born,Carroll, d'Inverno, Møller, Einstein, Falk, Fließbach, Martin, Misner, Sexl, Stephani, Straumann, Weinberg, S., Weyl, Woodhouse

[ehr] **Ehrenfest Paradox:** Berenda, Browne, Carmeli, Cavalleri, Cantoni, Davies, Ehrenfest, Grøn, Grünbaum, Held, Hrvoje, Herglotz, Ives (1939), Kelly, Logunov, Lorentz (1921), McCrea, Pachner, Pauli, Pecht, Phipps, Rodrigues, Roulet, Sama, Stachel, Strauss, Suresh, Vargas, Weber, Weinstein, D. H., Winterberg

[ex1] **Experimental confirmations of the special theory of relativity:** Alley, Bailey, Frisch, Grove, Häfele, Ives, Kaufmann, Kennedy, Michelson, Rossi

[ex2] **Experimental LI:** Chang (1988), Kelly, Maciel, Sherwin, Shamir, Spavieri, Rodrigues, Ruderfer (1970), Wesley, Winterberg (1986, 1987)

[he1] **Derivation of the Lorentz transformations on the basis of the LI and related considerations:**
Lorentz, Wiechert, Huntington, Lodge, Broad, Ives, Builder, Raab, Erlichson, Podlaha, Janossy, Grieder, Lorenzen, Rindler (1970), Mansouri I-III, Giannoni, Prokhovnik, Sjödin, Podlaha, Chang, Bridgman, Erlichson, Vargas, Ungar, Spavieri, Bell, Wilhelm, Brandes (1988, 1968), Renninger, Hoyer, Günther, Fritsch, Maciel, Selleri, Montanus, Herrmann, Schröder, Pierseaux, Niklowitz, Czerniawski, Levy, Duffy
A historical literature study comparable to [Hentschel] did not exist until now. This

has changed with the dissertation of Y. Piersevaux, which contains about 900 references to both interpretations. An above-average number of references can be found in [80], [104]. From the English-American language area: Erlichson [105], Prokhovnik [287] as well as Mansouri, Sexl [217]. Also [6] and especially numerous and not only on LI: Müller, G. O.

Those who want to be up-to-date can inform themselves in: Zeitschrift f. Naturforschung, American Journal of Physics and the Proceedings of Physical Interpretations of Relativity Theory VII, British Society for the Philosophy of Science London, 15-18 September 2000 and later years, M. C. Duffy (Chairman),http://www.physicsfoundations.org/ PIRT conferences in London, archives 2000 - 2008. There are also short biographies of dissident physicists (some of them Lorentzian): J. P. Wesley (ed.) Weiherdammstr. 24, D-78176 Blumberg

[he2] **Derivation of Lorentz transformations on the basis of the Einstein interpretation and related considerations:** Berzi, Einstein, Ives, Mittelstaedt, Pfarr, Raab, Robertson, Sen

[mag] **Magnetic reconnection:** Birk, Biskamp, Brown, Craig, de Gouveia dal Pino, Hirayama, Kliem, Kornack, Kuijpers, Lesch, Parnell, Priest, Ragot, Schopper, Somov

[neu] **Recent physical theoretical approaches to theories based on a distinguished inertial systems (ether):** Achuthan, Arminjon, Aspden, Cavalleri, Claybourne, Dingle (1967), Dirac, Duffy, Dudley, Einstein-Äther, Feinberg, Hunt, Ivert, Martin, Mitsopoulos, Mohorovicic, Mova, Morris, O'Rahilly, Peat, Pocci, Prodlaha(1977), Prokhovnik(1980), Rembielinski, Rowlands, Schaffner, Selleri, Shimmin, Simhony, Shupe, Whealton, Winterberg, Whittaker

[nic] **Non-Euclidean Geometry:** Efimov, Greenberg, Meschowski, Raschewski, Schreiber, Trudeau, Yaglom

[ph1] **Philosophical-fundamental scientific contributions to the question of the reality of Lorentz contractions.:** Atkinson, Dewan, Einstein, Evett, Jefimenko, Joseph, Kraus, MacGregor, Podlaha, Syrovatskii, Varicak, Winnie

[ph2] **Philosophical Space-Time Discussions:**
Anderson, Beauregard, Brown H., Büchel, Buth, Capek, De Ritis, Dessauer, Dieks, Dingler, Dörling, Earman, Ghosal, Grünbaum, Jammer, Janich, Maund, May, Montanus, Müller, A., Mundy, Robertson, Rietdijk, Petkov, Podlaha (1976), Roscoe, Schilpp, Schock, Tetens,Treder, Ungar, (s. a. [Hentschel], insbes. zu Reichenbach, H. und Grünbaum, A.)

[sp1] **Special relativity - introductions:** Bader, Sexl, Bergmann; Gerthsen, Westphal, Pohl, Joos, Lorentz

[sp2] **Special relativity, textbooks:** Einstein, French, Landau, Rindler, Ruder, Schröder

[zur] **To measure the one-way speed of light:** Brown, Chang, Cialdea, Clifton, Erlichson, Ohrstrom, Rindler, Ruderfer, Stedman, Tyapkin, Ungar, Vargas, Weinberg (1971) (Although these measurements did not achieve their true goal, they are, in rege, accuracy increases of [169].)

[zwi] **Twin paradox:** (vor allem [Schleichert], [Prokhovnik]) American, Dingle, Eisele, Fock, Genz, Ives, Langevin, Lowry, Marder, McCrea, Møller, Palacios, Prokhovnik, Schleichert, Terletzkii, Taylor

Lectures of the author, which are the basis of the 3rd to 5th edition of this book:

[vor1] Brandes, J., *Lorentzian approach to general relativity – Einsteins closed universe reinterpreted*, Relativistic Physics and some of its Applications: International Confer-

ence Athens 1997 and Proceedings Open Questions in Relativistic Physics, Edited by Franco Selleri, Bari, Italy, 1998

[vor2] Brandes, J., *Die Einstein'sche und lorentzianische Interpretation der speziellen und allgemeinen Relativitätstheorie,* Vortrag am Institut für Theoretische Physik, Prof. Dr. G. Mahler, Universität Stuttgart am 21.4. 98

[vor3] Brandes, J., *Experimental possibilities to prove Lorentzian interpretation*, Physical interpretations of relativity theory VI, British Society for the Philosophy of Science and M. C. Duffy, S.C.E.T. University of Sunderland, London: 11.-14. Sept. 1998,

[vor4] Brandes, J., *The Nimtz Experiment and the Lorentzian interpretation of special and general relativity,* Workshop on Superluminal(?) Velocities (W. Hehl, P. Mittelstaedt, G. Nimtz), Köln, Juni 1998. (Auf dieser Tagung hat sich die These von G. Nimtz, Information ließe sich mit Überlichtgeschwindigkeit übertragen (Mozart'sche Symphonie, 4.7c, s. [328], Seite 334), nicht durchsetzen können.)

[vor5] Brandes, J., *Die lorentzianische (und Einstein'sche) Interpretation der speziellen und allgemeinen Relativitätstheorie*, 63. Physikertagung, Deutsche Physikalische Gesellschaft, Frühjahrstagung Gravitation und Relativitätstheorie, Heidelberg, 15.-19.März 1999

[vor6] Brandes, J., *Die lorentzianische (und Einstein'sche) Interpretation der speziellen und allgemeinen Relativitätstheorie,* Tagung Aktuelle Entwicklungen in der Gravitationstheorie (C. Kiefer, Freiburg, V. Perlick, TU Berlin, G. Schäfer, Jena), Physikzentrum Bad Honnef 6.-8. September 1999

[vor7] Brandes, J., *Die lorentzianische Interpretation der Relativitätstheorie am Beispiel der Ehrenfestparadoxie,* Deutsche Physikalische Gesellschaft, Frühjahrstagung Didaktik der Physik, Ludwigsburg, 8.-10.3.1999

[vor8] Brandes, J., *Interpretationen der speziellen Relativitätstheorie – Born, Einstein, Lorentz,* 64. Physikertagung, Deutsche Physikalische Gesellschaft, Frühjahrstagung Didaktik der Physik, Dresden, 20.–24. März 2000, Tagungs-CD, Redaktion: V. Nordmeier, Münster

[vor9] Brandes, J., *Die lorentzianische Interpretation der Relativitätstheorie und die Winterberg'sche Hypothese eines Planckäthers,* 64. Physikertagung, Deutsche Physikalische Gesellschaft, Frühjahrstagung Gravitation und Relativitätstheorie, Dresden, 20.–24. März 2000

[vor10] Brandes, J., *Possible astronomical observations to prove Lorentzian interpretation of general relativity theory*, Physical Interpretations of Relativity Theory VII, British Society for the Philosophy of Science, London, 15-18 September 2000

[vor11] Brandes, J., *Schwarze Löcher als Ursprung ultrarelativistischer kosmischer Strahlung – ein Test der lorentzianischen Interpretation der Relativitätstheorie,* 66. Physikertagung, Deutsche Physikalische Gesellschaft, Frühjahrstagung Teilchenphysik, Leipzig, 18.-22. März 2002

[vor12] Brandes, J., *Die Dualität der klassischen und lorentzianischen Interpretation als didaktisches Hilfsmittel zur Einführung in die allgemeine Relativitätstheorie* in: Tagungs-CD Leipzig 2002 Didaktik der Physik Redaktion: V. Nordmeier, Münster

[vor13] Brandes, J., Doll, P. *Mögliche Beobachtung hoch energetischer Luftschauer aus der Richtung galaktischer Kerne,* 67. Physikertagung Deutsche Physikalische Gesellschaft, Frühjahrstagung Teilchenphysik Aachen, 10.–13. März 2003

[vor14] Brandes, J., *Die Zwillingsparadoxie – noch immer ungelöst?* in: V. Nordmeier, Berlin, Tagungs-CD: Didaktik der Physik – Augsburg 2003, Berlin: Lehmanns

[vor15] Brandes, J., Doll, P. *Teilchenemission aus galaktischen Kernen,* 68. Physikertagung Deutsche Physikalische Gesellschaft, Frühjahrstagung Teilchenphysik Mainz, 29.03 – 1.04 2004

[vor16] Brandes, J., *Fortsetzung: Die ungelöste Zwillingsparadoxie.* in: V. Nordmeier & A. Oberländer, Berlin Didaktik der Physik Düsseldorf 2004 Tagungs-CD Düsseldorf 2004 Didaktik der Physik

[vor17] Brandes, J., *Kosmische Strahlung und Gammabursts,* 69. Physikertagung, Deutsche Physikalische Gesellschaft, Frühjahrstagung Teilchenphysik, Berlin, 4.–9. März 2005

[vor18] Brandes, J., *Neue Konzepte in der Lehre der allgemeinen Relativitätstheorie (ART) und ihre Anwendung auf kosmische Strahlung und Gammabursts* in: Didaktik der Physik Berlin 2005 Nordmeier, V. & Oberländer, A. (Hrsg.), Berlin: Lehmanns Media 2005

[vor19] Brandes, J., *A cheap, rational and testable refinement of special and general relativity,* Physical Interpretations of Relativity Theory X, *British Society for the Philosophy of Science,* London, 8. –11. September 2006 http://www.physicsfoundations.org/ PIRT conferences in London, archives 2000 - 2008

[vor20] Brandes, J., *Quintessenz und Didaktik der allgemeinen Relativitätstheorie (ART)* in: Nordmeier, V. / Oberländer, A. (Hrsg.): Didaktik der Physik – Kassel 2006. Berlin: Lehmanns Media, 2006

[vor21] Brandes, J., *Ein neuartiger Mechanismus zur Umwandlung von Gravitationsenergie in Strahlungsenergie bei Gammabursts und Supernovae,* Deutsche Physikalische Gesellschaft, Frühjahrstagung Teilchenphysik, Heidelberg, 5.–9. März 2007

[vor22] Brandes, J., *Jets, supernova, gamma bursts – more light for theory?* Poster Cosmic Matter, Würzburg 24.9 – 29.9.2007

[vor23] Brandes, J., *Relativitätstheorie und philosophische Gegenargumente* in: Nordmeier, V. / Oberländer, A. / Grötzebauch, H. (Hrsg.): Didaktik der Physik – Regensburg 2007. Berlin: Lehmanns Media, 2007

[vor24] Brandes, J., *The smallness of the cosmological constant Λ,* Deutsche Physikalische Gesellschaft, Frühjahrstagung Teilchenphysik, Freiburg, 3.–7. März 2008 und München 9.–13. März 2009

[vor25] Brandes, J., *Relativitätstheorie und philosophische Gegenargumente; die Kleinheit der Kosmologischen Konstante* in: Nordmeier, V. / Grötzebauch, H. (Hrsg.): Didaktik der Physik – Berlin 2009. Berlin: Lehmanns Media, 2009.

[vor26] Brandes, J., *Relativitätstheorie und philosophische Gegenargumente II* (2010) in Internetzeitschrift (geplant): PhyDid B - Didaktik der Physik - Beiträge zur DPG-Frühjahrstagung,

The following lectures and contributions can be found on the Homepage of the author www.grt-li.de:

[vor27] 2012: „Ist die Klassische Allgemeine Relativitätstheorie unvollständig?"

[vor28] 2013: „GRT - well proven and also incomplete?"

[vor29] Counterarguments 2012: Einwand-Diffamierung-2012

[vor30] Counterarguments since 2013: Further_Counterarguments_since_2013

In the file Talks_Brandes_since_2014 on www.grt-li.de you can find the talks:

[jb.1] Fireballs of GRBs and Lorentz-Interpretation (LI) of GRT
[jb.2] GRT - well proven and also incomplete. Further arguments
[jb.3] Cosmology and Lorentz-Interpretation (LI) of GRT
[jb.4] (Cosmology and) Lorentz-Interpretation (LI) of GRT
[jb.5] Die didaktische Bedeutung der Lorentz-Interpretation der Allgemeinen Relativitätstheorie
[jb.6] Classical GRT and its Lorentz Interpretation
[jb.7] Interpretationen der Quantenmechanik und Lorentz-Interpretation
[jb.8] Das Zweikörperproblem der GRT
[jb.9] Lorentz interpretation and Kerr metric
[jb.10] First steps in calculating supermassive objects (black holes) using TOV equation
[jb.11] Supermassive objects (SMO's) calculated using the Tolman Oppenheimer Volkoff (TOV) equation and possible observation by gravitational waves (GW's) and by the event horizon telescope (EHT)
[jb.12] Possible experimental proof of Lorentz interpretation (LI) of GRT - further arguments Poster DPG 2019
[jb.13] Possible experimental proof of Lorentz interpretation (LI) of GRT – M87
[jb.14] Measuring results of M87 and Lorentz interpretation (LI) of GRT
[jb.15] Observations questioning classical GRT and preferring LI of GRT – EHT image of M87*, Spin, ALMA image of SGR A* and more
[jb.16] Poster-DPG-2021-SL-nicht-bestätigt
[jb 17] Homepage of the author: http://www.grt-li.de, please, look at first: Observations questioning classical GRT and preferring LI of GRT – EHT image of M87*, Spin, ALMA image of SGR A* and more, chapter 13.
[jb 18] Supermassive objects (SMO's) calculated using the Tolman Oppen-heimer Volkoff (TOV) equation and possible observation by gravitational waves (GW's) and by the event horizon telescope (EHT),
[jb 19] Measuring results of M87 and Lorentz interpretation (LI) of GRT, see homepage of the author: http://www.grt-li.de
[jb 20] Talk-DPG-2023-Questionable predictions of SgrA

26.4. The authors

Jürgen BRANDES, born 1939 in Berlin, physicist, diploma in experimental physics at DESY, Hamburg. Philosophically influenced by the value philosophy and ethics of Kurt Ports. Editor and one of the authors of "*Die Einstein'sche und Lorentzianische Interpretation der speziellen und allgemeinen Relativitätstheorie*" and author of "*Die relativistischen Paradoxien und Thesen zu Raum und Zeit*", the results of which are expanded upon in this edition. For the author, the Lorentz interpretation, in particular the realisation of what the *negative* Newtonian gravitational potential probably means, is more important than material successes. The path to the Lorentz interpretation began with the attempt to convince the philosopher PORT of the superiority of the relativistic space-time philosophy over Kant and Newton. The so far constant scientific recognition is the poor counter-arguments.
My homepage: www.grt-li.de

Ludwig NEIDHART, Dr. habil., philosopher and natural scientist, born 1963 in Bad Waldsee, studied theology and philosophy in Münster and Würzburg 1982-1978, mathematics and physics in Augsburg 1994-2002, doctorate in theology with an interdisciplinary doctoral thesis on "*Infinity at the Intersection of Mathematics and Theology*" 2005, habilitation for the subject of philosophy with a habilitation thesis on "*God and Time*" 2015, teaching as a private lecturer at the University of Augsburg at the Chair of Analytic Philosophy and Philosophy of Science. In addition to formal logic, his main research interests include the philosophical aspects of mathematics and the natural sciences, and in this context the concern to develop a philosophically appropriate interpretation of the theory of relativity based on a clearly understandable exposition has become very important to him. The Lorentzian interpretation approach proved to be groundbreaking for him, which is why he would like to make it known to a wider audience.

Jan CZERNIAWSKI, Dr. habil. Since 2020 Professor at the Jagiellonian University, Kraków, Poland. Born 1957 in Bychawa, Poland, master's degree in physics 1981 and philosophy 1988, doctorate 1995 and habilitation 2011 in philosophy at the Jagiellonian University. Since 1983, dealing with relativistic questions, recognized the dynamic interpretation through his own studies and subsequently found confirmation in the works of S. J. PROKHOVNIK and many other authors, especially H.A. LORENTZ. This became his own book: *"Understanding the Theory of Relativity"*, in Polish, 1993. His later field of work was radical Aristotelianism (i.e. his own version of T. KOTARBIŃSKI'S reism) and P. LORENZENS protophysics for additional support of the dynamic interpretation. The results were published, among others, in his book *"Motion, Space, Time"*, in Polish, 2009, https://ruj.uj.edu.pl/xmlui/handle/item/81681?locale-attribute=en. Currently he is occupied mainly with philosophical problems concerning quantum phenomena, especially quantum entanglement.

26.5. Index

A

absolute effect 194
absolute effects 117, 194, Fig. 9.5, ch. 24.1
absolute measurements 150
absolute simultaneity ch. 2.4, 25, 36, 58, Fig. 7.4, 71, 71, 79, 94, 96, 98, 99, 99, 100, 101, 180, Fig. 11.6.0, 188, 333, 336
absolute space geometry 100
absolute velocity 94
absolutely apparent 91
absolutely real 91, 92, 95
accelerated reference frame 314
acceleration field 151
acceleration objection 139, 140, 144, 146, 157
acceleration phase 151, 156
accretion 285, 289
accretion disc 286, 328, Fig. 21.6
accretion disk Fig. 21.8
addition theorem 35, 38, 39, 40, 43, 64, 67
addition theorem of the velocities 35, 38
adiabatic clock transport 76, 78
affine 75
AGN 252
aircraft 44, 45
ALMA observation of Sgr A* 287, 288
ALMA-Beobachtung 349, 365
alternative 2nd postulate 195
Andromeda Nebula 243
angle of rotation 123, 213
angular momentum 286, 288, 289, 306, 322, 325
anisotrop 250
anisotropy 97
annihilation frequency 280, 281
antinomian formulae 146
antinomic prediction 155
antinomy 119, 128, 146, 147, 153, 338
Antinomy 146
aphelion 204
astronomical observations 24, 100, 202, 256, 305, 322, 327, 328
AUDRETSCH 216
AUGUSTINUS 242, 260

B

background geometry 101, 102
background radiation 98, 100, 252, 253, 254
ballistic theories 93
baryon number 277
BASCHEK 256
being real 181
BELL 85, 317
Bell's spaceship paradox 120, 127, 132, 194, 315
BERENDA 130
BERENDS 4, 68
BERGMANN-SCHÄFER 313
BERGSTRÖM 330
BERRY 256
big bang 98, 248, 249, 251, 254, 258, 260, 263, 279, 330, cp. 14.8.2
bimetric 101, 316
Birkhoff theorem 217, 299, 301
Birkhoff-Theorem 225, 226
black hole 19, 24, 27, 259, 266, 275, 276, 286, 288, 289, 292, 305, 321, 323, 324, 325, 326, 327, 328, 329, 337, Fig. 21.6
black holes cp. 21
black holes, inner SM und TOV 293
block universe 98, 99
bodies of the theory of relativity 177
BORN 69, 113, 308, 309
Born-Einstein'sche These 309
boundary layer 286, 287, 288, 289, 324, 325, Fig. 21.8
BRANDES 312, 366
BROWN, 187
bucket experiment 59, 187
BURGSTALLER, S. 4

C

calculation of supermassive objects using TOV 290
calculus of variations 218
CAPEK 179
Casimir effect 312
cavity 301
channel beams 45
circle segments 194
circles of contact 228
circular disc 116, 127, 128, 130, 131, 229, 238
circular ring 113, 114, 116, 117, 130, 194, 309, 310, 311, 314
circularly accelerated 113
clock dilation 24
clock paradox 137, 144, 149, Fig. 10.8, Fig. 10.10
clock synchronisation 33
co-moving coordinate system ch. 19.3
comparison Einstein (EI) and Lorentz (LI) interpretations of general relativity 305
comparison of the Einstein and Lorentz interpretations 194
concept of force 188
conservation 38
conservation of energy 7, 19, 22, 25, 35, 42, 267, 270, 272, 276, 277, 279, 280, 285, 301, 305, 316, 320, *ch. 4.5*
conservation of energy for light cp. 21.1
conservation of energy RWM 301
conservation of energy RWM ch. 22
conservation of energy SM ch. 21
conservation of Mass 39
conservation of momentum 38, 39
constancy of the speed of light 30, 36, 47, 62, 338
constituent and gravitational mass 28
conventional 86
conventionality 86
coordinate time 153, 217, 222, 246, 270, 275, 283, 316, 317
cosmic microwave background (CMB) 329
cosmic time 98, 99
cosmological constant 225, 226, 242, 245, 254, 256, 258, 304, 316, 322, 334, 335
cosmological principle Fig. 19.1, Fig. 19.9, 243, 244, ch. 19.3, 244, 246, 246, 247, 249, 254, 260, 335, 336
cosmological simultaneity ch. 19.8.4, 19, 24, 241, 242, 256, 258, 260, 262, 263, cp. 19.8.4, 263, 265, 304, 329, 329
cosmological time 262, 263
cosmology 70, 306, 335
covariance 22, 220
covariance principle 22, 267
critics of the Lorentz interpretation 316
curvature 24, 101, 199, 200, 202, 208, 210, 214, 216, 218, 220, 221, 224, 225, 228, 229, 230, 233, 239, 240, 247, 257, 260, 316, 332
curved space *ch. 18*
cylinder coordinates Fig. 18.5
CZERNIAWSKI 7, 74, 136, 367

D

D'INVERNO 273, 297
dark energy 335
dark matter 255, 279, 322, 334, 335
DE GOURMONT 194
degenerate object 325
degenerative pressure 295
density fluctuation 252
derivation of the Lorentz transformations 60, ch. 6
DESCARTES 196
DESSAUER 25
differential geometry 228
dimension concept Fig. 11.1
DINGLER 181
DIRAC 188
distance
(s. a. metric) 230
distance in plane Fig. 16.7
distance of spherical coordinates Fig. 18.4
distinguished inertial frame 48, 63, 189, 316
distinguished inertial system 63, 149, 185, 190, 192

distinguished IS 149
distinguished or excellent inertial system 336
Doppler effect 45, 90, 265
Doppler formula 263, (19.9)
DUFFY 188
dust model 254, 259
dust star 225, 226, 259, 263, 270, 297, 299, 300, 302, 324, 329, 330
dynamic mass 42, 46, 115

E

edge of the universe 247, 257
edge-on 325, 327
EHRENFEST 127
Ehrenfest paradox 113, 117, 120, 127, 129, 130, 131, 132, 136, 194, 309, Fig. 10.6
Ehrenfest Paradox 131, 317
Ehrenfest-paradox Fig. 10.5
Ehrenfest-Paradoxie 363
EINSTEIN 4, 7, 21, 25, 43, 48, 58, 59, 62, 68, 69, 72, 74, 107, 108, 117, 129, 131, 171, 181, 183, 186, 188, 196, 197, 204, 214, 215, 216, 225, 227, 267, 272, 279, 304, 308, 310, 314, 315, 319, 331, 333, 335, 337, 339
Einstein and Lorentz interpretation *ch. 6.6, SM*
RWM *cp. 19.8*
Einstein- and Lorentz-Interpretation ch. 2, ch. 23
Einstein clock synchronisation 62, 147
Einstein contraction 86, 87, 88, 89, 90
Einstein convention 33, 34, 35, 55, 67, 336
Einstein convention of clock synchronisation 34
Einstein definition of simultaneity 56, 58
Einstein equation 99, 101, 220, 225, 256
Einstein interpretation 19, 22, 25, 35, 48, 53, 55, 58, 60, 63, 66, 71, 72, 107, 108, 109, 129, 147, 175, 179, 180, 181, 190, 191, 192, 199, 214, 221, 248, 250, 256, 257, 260, 264, 305, 314, 335, 336
Einstein Interpretation 21
Einstein law of conservation of energy 319
Einstein synchronisation 57
Einstein's postulates 74, 82, 83, 332, 333
Einstein's procedure of synchronization 84, 86
Einstein's second postulate 93
Einstein's synchronization procedure 86
Einsteinians 94, 95, 96
Einstein's clock synchronisation 260
EINSTEIN's thought experiment 107, Fig. 9.1
electric and magnetic field strength 190
elementary particle 44, 46, 316
elementary particles 37, 63, 64, 65, 164, 194, 308, 327
ellipsoid 177
elliptical geometry 241
elliptical space 247, 259
empty, matter-free space *ch. 17*
equality of heavy and inert mass 338
equation of state 289, 290, 292, 293, 323, 336
equivalence principle 22, 271, 272, 279, 280, 281, 282, 283, 285, 324, 338
escape velocity 248, 251, 263
ether 69, 71, 72, 92, 93, 95, 96, 100, 102, 103, 186, 187, 188, 189, 196, 256, 305, 308, 313, 314, 315, 333
ether theories 93, 187
ether theory 72
event horizon 301, 325, 326
exact definition of the LI 195
excellent inertial system 49, 53, 54, 62, 63, 68, 71, 110, 113, 128, 133, 135, 147, 148, 172, 180, 184, 188, 189, 190, 191, 192, 194, 195, 196, 197, 269
excellent IS 149
expanding dust star Fig. 22.1
expanding space 199, 242, 256, 260, 263, 264, 270, 301, 329
expanding universe 206, 241, 242, 244, 249, 256, 259, 263, 297, 301, 324, Fig. 19.8
expansion of space 335, ch. 19.8.4
expansion of the space 262

expansion of the universe 225, 242, 248, 249, 251, 254, 256, 258, 324, 325, 330, 334, 335, ch. 19.8.5
exploding dust star 297

F

face-on 325, 327
FALK 146, 151
Fermat's Principle in General Relativity 317
Fermi gas 290, 291
finite and closed universe 227, 247
finite closed space 242, 259
finite universe 241, 243
finite, closed universe 227
first postulate 83
five-dimensional space 161
FIZEAU 121
flare 286
FLIESSBACH 202, 252
four vector 191
four-dimensional distance 109, 168
four-dimensional energy-momentum vector 282
four-dimensional manifold 161, 162, 172
four-dimensional pseudo-Euclidean space 93
four-dimensional space 62, 109, 110, 117, 159, 160, 162, 163, 165, 166, 168, 169, 172, 175, 176, 177, 189, 200, 209, 233, 234, 235, 308, 336, 337, 338, 339
four-dimensional space-time continuum 62, 109, 110, 117, 163, 165, 168, 172, 175, 176, 177, 189, 200, 308, 336, 338, 339
four-dimensional space-time-continuum 159
four-vector 100, 189, 191, 196
four-vectors 191, 192
free fall 25, 27, 112, 272, 273, 277, 278, 279, 280, 285, 297, 299, 316, 318, 323, 338
FRENCH 131, 171
Friedmann cosmos 254, 255, 259, 301, 335, 336
fundamental frames 98

G

galaxy distribution 245
Galilean geometry 102, 169
Galilean inertial space 225
Galilean principle of relativity 59
Galilean transformation 33, 60, 81, 189, 333, 337
Galilean transformations 33, 43, 77, 79, 82, 83, 84, 94, 96, 97, 163, 168, 169, 171
Galilei transformation 30
Galilei transformations Fig. 11.3
Galileo space 171
gamma burst 323, Fig. 24.7
garage problem 120
GAUSS 224, 240
Gaussian or curvilinear coordinate system 212
general relativistic solution 151, 153, 156
generalised principle of relativity 220
GENZEL 24
geodesic 28
geodesic movement 318
geometrical ether 100, 102
gläubige Physiker 196
global features 239
Gödel's solution 99
GOOBAR 330
gravitational collapse 223, 259, 289, 293, cp. 19.8.3
gravitational force 28, 202, 279, 289, 293, 318
gravitational mass 28, 283, 299, 302, 323
gravitational pressure 259, 293, 295, 296
gravitational redshift 24, 28, 250, 257, 271, 329
great circle 200, 231, 240, 250, 302
GRØN 129, 130
GRÜNBAUM 181

H

H.E.S.S. 19
HAERING 185
HÄFELE and KEATING 44

HÄFELE und KEATING 137, 205
HAMILTON 103
HECKMANN 258
HERING 132, 308, 312, 314
HERRMANN 186
Higgs mechanisms 279
HILBERT 315
Hilbert-Einstein equation 220, 223
Hilbert-Einstein equations 247, 315, 336, 337
Hilbert-Einstein field equations 220, 223, 225
hollow cylinder 113, 115
hollow sphere 225, 226, 297, 299
HOLST 69
homogeneity 60, 244, 252, 256
homogeneous isotropic universe 261
Hubble constant 251, 256, 260, 261, 324, Fig. 19.9
Hubble parameter 324
Hubble Parameter 324
Hubble’s law Fig. 19.5
Hubble-Konstante K
Hubble's law 199, 251, 252, 254, 258, 260, 304, 324
hyperbolic 260
hyperbolic acceleration 154
hyperbolic angle function 240
hyperbolic geometry Fig. 18.12
hyperbolic geometry 240
hyperbolic geometry 224
hyperbolic geometry 241
hyperbolic infinite 254, 257
hyperbolic space 199, 241, 246, 255, 258, 259, 262, 309, 310, ch. 14.7
hypersurfaces of constant absolute time 101

I

ideal inelastic collision Fig. 4.2
inertial system requirement 136, 138, 146, 147, 148, 149, 196, 267
inertial system with c 173
inertial transformations 58
infinitely expanding universe 241
inner curvature 229, 230, 232, 234, 239, Fig. 18.3
inner SM 283, 284, 289, 293, 294
inspection shaft 123, 124, 125
interferometer 96
Interferometer 47, 49, Fig. 5.3
Interpretation BORN, EINSTEIN, LORENTZ 308
IS requirement 149, 194
isometry problem 212
isotropic coordinates Fig. 16.11
isotropy 60, 244, 252, 256, 329
IVES 45, 128, 129, 138
IVES-STILWELL 45, 49, 51, 52, 53

J

JAMMER 188, 224
JANICH 59
JANOSSY 69
JOB 186

K

KANT 7
KAUFMANN 46
KENNEDY 48
KENNEDY-THORNDIKE 47, 48, 49, 50, 51, 52, 53
kinematic interpretation Fig. 24.1
kinematic transformation 76, 333
kinematical transformations 74, 79, 82, 83, 84
KLEIN 315
kosmologische Konstante 226
KOTARBIŃSKI 367
KRAUS 117
Kugelfläche 233, 234, 236

L

LANDAU 259, 279, 297, 309, 317
LANGEVIN 137
law of addition of velocities 94
laws of fall 59
LEIBNIZ 7
Leibnizian background geometry 103
Leibnizian geometry 102, 103
Leibnizian space-time 102
length contraction 35, 36, 37, 44, 48, 49, 54, 56, 58, 68, 71, 85, 86, 87, 89,

90, 91, 94, 95, 101, 103, 107, 113, 119, 128, 154, 197, 336, 338
length of rotation 108
lid paradox 123, 124, 125, 332
LIEBSCHER 169, 170
lifetime 37, 38, 44, 63, 64
LIFSCHITZ 259, 279, 297, 309, 317
light curvature 224
light deflection Fig. 16.1
LIGHTMAN 290, 297
limiting case v→ c 175
limiting case, *g* infinite 155
linearly accelerated 111
lines of free fall 213
LISLE 103
LOBATSCHEWSKI 240
local inertial frame 271, 273, 274, 279, 280, 281, 285
local inertial system 273, 285
local system 313
LORENTZ 48, 51, 53, 62, 69, 74, 100, 102, 129, 308, 312, 315, 367
Lorentz contraction 22, 24, 48, 51, 52, 69, 72, 85, 86, 87, 88, 89, 90, 92, 107, 111, 112, 113, 115, 116, 117, 120, 121, 127, 128, 129, 130, 131, 132, 134, 177, 187, 194, 310, 311, 315, 332
Lorentz contraction of length and distance Fig. 9.3, Fig. 9.4
Lorentz contraction of lengths and distances Fig. 9.2
Lorentz interpretation 7, 19, 20, 21, 22, 25, 35, 48, 53, 55, 58, 60, 62, 63, 64, 66, 68, 69, 71, 72, 92, 93, 98, 99, 109, 110, 128, 129, 130, 136, 137, 147, 180, 188, 189, 190, 191, 192, 193, 199, 200, 214, 221, 247, 257, 258, 265, 305, 306, 307, 309, 314, 326, 335, 336, 337,338, 366, Fig. 2.1, ch. 20
  Definition ch. 2, 6.4, 6.6, 12, 13, 20
  RWM ch. 19.5, 19.8
  statements ch. 7.4
Lorentz interpretation, dust star ch. 22.2
Lorentz kinematics 74
Lorentz spacetime 99, 100
Lorentz tansformation 32
Lorentz theory 59
Lorentz transformation 81
Lorentz transformations 25, 30, 31, 32, 33, 34, 35, 36, 38, 40, 42, 43, 44, 45, 53, 54, 55, 56, 57, 58, 59, 61, 62, 63, 71, 79, 83, 97, 99, 100, 116, 119, 120, 123, 124, 125, 126, 128, 130, 137, 139, 140, 146, 147, 158, 163, 165, 166, 167, 168, 169, 170, 171, 172, 173, 174, 175, 176, 178, 184, 189, 267, 312
Lorentz-FitzGerald contraction hypothesis 83
Lorentzian principle of relativity 59
Lorentzians 93, 94, 95, 102
Lorentz-Interpretation 58, 191, 366
LORENZEN 367
lot 236, 237
Luftballonbeispiel 251
LUTHER 316

M

Mach's thesis 188
machsche These 343
magnetar 323
magnetic reconnection 286, 322
manifold 159
MANSOURI 71, 72
mass increase 85, 89
mass-energy equivalence 92
MATLAB implementation 291
Maxwellian equations 22
Maxwell's equations 43, 187, 192
MAY 184
MCCREA 138
MCCULLOCH 19
metagalaxy 249, 260, 263, 265, 324, 325, 330
MICHELSON 48
MICHELSON-MORLEY 47, 48, 49, 50, 51, 52, 53, 185, 186, 192, 312, 337
MICHELSON-MORLEY experiment 314
microwave radiation (CMB) 304
midpoint plane 211, 212, 213
Milky Way system 243
MINKOWSKI 74, 165, 166, 170, 172
Minkowski diagram 87, 169, 171, 179
Minkowski diagramm Fig. 11.2

Minkowski space 7, 21, 62, 71, 165, 168, 169, 170, 171, 172, 173, 174, 178, 179, 180, 181, 182, 184, 185, 187, 193, 194, 196, 200, 206, 211, 214, 217, 221, 224, 225, 243, 298, 299, 332, 337, ch. 11.3, G
Minkowski space, alternative interpretation ch. 11.4
Minkowski's spacetime 99
Minkowskian geometry 101, 102
Minkowski-space Fig. 11.6.3
MISNER 275, 297
missing matter 335
MITTELSTAEDT 68
modified equivalence principle 281
MØLLER 130, 151, 153
MOORE 319
MORLEY 48
MÜLLER, A. 163, 165
Muon 37

N

negative gravitational potential 306, 319, 321
negative mass 338
negative Masse 7, 306
negative Newtonian gravitational potential 19, 20, 318, 319, 332, 366
negative proper time 155
NEIDHART 7, 366
neutron star 178, 259, 291, 323, 337
NEWTON 28, 99
Newtonian bucket effect 194
Newtonian bucket experiment 59, 107
Newtonian limit case 319
Newtonian limiting case 7, 257, 280, 281
non-reality thesis 85, 91
non-relativistic limiting case 296
nuclear fuel 223, 259
nuclear fusion 259, 285
numerical example 32, 157, 168, 234, 309
Numerical example 33, 35, 37, 38, 145, 148, 165, 208, 209, 228, 231, 234, 236, 238, 265, 323

O

Occam's razor 317
Occam's razor 313, 314
one-dimensional and multidimensional 159
one-way speed of light 35, 52, 66, 180, 192, 194, 196, 331
one-way velocity of light 48, 94
opposite 183, 184
oppositeness 194
outer curvature 232

P

pair instability supernovae 323
PAIS 304
paradigm 20, 306
paradox 119
paradoxes ch. 10
parallel 31, 128, 211, 233, 235, 240, 241, 255
particle picture 277
PAULI 129, 272
Pauli principle 327
perception 165, 166
perihelion rotation 202, 204, 218
PETKOV 136, 271
phase difference 66, 68
phenomenology 163
phenomenology 233
philosophical objections. 19
PHIPPS 130, 131
PIERSEAUX 59
pion 44
Poincaré 53, 160, 224, 257, 304
Poincaré model Fig. 18.12
Poincaré thesis 199, 224
Poincaréan model 338
Poincaréan thesis 338
poincarésche 241
polynomial division 294
PORT 366
potential energy 27
POUND 271
principle of constancy of the speed of light 34, 60, 66
principle of economy 313, 314
principle of equivalence 267, 338

principle of methodical order 100
principle of relativity 21, 30, 43, 48, 59, 60, 61, 62, 63, 68, 69, 71, 72, 79, 82, 83, 84, 86, 90, 93, 94, 96, 97, 98, 100, 102, 110, 111, 117, 119, 128, 133, 146, 172, 180, 181, 182, 186, 187, 190, 191, 192, 193, 200, 314, 315, 333, 336, 337, 338, 339
principle of Relativity (RP) of LI 195
principle of relativity, two different one 69
principle of the constancy of the speed of light 21, 35, 44, 51, 60, 61, 62, 190, 193, 262, 263, 271, 312, 333
principle of the constancy of the speed of light of the LI 195
privileged reference frame 92, 93
problem of time in quantum gravity 103
projection 228, 238, 247
Projection 161
projections 233
PROKHOVNIK 187, 367
propagation delay Fig. 16.4
proper time 55, 91, 92, 95, 138, 140, 143, 144, 145, 148, 152, 153, 154, 155, 157, 158, 215, 216, 217, 222, 246, 247, 262, 263, 264, 266, 270, 273, 275, 276, 281, 283, 284, 298, 338
protophysics 59, 100, 101
pseudo-Riemannian space 22, 200
pseudo-rotations 84
pseudosphere Fig. 18.11
pseudotensor 267
psychische Vorgänge 163
PYTHAGORAS 50, 168, 170, 210, 230, 231

Q

quantum ether 186, 187, 188, 197, 226, 305, 312
quantum field theory 186, 277, 330
quantum gravity 103, 105, 277, 279, 314, 323, 324
quantum mechanics 277
quark star 291
quasar 252

R

Radar signals
Venus Fig. 16.3
radial motion 301, 321, 322
RASCHEWSKI 167, 171
real and apparent quantities 110
Real space *ch. 11.6.3*
reale Welt 169
REBHAN 283
reciprocity principle 57, 58
recombination radiation 329
red shift 213, 252, 254, 277
redshift 213, 214, 251, 264, 265, 271, 272, 324
Reference- or coordinate system 30
REICHENBACH 86, 181
relative and philosophical simultaneity 178
relative velocity 32
relativistic law of gravity 319
relativistic limiting case 296
relativistic virial theorem 319
relativity of simultaneity ch. 4.3, ch. 4.4, 7, 26, 33, 36, 79, 89, 107, 108, 119, 121, 126, 132, 146, 181, 182, 182, 313
relativity of simultaneity relativistisch Fig. 7.4
relativity principle 110
remote effects 188
REPKA 271
rest length 86, 88, 91, 92, 95, 107, 108, 110, 111, 115, 196, 338
rest mass 27, 28, 42, 46, 85, 91, 92, 95, 115, 161, 162, 252, 272, 276, 277, 279, 281, 282, 285, 316, 319, 320, 323
ribbon 113
Riemannian 22, 24, 101, 200, 267, 269
Riemannian space 24, 269
RINDLER 27, 28, 271, 276, 285
RITCHIE et al. 323
RITZ 93
Robertson-Walker metric 24, 243, 246, 247, 249, 250, 251, 254, 257, 258, 259, 263, 264, 294
Robertson-Walker-metric ch. 19.4
Robertson-Walker-Metrik 330, Fig. 19.4

RÖMER 312
rotating circular ring 115, Fig. 24.4
rotating disc Fig. 10.6
rotating reference frame 314
rotating reference systems 309
rotating wheel Fig. 24.5
RUPPEL 146, 151

S

saddle surface 229, 230, 245
Sagnac effect 96, 98, 117, 120, 136, 194
Sagnac-Effekt 132, Fig. 10.7.1
Saphiro effect 218
scale changes Fig. 20.1, Fig. 20.2
scale contraction 24
scale distortion 239, Fig. 18.9
Scharnhorst effect 312
SCHILLER 159, 178
SCHREIBER 241
SCHRÖDER, U. E. 60, 62
SCHRÖDER, W. 315
Schwarzschild metric 24, 199, 200, 202, 204, 206, 207, 208, 209, 210, 211, 212, 213, 214, 215, 216, 217, 218, 219, 220, 221, 222, 223, 224, 226, 227, 242, 243, 246, 247, 249, 250, 257, 259, 269, 279, 280, 283, 284, 294, 299, 323, 336, 339
Schwarzschild radius 277
Schwarzschild radius for the LI 302
Schwarzschild-metric experiment Fig. 16.1 - Fig. 16.4
Schwarzschild-Metrik 284
SELLERI 134, 136
Selleri transformations 58
sensation 164
SEXL 68, 70, 71, 72, 215, 255, 317
SHAMIR 324
shrinking Fig. 16.9
simulation supernova 323
SINGH 297
singularity 222, 284, 293, 322
singularity SM *ch. 19*
slowing down of clocks 205
smallness of the cosmological constant 265, 270, 305, 322, 330, 335
SMO supermassives Objekt
SOMMER 315
SOMMERFELD 166
spacetime 93
space-time interval 191
spatialization of time 84
special-relativistic limiting case 19, 199, 246, 256, 260, 261, 262, 263, 264, 265
special-relativistic simultaneity 24, 265
spherical coordinates Fig. 18.4
Spherical coordinates Fig. 16.6
spherical geometry 219, 224, 240, 241, 255, 337
spherical surface Fig. 18.6
spin 324
Standard Model 252
star collapse 259
STEPHANI 255, 260, 271, 275, 279, 297, 298, 302, 317
stereographic projection 240
STILWELL 45
stress 129
stress forces 111
superclusters 243
supermassive object Fig. 21.10, Fig. 21.11
supermassive objects (SMO) 291
supermassive stellar object Fig. 21.7
supermassive stellar objects (SMO) 285
synchronisation effect 147, 148

T

temporalization of space 84
tension 112, 128, 129, 131
tension forces 113
tensor 191, 192
Tensor 189, 191
TETENS 186
Theory of LORENTZ 72
thermalisation 326, 327
Thomas precession 119, 138
THORNDIKE 48
THORNE 20, 72, 275, 297, 306
thought experiment 63, 66, 108, 112, 116, 138, 140, 147, 152, 185, 196, 215, 219, 314, 318, Fig. 7.1
three-dimensional spherical surface Fig. 18.8

three-plus-one-dimensional 162, 163, 165, 166, 172
THÜRAUF 288
time dilation 22, 26, 35, 37, 44, 45, 48, 49, 51, 52, 54, 56, 58, 63, 64, 68, 69, 71, 85, 89, 90, 94, 96, 100, 101, 103, 116, 119, 134, 137, 139, 141, 142, 145, 146, 147, 153, 154, 187, 197, 206, 213, 218, 262, 329, 336, 338
TOLMAN 27, 28, 322
Tolman-Oppenheimer-Volkhoff equation 284
Tolman-Oppenheimer-Volkhoff-Gleichung 290
torus Fig. 16.12
total energy 27
total energy free fall Fig. 21.4, Fig. 21.5
traktrix 240
transit time 48, 122, 204, 205, 218
TREBIN 283
TREDER 68, 69, 315
triangulation 224
twin paradox 85
twin paradox 19, 26, 91, 119, 120, 137, 138, 146, 149, 151, 157, 194, 332, 338, 339, Fig. 10.8
two principles of relativity 59
two-way speed of light 35, 35, 180, 196
two-way velocity of light 48

U

UHECR 19, 21, 305
ultra-high-energy cosmic ray 305
Umkehrphase 157
uniformity of the clock transport 76
UNSÖLD 256
URBANTKE 72

V

V. LAUE 69
vacuum 186
vacuum fluctuation 186
vanishing masses 225
variation principle 317
velocity addition Fig. 5.4
vierdimensionale 162
view into space Fig. 19.6
VIRCHOW 163

W

wave energy 277, 323
wave picture 277
weight force 318
WEYL 210, 215, 218, 220, 221
WHEELER 255, 275, 297
WIECHERT 315
WIKIPEDIA 157, 315
WINTERBERG 315
WOOD 121
WOODHOUSE 318
Wood's paradox Fig. 10.2
Wood's Paradox 121
world point 166, 170
world radius 245, 246, 247, 254, 260, 297, 301, Fig. 19.7
WÜNSCH
DANIELA 315

X

X-ray 252

Z

zero-point oscillation 186
ZIPPELIUS 283

Made in United States
Troutdale, OR
12/30/2023

16559144R00229